Practical Business Analytics Using SAS

A Hands-on Guide

■■■

Venkat Reddy Konasani

Shailendra Kadre

Apress®

Practical Business Analytics Using SAS: A Hands-on Guide

ISBN-13 (pbk): 978-1-4842-0044-5

ISBN-13 (electronic): 978-1-4842-0043-8

Managing Director: Welmoed Spahr
Lead Editor: Jeff Olson
Developmental Editor: Anne Marie Walker
Editorial Board: Steve Anglin, Mark Beckner, Gary Cornell, Louise Corrigan, James DeWolf, Jonathan Gennick, Robert Hutchinson, Michelle Lowman, James Markham, Matthew Moodie, Jeff Olson, Jeffrey Pepper, Douglas Pundick, Ben Renow-Clarke, Gwenan Spearing, Matt Wade, Steve Weiss
Coordinating Editor: Kevin Walter
Copy Editor: Kim Wimpsett
Compositor: SPi Global
Indexer: SPi Global
Artist: SPi Global
Cover Designer: Anna Ishchenko

Distributed to the book trade worldwide by Springer Science+Business Media New York, 233 Spring Street, 6th Floor, New York, NY 10013. Phone 1-800-SPRINGER, fax (201) 348-4505, e-mail orders-ny@springer-sbm.com, or visit www.springeronline.com. Apress Media, LLC is a California LLC and the sole member (owner) is Springer Science + Business Media Finance Inc (SSBM Finance Inc). SSBM Finance Inc is a Delaware corporation.

For information on translations, please e-mail rights@apress.com, or visit www.apress.com.

Apress and friends of ED books may be purchased in bulk for academic, corporate, or promotional use. eBook versions and licenses are also available for most titles. For more information, reference our Special Bulk Sales–eBook Licensing web page at www.apress.com/bulk-sales.

Any source code or other supplementary material referenced by the author in this text is available to readers at www.apress.com. For detailed information about how to locate your book's source code, go to www.apress.com/source-code/.

To
My mother, Atchamma Konasani
My wife, Naga Rani Konasani

—Venkat

To my entire family:
Father, Sharad Madhav Kadre
Mother, Shakuntala Sharad Kadre
Wife, Meenakshi Kadre
Daughter, Neha Kadre
Son, Vivek Kadre

—Shailendra

Contents at a Glance

Contents

About the Authors

Venkat Reddy Konasani is a postgraduate in applied statistics and informatics technology from the Indian Institute of Technology (IIT) in Mumbai. He has more than eight years of experience in predictive modeling, big data analytics, and several analytics software programs and tools including SAS and R. His major areas of expertise include credit risk model building, customer experience management, social media analytics, machine learning, NLP, text mining, and big data analytics. He has been associated with several prestigious analytics projects with reputed organizations and for international clients. Some of his major projects include an analytics plug-in for a GIS tool, text mining and Twitter sentiment on social media data, football game prediction model building, customer experience management, credit risk model building for consumer credit cards and loans, and visualizations of consumer credit risk metrics using Tableau. He has expertise in predictive modeling, social media analytics, and data analytics that he has applied to projects for international companies.

Shailendra Kadre is a senior IT and management consultant from Bangalore, India. He is the author of the 2011 Apress book *Going Corporate: A Geek's Guide*, which covers IT operations management and the business aspects of IT. He has written several research papers and articles on technology, IT portfolio management, and project and program management.

Currently, he works with Hewlett-Packard India as a solutions consultant. He has more than 20 years of experience in technology, program management, pre-sales, enterprise sales, and business analytics. His current interests include business analytics and enterprise printing solutions.

Shailendra holds a master's degree in mechanical engineering from the Indian Institute of Technology (IIT) in Delhi. He resides in Bangalore, India, with his wife, Meenakshi; daughter, Neha; and son, Vivek. He can be contacted at shailendrakadre@gmail.com.

Acknowledgments

The journey of this project was comparatively smooth, much smoother than my first book, *Going Corporate*. Thanks to Jeff Olson, an executive editor at Apress, and the entire Apress team. Jeff, also editor of my first book, picked up the idea of this book at a nascent stage.

We would like to thank development editor Anne Marie Walker, who did a great job of hand-holding while going through the manuscript. She helped improve the book in numerous ways. Kevin Walter, the coordinating editor of this book, did a wonderful job of project management and problem solving. He was the person through which we got all our work done within Apress. He created deadlines for completing each and every task in this project, and his meticulous follow-ups and facilitation made us complete this book right on schedule.

We are particularly thankful to Kim Wimpsett, the copy editor, for her patience in correcting mistakes and great assistance on the language part. Her suggestions for improving the overall presentation of the book were very useful. Dhaneesh Kumar was patient enough to take on all the final formatting and indexing work that brought this book into its current form. Thanks to him as well.

Together we formed a great international team that worked 24/7 on this project. When everyone was asleep in the United States, I and Venkat were working from Bangalore, and vice versa. It was a great experience indeed! And after such a long duration of working together, Apress is now like our immediate family. We both are looking forward to doing many more projects with Apress.

No book can succeed without the blessings of its readers. We have tried to keep this book as simple and example-oriented as possible. The needs of working professionals in the analytics industry were at the top of our minds while writing each chapter. At any time, we welcome you to interact with us via shailendrakadre@gmail.com and 21.venkat@gmail.com. We are active on Facebook and LinkedIn as well.

Any project of this size invariably creates a number of hardships for the family members of the authors and the development team. This book was no exception. Both I and Venkat owe the credit of completing this book on time to our respective family members. Without their help and support, it would not have been possible. We cheer them all. At the same time, we both feel lucky to have gotten an opportunity to present our work to readers worldwide.

Last but not least, there are relatively few people on this planet who make a living from writing, especially those who write professional books. This book was definitely not written for money. We firmly believe that whatever we are today is by the virtue of society. Writing is a beautiful and noble way to return something to our society, and we hope to continue with it in the future as well.

—Shailendra Kadre

I am thankful to everyone mentioned by Shailendra in the previous paragraphs. But I would like to mention personally a few people who worked exclusively with me in creating this book.

Sandeep Sunkara, a close friend of mine from IIT Bombay, patiently went through every page of this book. His expert technical suggestions were particularly useful. I am thankful to Swapnil Kitcha, now working with Fidelity as a specialist, for validating all the SAS code used in this book.

■ ACKNOWLEDGMENTS

The acknowledgments wouldn't be complete without a special mention of Gopal Prasad Malakar, who has been my analytics guru throughout my career. I would also like to mention Shailendra Kadre, the coauthor of this book, who envisioned, steered, influenced, and encouraged me throughout this work.

I sincerely thank the numerous students of my analytics training programs, who always encouraged me to document my lectures and who shaped my ideas, as they appear in this book.

This book would not have been possible without the sacrifices made by my wife, little son, and other family members, who saw me working late hours and were patient with me. Each one of them should get equal credit for bringing this book to fruition.

—Venkat Reddy Konasani

Preface

Coauthor Venkat Reddy Konasani and I, Shailendra Kadre, both work for Hewlett-Packard in Bangalore, India. He and I have been friends for a while, long before he started working for Hewlett-Packard. I had been becoming more and more interested in business analytics, using it in my work and thinking about doing a book on the subject. I was aware that he was a trainer in business analytics, and I attended a few of his training sessions. I was really impressed with his effective, simple presentation style, not to mention the content of his lectures. Here, I realized, was the perfect partner for the book I had in mind. It didn't take long for us to decide to join forces to write this book.

But before starting work on this volume, we spent a couple years looking at books on business analytics and predictive modeling. Some of them were really good. But most of them were too intense and deep on the theory and mathematics of statistical algorithms, which are an integral part of this subject. Some people like books that take that tack, but most practitioners—even those in the industry—don't have the deep background in the math required or the interest in learning it. Working professionals, particularly newcomers to the field of business analytics, are not very comfortable with the deep theoretical treatment of statistical algorithms generally provided in most of the books available on analytics. The market need we discerned, therefore, was to simplify the presentation of algorithms for professionals who don't need to know the details to succeed in their work. Besides, once introduced to the subject, one can always refer to the advanced texts on statistics if such academic rigor is required.

The good news is that today's analytics software, like SAS, is designed to do most of the math. Thus, we strongly felt there was a need for a book like this one, which takes the power of the software into account and, at the same time, simplifies the mathematical concepts involved in the process. With this motivation in mind, we started our work and strongly feel we have been successful in showing you how to use SAS to perform common analytical procedures while providing the basic knowledge of statistics required. The book keeps the theoretical part as simple as possible yet uses numerous real business scenarios to explain the concepts and the way they are used in the industry. Venkat's working experience with the world's leading banks and his vast experience working with students as an analytics trainer has come in handy in designing the case studies and examples used in this book.

As usual, readers are the best judge, and we invite you to join us in an active dialogue on this point so that subsequent editions can be improved. Please feel free to reach out to us at shailendrakadre@gmail.com and 21.venkat@gmail.com.

Whenever we look at books on technical subjects, we ask ourselves particular questions to decide whether the book is worth the investment of money, time, and effort. We think you probably do the same, so here are answers to the questions you may be asking.

What Is the Authors' Training and Experience?

Both of us are working professionals in the field of business analytics with years of experience in real-life business problems. We currently work for HP India in Bangalore. We are postgraduates in either engineering or statistics from India's top technology institutes—the Indian Institute of Technology (IIT). This is Shailendra's second book with Apress, after *Going Corporate: A Geek's Guide*, published in 2011.

What Does the Book Cover?

This book is basically a first course in analytics using SAS. You will get the foundational knowledge needed to be successful in the field. It has 13 chapters, each uniquely designed to develop a step-by-step understanding of the subject. Numerous real-life examples are provided with data and SAS code so that you, after completing this book, feel comfortable working on actual business challenges. Here's a rundown of the content in each chapter:

- Chapter 1 starts with the basics of business analytics and presents some use cases to build a background for the upcoming chapters. We cover some of the most widely used analytical techniques. Next we move on to analytical tools including the most widely used analytics tool in technology today: SAS.

- Chapters 2, 3, and 4 impart the basic code-level knowledge you need in order to work with the SAS software. In these chapters, we discuss the basics of SAS, followed by data handling in SAS, and, finally, important SAS functions and procedures. These chapters give you a foundation in the SAS software, which will be required throughout this book, while working with real-life business scenarios on business analytics.

- Chapters 5 and 6 cover the basics of statistics and simple descriptive statistical techniques. They are the foundation-building chapters in statistics.

- Chapter 7 proceeds—since you have a solid foundation on statistics and SAS—to the life-cycle steps in analytics. Any analytical problem solving starts with data exploration, validation, and data cleaning. In this chapter, we document some of the most creative data-cleaning techniques used in the industry.

- Chapter 8 is all about testing. Testing hypotheses is a concept that is linked to many other statistical topics. In this chapter, we explain how to test hypotheses in simple terms.

- Chapter 9 begins the journey into advanced topics. The first steps in predictive modeling are correlation and regression. We explain these concepts with abundant real-life examples.

- Chapter 10 shows that regression is a vast subject; it requires more than one chapter to explain. We discuss multiple regressions in this chapter, as well as concepts such as multicollinearity and adjusted R-square.

- Chapter 11 explains why logistic regression is a commonly used predictive modeling technique. In this chapter, we discuss model building using logistic regression.

- Chapter 12 demonstrates time-series analysis, along with its applications, as well as how to implement the ARIMA technique.

- Chapter 13 is all about big data. Considering the exponential growth and popularity of the big data analytics domain, we felt it was appropriate to end this book with the basics of big data. This final chapter introduces big data with real-time data and examples that demonstrate the code and use of Hadoop.

What Is Not Covered in the Book?

While this book prepares you well in the basics of analytics and SAS, the following topics are left for you to explore in more advanced resources. This list is not an exhaustive one.

- Advanced SAS programing techniques such as macros

- Advanced analytical concepts such as optimization, advanced operation research techniques, machine learning, and decision trees

- Advanced big data concepts such as MapReduce coding

What Are the Unique Features of the Book?

A simple style of presentation and hands-on, real-life examples are two things we are particularly proud of. The following are some unique design features of this book:

- It contains numerous examples right from the first chapter until the end.

- Every concept is explained using a business scenario or case study.

- It offers the right mix of theory and hands-on labs.

- It is written by industry professionals who are currently working in the field of analytics on real-life problems for paying customers.

- It simplifies complex statistical concepts.

- It offers SAS code and a sufficient number of data sets.

- The book is self-sufficient. You don't need any other resource or reference to grasp the concepts.

Does the Book Contain Hands-on Exercises?

Yes! Every chapter has demos and hands-on examples. The data sets used in each case study are available to you through the book's web page on Apress.com. You can download the data and code files to practice.

What Industry Examples Does the Book Cover?

The examples in this book are not limited to any particular domain. We have tried to cover as wide a spectrum of industries as possible. The following are some specifics:

- Analysis of loan defaulter data in the banking domain

- Examples from retailers such as Walmart

- Applications of analytics in social media analytics

- Analysis of call volumes data in a contact center

- Analysis of smartphone sales predictions

- Analysis of associations between customer profile variables

- Forecasting stock prices

- Testing machine quality in a manufacturing plant
- Getting insights into weblog data
- Predicting risk associated with each customer in loan data
- And many more

What Will You Learn in This Book?

This book will give you a solid, hands-on foundation in business analytics and SAS. You can expect to learn the following:

- How to apply analytics to real-life business problems
- How to build analytics models and draw useful inferences from them
- How to write SAS programs and how to handle data in SAS
- How to perform basic descriptive analysis to gain insights from the data
- How to explore, validate, and clean data in a systematic way
- How to find associations between variables
- How to test an assumption and accept or reject it statistically
- How to predict a dependent variable using several independent factors
- How to forecast future values using historical data
- How to find the probability of winning versus losing
- How to handle big data
- And much more

What Are the Prerequisites for Reading This Book?

This book doesn't require a graduate-level degree in math or statistical background. No prior knowledge in statistics is in fact required. Studying math at the high-school level should be adequate to grasp most of the concepts explained in this book. Basic knowledge of computers and spreadsheets is required, though. An elementary knowledge of databases such as Microsoft Access and SQL will also help. A basic knowledge in computer programming is a plus.

Who Is This Book For?

This book is for working professionals, new graduates, students, and analytics enthusiasts. While designing this book, we envisioned our audience as follows:

- Anybody who wants to get hands-on with analytics
- Professionals who want to upgrade their analytical skills
- Anybody who wants to get started with SAS

- Graduate and postgraduate statistics and business analytics students
- Any new college graduate who wants to enter the world of analytics in domains such as retail, banking, insurance, and services
- Lecturers or teachers who want to teach a basic analytics course to students at the graduate or postgraduate level
- Individuals with nonstatistical backgrounds
- Anyone who wants to get a basic idea of big data analytics

Can I Use This Book for an Analytics Course?

You bet! The goal of this book is to impart the concepts rather than just giving information. All the concepts covered are self-contained and don't require any further references. This book can be easily converted into a semester's course as an introduction to SAS, analytics, and predictive modeling. It can be part of any undergraduate or postgraduate program that aims to introduce business analytics to students.

The following is the suggested class plan for a one-semester course. It consists of 30 classroom theory and 12 lab sessions. Course instructors can adapt it to suit their specific requirements.

Chapter	Contents	Theory Classes	Labs
Chapter 1	Introduction to Business Analytics and Data Analysis Tools	2	0
Chapter 2	SAS Introduction	2	1
Chapter 3	Data Handling Using SAS	2	1
Chapter 4	Important SAS Functions and Procs	2	1
Chapter 5	Introduction to Statistical Analysis	2	1
Chapter 6	Basic Descriptive Statistics and Reporting in SAS	3	1
Chapter 7	Data Exploration, Validation, and Data Sanitization	3	1
Chapter 8	Testing a Hypotheses	3	1
Chapter 9	Correlation and Linear Regression	3	1
Chapter 10	Multiple Regression Analysis	2	1
Chapter 11	Logistic Regression	2	1
Chapter 12	Time-Series Analysis and Forecasting	2	1
Chapter 13	Introducing Big Data Analytics	2	1
	Total	30	12

## PART 1

Basics of SAS Programming for Analytics

CHAPTER 1

■ ■ ■

Introduction to Business Analytics and Data Analysis Tools

There is an ever-increasing need for advanced information and decision support systems in today's fierce global competitive environment. The profitability and the overall business can be managed better with access to predictive tools—to predict, even approximately, the market prices of raw materials used in production, for instance. Business analytics involves, among others, quantitative techniques, statistics, information technology (IT), data and analysis tools, and econometrics models. It can positively push business performance beyond executive experience or plain intuition.

Business analytics (or advanced analytics for that matter) can include nonfinancial variables as well, instead of traditional parameters that may be based only on financial performance. Business analytics can effectively help businesses, for example, in detecting credit card fraud, identifying potential customers, analyzing or predicting profitability per customer, helping telecom companies launch the most profitable mobile phone plans, and floating insurance policies that can be targeted to a designated segment of customers. In fact, advanced analytical techniques are already being used effectively in all these fields and many more.

This chapter covers the basics that are required to comprehend all the analytical techniques used in this book.

Business Analytics, the Science of Data-Driven Decision Making

Many analytical techniques are data intensive and require business decision makers to have an understanding of statistical and various other analytical tools. These techniques invariably require some level of IT and database knowledge. Organizations using business analytics techniques in decision making also need to develop and implement a data-driven approach in their day-to-day operations, planning, and strategy making. However, in a large number of cases, businesses have no other choice but to implement a data-driven decision-making approach because of fierce competition and cost-cutting pressures. This makes business analytics a lucrative and rewarding career choice. This may be the right time for you to enter this field because the business analytics culture is still in its nascent stage in most organizations around the world and is on the verge of exploding with respect to growing opportunities.

Business Analytics Defined

Business analytics is all about data, methodologies, IT, applications, mathematical, and statistical techniques and skills required to get new business insights and understand business performance. It uses iterative and methodical exploration of past data to support business decisions.

Business analytics aims to increase profitability, reduce warranty expenditures, acquire new customers, retain customers, upsell or cross-sell, monitor the supply chain, improve operations, or simply reduce the response time to customer complaints, among others. The applications of business analytics are numerous and across industry verticals, including manufacturing, finance, telecom, and retail. The global banking and financial industry traditionally has been one of the most active users of analytics techniques. The typical applications in the finance vertical are detecting credit card fraud, identifying loan defaulters, acquiring new customers, identifying responders to e-mail campaigns, predicting relationship value or profitability of customers, and designing new financial and insurance products. All these processes use a huge amount of data and fairly involved statistical calculations and interpretations.

Any application of business analytics involves a considerable amount of effort in defining the problem and the methodology to solve it, data collection, data cleansing, model building, model validation, and the interpretation of results. It is an iterative process, and the models might need to be built several times before they are finally accepted. Even an established model needs to be revisited/rebuilt periodically for changes in the input data or changes in the business conditions (and assumptions) that were used in the original model building.

Any meaningful decision support system that uses data analytics thus requires development and implementation of a strong data-driven culture within the organization and all the external entities that support it.

Let's take an example of a popular retail web site that aims to promote an upmarket product. To do that, the retail web site wants to know which segment of customers it needs to target to maximize product sales with minimum promotional dollars. To do this, the web site needs to collect and analyze customer data. The web site may also want to know how many customers visited it and at what time; their gender, income bracket, and demographic data; which sections of the web site they visited and in what frequency; their buying and surfing patterns; the web browser they used; the search strings they used to get into the web site; and other such information.

If analyzed properly, this data presents an enormous opportunity to garner useful business insights about customers, thereby providing a chance to cut promotional costs and improve overall sales. Business analytics techniques are capable of working with multiple and a variety of data sources to build the models that can derive rich business insights that were not possible before. This derived rich fact base can be used to improve customer experiences, streamline operations, and thereby improve overall profitability. In the previous example, it is possible, by applying business analytics techniques, to target the product to a segment of customers who are most likely to buy it, thereby minimizing the promotional costs.

Conventional business performance parameters are based mainly on finance-based indicators such as top-line revenue and bottom-line profit. But there is more to the performance of a company than just financial parameters. Measures such as operational efficiency, employee motivation, average employee salary, working conditions, and so on, may be equally important. Hence, the numbers of parameters that are used to measure or predict the performance of a company have been increased here. These parameters will increase the amount of data and the complexity of analyzing it. This is just one example. The sheer volume of data and number of variables that need to be handled in order to analyze consumer behavior on a social media web site, for instance, is immense. In such a situation, conventional wisdom and reporting tools may fail. Advanced analytics predictive modeling techniques help in such instances.

The subsequent chapters in this book will deal with data analytics. Statistical and quantitative techniques used in advanced analytics, along with IT, provide business insights while handling a vast amount of data that was not possible until a few years ago. Today's powerful computing machines and software (such as SAS) take care of all the laborious tasks of analytics algorithms coding and frees the analyst to work on the important tasks of interpretation and applying the results to gain business insights.

Is Advanced Analytics the Solution for You?

Anyone who is in a competitive business environment and faces challenges such as the following, or almost any problem for which data is available, might be a potential candidate for applying advanced analytics techniques:

- Consumer buying pattern analysis
- Improving overall customer satisfaction
- Predicting the lead times in supply chain
- Warrant costs optimization
- Right sizing or the optimization of the sales force
- Price and promotion modeling
- Predicting customer response to sales promotion
- Credit risk analysis
- Fraud identification
- Identifying potential loan defaulters
- Drug discovery
- Clinical data analysis
- Web site analytics
- Text analysis (for instance on Twitter)
- Social media analytics
- Identifying genes responsible for a particular disease

As discussed, business analytics is a culture that needs to be developed, implemented, and finally integrated as a way of life in any organization with regard to decision making. Many organizations around the world have already experienced and realized the potential of this culture and are successfully optimizing their resources by applying these techniques.

EXAMPLE

Trade-offs such as sales volumes versus price points and the costs of carrying inventory versus the chances of stocks not available on demand are always part of day-to-day decision making for managers. Many of these business decisions are highly subjective or based on available data that is not that relevant.

In one such example, a company's analysis found that the driving force of customer sentiments on key social media sites is not its TV commercials but the interaction with the company's call centers. The quality of service provided by the company, and the quality of call center interaction, was largely affecting the brand impact. Based on these insights, the company decided to divert part of the spending on TV commercials toward improving the call center satisfaction levels. The results were clearly visible; customer satisfaction surveys improved considerably, and there was a significant increase in customer base and revenues.

Simulation, Modeling, and Optimization

This section (and the chapter, by and large) explains the terminology and basics to build a background for the coming chapters, which will be more focused and technical in nature.

Simulation

There are various types of simulations. In the context of analytics, computer simulations, an oft-used term, is more relevant. Some real-world systems or scenarios might be complex and difficult to comprehend or predict. Predicting a snow storm or predicting stock prices are classic examples. They depend upon several variables or factors, which are practically impossible to predict. Daily stock prices, for instance, may be affected by current political conditions, major events during the day, international business environment, dollar prices, or simply the overall mood in the market.

There can be various levels of simulations, from simple programs that are a few hundred of lines of code which are complex and millions of lines of code. Computer simulations used in atmospheric sciences are another classic example where complex computer systems and software are used for weather prediction and forecasting.

Computer simulations use various statistical models in analytics.

Modeling

A modeling is merely the mathematical logic and concepts that go into a computer program. These models, along with the associated data, represent the real-world systems. These models can be used to study the effect of different components and predict system behavior. As discussed earlier, the accuracy with which a model represents a real-world system may vary and depends upon the business needs and resources available. For instance, 90 percent accuracy in prediction might be acceptable in banking applications such as the identification of loan defaulters, but in systems that involve human life—for instance, reliability models in aerospace applications—accuracy of 100 percent, or as close to it, is desired.

Optimization

Optimization is a term related to computer simulations. The sole objective of some computer simulations may be simply to ensure optimization, which in simple terms can be explained as minimization or maximization of a mathematical function, subject to a given set of constrains. In optimization problems, a set of variables might need to be selected from a range of available alternatives to minimize or maximize a mathematical function while working with constraints. Although optimization is discussed here in its most simple form, there is much more to it.

An instance of a simple optimization problem is maximizing the working time of a machine, while keeping the maintenance costs below a certain level. If enough data is available, this kind of problem may be solved using advanced analytics techniques. Another instance of a practical optimization problem is chemical process factories, where an engineer may need to adjust a given set of process parameters in order to get maximum output of a chemical reaction plant, while also keeping the costs within budget. Advanced analytical techniques can be an alternative here as well.

Data Warehousing and Data Mining

Creating a data warehouse can be considered one of the most important basics. It can give a jump start to any business analytics project. Consider an example of a multilocational business organization with sales offices and manufacturing plants spread across the country. Today, in almost all large establishments, some amount of business process automation using homegrown or packaged applications such as SAP is expected. Some processes can be local, and their transaction data might be maintained at the branch level. It may not be possible to provide the head office with quarterly sales reports across the products and locations, unless all the relevant data is readily available to the reporting engine. This task is easier if the company links its branch-level data sources and makes them available in a central database. The data may need cleansing and transformations before being loaded in the central database. This is done in order to make the raw data more meaningful for further analysis and reporting. This central database is often called a *data warehouse*.

The previous instance was just one example of a data warehouse. We live in a data age. Terabytes and petabytes of data are being continuously generated by the World Wide Web, sales transactions, product description literatures, hospital records, population surveys, remote-sensing data by satellites, engineering analysis results, multimedia and videos, and voice and data communications networks. The list is endless. The sources of data in a data warehouse can be multiple and heterogeneous. Interesting patterns and useful knowledge can be discovered by analyzing this vast amount of available data. This process of knowledge discovery is termed *data mining* (Figure 1-1). The sources of data for a data mining project may be multiple, such as a single large company-wide data warehouse or a combination of data warehouses, flat files, Internet, commercial information repositories, social media web sites, and several other such sources.

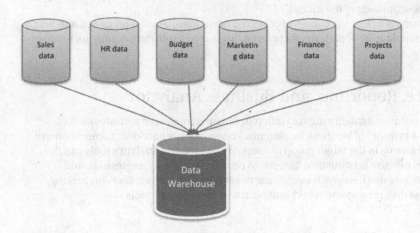

Figure 1-1. Data mining

What Can Be Discovered Using Data Mining?

There are a few defined types of pattern discoveries in data mining. Consider the familiar example of the bank and credit card. Bank managers are sometimes interested only in summaries of a few general features in a target class. For instance, a bank manager might be interested in credit card defaulters who regularly miss payment deadlines by 90 days or more. This kind of abstraction is called *characterization*.

In the same credit card example, the bank manager might want to compare the features of clients who pay on time versus clients who regularly default beyond 90 days. This is a comparative study, termed as *discrimination*, between two target groups.

In yet another type of abstraction called *association analysis*, the same bank manager might be interested in knowing how many new credit card customers also took personal loans. The bank may also be interested in building a model that can be used as a support tool to accept or reject the new credit card applications. For this purpose, the bank might want to classify the clients as "very safe," "medium risky," and "highly risky," as one of the steps. It might be done after a thorough analysis of a large number of client attributes.

The bank might also be interested in predicting which customers can be potential loan defaulters, again based on an established model, which consumes a large number of attributes pertaining to its clients. This is *predictive analysis*.

To open new branches or ATMs, the bank might be interested in knowing the concentration of customers by geographical location. This abstraction, called *clustering*, is similar to classification, but the names of classes and subclasses are not known as the analysis is begun. The class names (the geography names with sizable concentration) are known only after the analysis is complete. While doing a cluster analysis or classification on a given client attribute, there may be some values that do not fit in with any class or cluster. These exclusions or surprises are *outliers*.

Outlier values might not be allowed in some model-building activities because they tend to bias the result in a particular direction, which may not be a true interpretation of the given data set. Such outliers are common while dealing with $ values in data sets. *Deviation analysis* deals with finding the differences between the expected and actual values.

For example, it might be interesting to know the deviation with which a model predicts the credit card loan defaulters. Such an analysis is possible when both the model-predicted values and actual values are present. It is, in fact, periodically done to ascertain the effectiveness of models. If the deviation is not acceptable, it might warrant the rebuilding of the model.

Deviation analysis also attempts to find the causes of observed deviations between predicted and actual values. This is by no means a complete list of patterns that can be discovered using data mining techniques. The scope is much wider.

Business Intelligence, Reporting, and Business Analytics

Business intelligence (BI) and business analytics are two different but interconnected techniques. As reported in one of SAS's blogs, a majority of business intelligence systems aim at providing comprehensive reporting capabilities and dashboards to the target group of users. While business analytics tools can do reporting and dashboards, they can also do statistical analysis to provide forecasting, regression, and modeling. SAS business analytics equips users with everything needed for data-driven decision making, which includes information and data management and statistical and presentation tools.

Analytics Techniques Used in the Industry

The previous few sections introduced the uses of data mining or business analytics. This section will examine the terminology in detail. Only the frequently used terms in the industry are discussed here.

Then the chapter will introduce and give examples of many of these analytics techniques and applications. Some of the more frequently used techniques will be covered in detail in later chapters.

Regression Modeling and Analysis

To understand regression and predictive modeling, consider the same example of a bank trying to aggressively increase its customer base for some of its credit card offerings. The credit card manager wants to attract new customers who will not default on credit card loans. The bank manager might want to build a model from a similar set of past customer data that resembles the set of target customers closely. This model will be used to assign a credit score to the new customers, which in turn will be used to decide whether to issue a credit card to a potential customer. There might be several other considerations aside from the calculated credit score before a final decision is made to allocate the card.

The bank manager might want to view and analyze several variables related to each of the potential clients in order to calculate their credit score, which is dependent on variables such as the customer's age, income group, profession, number of existing loans, and so on. The credit score here is a dependent variable, and other customer variables are independent variables. With the help of past customer data and a set of suitable statistical techniques, the manager will attempt to build a model that will establish a relationship between the dependent variable (the credit score in this case) and a lot of independent variables about the customers, such as monthly income, house and car ownership status, education, current loans already taken, information on existing credit cards, credit score and the past loan default history from the federal data bureaus, and so on. There may be up to 500 such independent variables that are collected from a variety of sources, such as credit card application, federal data, and customers' data and credit history available with the bank. All such variables might not be useful in building the model. The number of independent variables can be reduced to a more manageable number, for instance 50 or less, by applying some empirical and scientific techniques. Once the relationship between independent and dependent variables is established using available data, the model needs validation on a different but similar set of customer data. Then it can be used to predict the credit scores of the potential customers. A prediction accuracy of 90 percent to 95 percent may be considered good in banking and financial applications; an accuracy of 75 percent is must. This kind of predictive model needs a periodic validation and may be rebuilt. It is mandatory in some financial institutions to revalidate the model at least once a year with renewed conditions and data.

In recent times, revenues for new movies depend largely on the buzz created by that movie on social media in its first weekend of release. In an experiment, data for 37 movies was collected. The data was the number of tweets on a movie and the corresponding tickets sold. The graph in Figure 1-2 shows the number of tweets on the x-axis and number of tickets sold on the y-axis for a particular movie. The question to be answered was, If a new movie gets 50,000 tweets (for instance), how many tickets are expected to be sold in the first week? A regression model was used to predict the number of tickets (y) based on number of tweets (x) (Figure 1-3).

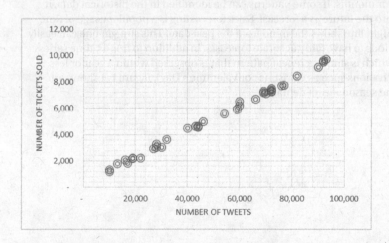

Figure 1-2. *Number of Tickets Sold vs. Number of Tweets—a data collection for sample movies*

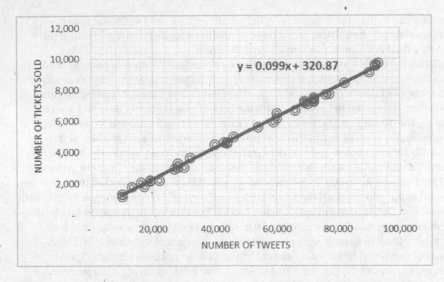

Figure 1-3. *The regression model for Number of Tickets Sold vs. Number of Tweets—prediction using regression model*

Using the previous regression predictive model equation, the number of tickets was estimated to be 5,271 for a movie that had 50,000 tweets in the first week of release.

Time Series Forecasting

Time series forecasting is a simple form of forecasting technique, wherein some data points are available over regular time intervals of days, weeks, or months. If some patterns can be identified in the historical data, it is possible to project those patterns into the future as a forecast. Sales forecasting is a popular usage of time series forecasting. In Figure 1-3, a straight line shows the trend from the past data. This straight line can easily be extended into a few more time periods to have fairly accurate forecasts. In addition to trends, time series forecasts can also show seasonality, which is simply a repeat pattern that is observed within a year or less (such as more sales of gift items on occasions such as Christmas or Valentine's Day). Figure 1-4 shows an actual sales forecast, the trend, and the seasonality of demand.

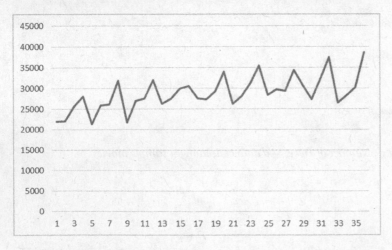

Figure 1-4. *A time series forecast showing the seasonality of demand*

Figure 1-4 shows the average monthly sales of an apparel showroom for three years. There is a stock clearance sale every four months, with huge discounts on all brands. The peak in every fourth month is apparent from the figure.

Time series analysis can also be used in the example of bank and credit card to forecast losses or profits in future, given the same data for a historical period of 24 months, for instance. Time series forecasts are also used in weather and stock market analysis.

Other examples of time series data include representations of yearly sales volumes, mortgage interest rate variations over time, and data representations in statistical quality control measurements such as accuracy of an industrial lathe machine for a period of one month. In these representations, the time component is taken on the x-axis, and the variable, like sales volume, is on the y-axis. Some of these trends may follow a steady straight-line increase or a decline over a period of time. Others may be cyclic or random in nature. While applying time series forecasting techniques, it is usually assumed that the past trend will continue for a reasonable time in the future. This future forecasting of the trend may be useful in many business situations, such as stocks procurement, planning of cash flow, and so on.

Conjoint Analysis

Conjoint analysis is a statistical technique, mostly used in market research, to determine what product (or service), features, or pricing would be attractive to most of the customers in order to affect their buying decision positively.

In conjoint studies, target responders are shown a product with different features and pricing levels. Their preferences, likes, and dislikes are recorded for the alternative product profiles. Researchers then apply statistical techniques to determine the contribution of each of these product features to overall likeability or a potential buying decision. Based on these studies, a marketing model can be made that can estimate the profitability, market share, and potential revenue that can be realized from different product designs, pricing, or their combinations.

It is an established fact that some mobile phones sell more because of their ease of use and other user-friendly features. While designing the user interface of a new phone, for example, a set of target users is shown a carefully controlled set of different phone models, each having some different and unique feature yet very close to each other in terms of the overall functionality. Each user interface may have a different set of background colors; the placement of commonly used functions may also be different for each phone. Some phones might also offer unique features such as dual SIM. The responders are then asked to rate the models and the controlled set of functionalities available in each variation. Based on a conjoint analysis of this data, it may be possible to decide which features will be received well in the marketplace. The analysis may also help determine the price points of the new model in various markets across the globe.

Cluster Analysis

The intent of any cluster analysis exercise is to split the existing data or observations into similar and discrete groups. Each observation is divided groupwise in classification type of problems, while in cluster analysis, the aim is to determine the number and composition of groups that may exist in a given data or observation set.

For example, the customers could be grouped into some distinct groups in order to target them with different pricing strategies and customized products and services. These distinct customer groups (Figure 1-5) may include frequent customers, occasional customers, high net worth customers, and so on. The number of such groups is unknown when beginning the analysis but is determined as a result of analysis.

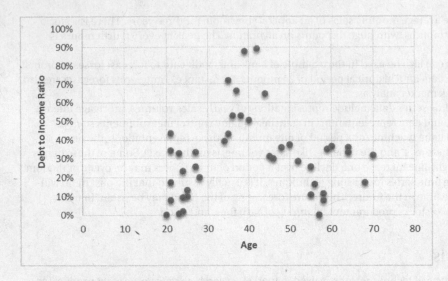

Figure 1-5. *A cluster analysis plot*

The graph in Figure 1-6 shows the debt to income ratio versus age. Customer segments that are similar in nature can be identified using cluster analysis.

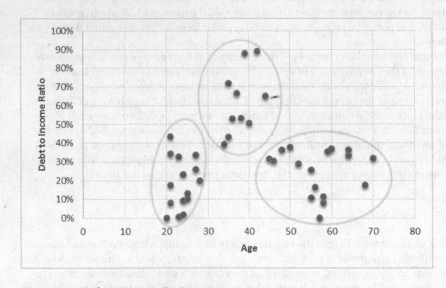

Figure 1-6. *Debt to Income Ratio vs. Age*

The debt to income ratio in Figure 1-6 is low for age groups 20 to 30. The 30-to-45 age group segment has a higher debt to income ratio. The three groups need to be treated differently instead of as one single population, depending on the business objective.

Segmentation

Segmentation is similar to classification, where the criteria to divide observations into distinct groups needs to be found. The number of groups may be apparent even at the beginning of the analysis, while the aim of cluster analysis is to identify areas with concentrations different than other groups. Hence, clustering is discovering the presence of boundaries between different groups, while segmentation uses boundaries or some distinct criterion to form the groups.

Clustering is about dividing the population into different groups based on all the factors available. Segmentation is also dividing the population into different groups but based on predefined criteria such as maximizing the profit variable, minimizing the defects, and so on. Segmentation is widely used in marketing to create the right campaign for the customer segment that yields maximum leads.

Principal Components and Factor Analysis

These statistical methodologies are used to reduce the number of variables or dimensions in a model building exercise. These are usually independent variables. Principal component analysis is a method of combining a large number of variables into a small number of subsets, while factor analysis is a methodology used to determine the structure or underlying relationship by calculating the hidden factors that determine the variable relationships.

Some analysis studies may start with a large number of variables, but because of practical constraints such as data handling, data collection time, budgets, computing resources available, and so on, it may be necessary to drastically reduce the number of variables that will appear on the final data model. Only those independent variables that make most sense to the business need to be retained.

There might also be interdependency between some variables. For example, income levels of individuals in a typical analysis might be closely related to the monthly dollars they spend. The more the income, the more the monthly spend. In such a case, it is better to keep only one variable for the analysis and remove the monthly spend from the final analysis.

The regression modeling section discussed using 500 variables as a starting point to determine the credit score of potential customers. The principal component analysis can be one of the methods to reduce the number of variables to a manageable level of 40 variables (for example), which will finally appear in the final data model.

Correspondence Analysis

Correspondence analysis is similar to principal component analysis but applies to nonquantitative or categorical data such as gender, status of pass or fail, color of objects, and field of specialization. It especially applies to cross-tabulation. Correspondence analysis provides a way to graphically represent the structure of cross-tabulations with each row and column represented as a point.

Survival Analytics

Survival analytics is typically used when variables such as time of death, duration of a hospital stay, and time to complete a doctoral thesis need to be predicted. It basically deals with the time to event data. For a more detailed treatment of this topic, please refer to www.amstat.org/chapters/northeasternillinois/pastevents/presentations/summer05_Ibrahim_J.pdf.

Some Practical Applications of Business Analytics

The following sections discuss a couple of examples on the practical usage of application of business analytics in the real world. Predicting customer behavior towards some product features, or application of business analytics in the supply chain to predict the constraints, such as raw material lead times, are very common examples. Applications of analytics are very popular in retail and predicting trends on social media as well.

Customer Analytics

Predicting consumer behavior is the key to all marketing campaigns. Market segmentation, customer relationship management, pricing, product design, promotion, and direct marketing campaigns can benefit to a large extent if consumer behavior can be predicted with reasonable accuracy. Companies with direct interaction with customers collect and analyze a lot of consumer-related data to get valuable business insights that may be useful in positively affecting sales and marketing strategies. Retailers such as Amazon and Walmart have a vast amount of transactional data available at their disposal, and it contains information about every product and customer on a daily basis. These companies use business analytics techniques effectively for marketing, pricing policies, and campaign designs, which enable them to reach the right customers with the right products. They understand customer needs better using analytics. They can swap better-selling products at the cost of less-efficient ones. Many companies are also tapping the power of social media to get meaningful data, which can be used to analyze consumer behavior. The results of this analytics can also be used to design more personalized direct marketing campaigns.

Operational Analytics

Several companies use operational analytics to improve existing operations. It is now possible to look into business processes in real time for any given time frame, with companies having enterprise resource planning (ERP) systems such as SAP, which give an integrated operational view of the business. Drilling down into history to re-create the events is also possible. With proper analytics tools, this data is used to analyze root cases, uncover trends, and prevent disasters. Operational analytics can be used to predict lead times of shipments and other constraints in supply chains. Some software can present a graphical view of supply chain, which can depict any possible constraints in events such as shipments and production delays.

Social Media Analytics

Millions of consumers use social media at any given time. Whenever a new mobile phone or a movie, for instance, is launched in the market, millions of people tweet about it almost instantly, write about their feelings on Facebook, and give their opinions in the numerous blogs on the World Wide Web. This data, if tapped properly, can be an important source to uncover user attitudes, sentiments, opinions, and trends. Online reputation and future revenue predictions for brands, products, and effectiveness of ad campaigns can be determined by applying proper analytical techniques on these instant, vast, and valuable sources of data. In fact, many players in the analytics software market such as IBM and SAS claim to have products to achieve this.

Social media analytics is simply text mining or text analytics in some sense. Unstructured text data is available on social media web sites, which can be challenging to analyze using traditional analytics techniques. (Describing text analytics techniques is out of scope for this book.)

Some companies are now using consumer sentiment analysis on key social media web sites such as Twitter and Facebook to predict revenues from new movie launches or any new products introduced in the market.

Data Used in Analytics

The data used in analytics can be broadly divided into two types: qualitative and quantitative. The qualitative, discrete, or categorical data is expressed in terms of natural languages. Color, days of a week, street name, city name, and so on, fall under this type of data. Measurements that are explained with the help of numbers are quantitative data, or a continuous form of data. Distance between two cities expressed in miles, height of a person measured in inches, and so on, are forms of continuous data.

This data can come from a variety of sources that can be internal or external. Internal sources include customer transactions, company databases and data warehouses, e-mails, product databases, and the like. External data sources can be professional credit bureaus, federal databases, and other commercially available databases. In some cases, such as engineering analysis, a company may like developing its own data to solve an uncommon problem.

Selecting the data for building a business analytics problem requires a thorough understanding of the overall business and the problem to be resolved. The past sections discussed that an analytics model uses data combined with the statistical techniques used to analyze it. Hence, the accuracy of any model is largely dependent upon the quality of underlying data and statistical methods used to analyze it.

Obtaining data in a usable format is the first step in any model-building process. You need to first understand the format and content of the raw data made available for building a model. Raw data may require extraction from its base sources such as a flat file or a data warehouse. It may be available in multiple sources and in a variety of formats. The format of the raw data may warrant separation of desired field values, which otherwise appear to be junk or have little meaning in its raw form. The data may require a cleansing step as well, before an attempt is made to process it further. For example, a gender field may have only two values of male and female. Any other value in this field may be considered as junk. However, it may vary depending upon the application. In the same way, a negative value in an amounts field may not be acceptable.

In some cases, the size of available data may be so large that it may require sampling to reduce it to a manageable form for analysis purposes. A sample is a subset from the available data, which for all practical purposes represents all the characteristics of the original population. The data sourcing, extraction, transformation, and cleansing may eat up to 70 percent of total hours made available to a business analytics project.

Big Data vs. Conventional Business Analytics

Conventional analytical tools and techniques are inadequate to handle data that is unstructured (like text data), that is too large in size, or that is growing rapidly like social media data. A cluster analysis on a 200MB file with 1 million customer records is manageable, but the same cluster analysis on 1000GB of Facebook customer profile information will take a considerable amount of time if conventional tools and techniques are used. Facebook as well as entities like Google and Walmart generate data in petabytes every day. Distributed computing methodologies might need to be used in order to carry out such analysis.

Introduction to Big Data

The SAS web site defines *big data* as follows:

> *Big data is a popular term used to describe the exponential growth and availability of data, both structured and unstructured. And big data may be as important to business— and society—as the Internet has become. Why? More data may lead to more accurate analyses.*

15

It further states, "More accurate analyses may lead to more confident decision making. And better decisions can mean greater operational efficiencies, cost reductions, and reduced risk."

This definition refers to big data as a popular term that is used to describe data sets of large volumes that are beyond the limits of commonly used desktop database and analytical applications. Sometimes even server-class databases and analytical applications fail to manage this kind of data set.

Wikipedia describes big data sizes as a constantly moving target, ranging from a few dozen terabytes to some petabytes (as in 2012) in a single data set. The size of big data may vary from one organization to the other, depending on the capabilities of software that are commonly used to process the data set in its domain. For some organizations, only a few hundred gigabytes of data may require reconsideration using their data processing and analysis systems, while some may feel quite at home with even hundreds of terabytes of data.

CONSIDER A FEW EXAMPLES

Cern's Large Hydron Collider experiments deal with 500 quintillion bytes of data per day, which is 200 times more than all other sources combined in the world.

In Business

eBay.com uses a 40 petabytes Hadoop cluster to support its merchandising, search, and consumer recommendations.

Amazon.com deals with some of the world's largest Linux databases, which measure up to 24 terabytes.

Walmart's daily consumer transactions run into 1 million per hour. Its databases sizes are estimated to be 2.5 petabytes.

All this, of course, is extremely large data. It is almost impossible for conventional database and business applications to handle it.

The industry has two more definitions for big data.

- Big data is a collection of data sets so large and complex that it becomes difficult to process using on-hand database management tools or traditional data processing applications.

- Big data is the data whose scale, diversity, and complexity requires new architecture, techniques, algorithms, and analytics to manage it and extract value and hidden knowledge from it.

In simple terms, big data cannot be handled by conventional data-handling tools, and big data analysis cannot be performed using conventional analytical tools. Big data tools that use distributed computing techniques are needed.

Big Data Is Not Just About Size

Gartner defines the three v's of big data as volume, velocity, and variety. So far, only the volume aspect of big data has been discussed. In this context, the speed with which the data is getting created is also important. Consider the familiar example of the Cern Hydron Collider experiments; it annually generates 150 million petabytes of data, which is about 1.36EB (1EB = 1073741824GB) per day. Per-hour transactions for Walmart are more than 1 million.

The third v is variety. This dimension refers to the type of formats in which the data gets generated. It can be structured, numeric or non-numeric, text, e-mail, customer transactions, audio, and video, to name just a few.

In addition to these three v's, some like to include veracity while defining big data. Veracity includes the biases, noise, and deviation that is inherent in most big data sets. It is more common to the data generated from social media web sites. The SAS web site also counts on data complexity as one of the factors for defining big data.

Gartner's definition of the three v's has almost become an industry standard when it comes to defining big data.

Sources of Big Data

Some of the big data sources have already been discussed in the earlier sections. Advanced science studies in environmental sciences, genomics, microbiology, quantum physics, and so on, are the sources of data sets that may be classified in the category of big data. Scientists are often struck by the sheer volume of data sets they need to analyze for their research work. They need to continuously innovate ways and means to store, process, and analyze such data.

Daily customer transactions with retailers such as Amazon, Walmart, and eBay also generate large volumes of data at amazing rates. This kind of data mainly falls under the category of structured data. Unstructured text data such as product descriptions, book reviews, and so on, is also involved. Healthcare systems also add hundreds of terabytes of data to data centers annually in the form of patient records and case documentations. Global consumer transactions processed daily by credit card companies such as Visa, American Express, and MasterCard may also be classified as sources of big data.

The United States and other governments also are big sources of data generation. They need the power of some of the world's most powerful supercomputers to meaningfully process the data in reasonable time frames. Research projects in fields such as economics and population studies, conducted by the World Bank, UN, and IMF, also consume large amounts of data.

More recently, social media sites such as Facebook, Twitter, and LinkedIn are presenting some great opportunities in the field of big data analysis. These sites are now among some of the biggest data generation sources in the world. They are mainly the sources of unstructured data. Data forms included here are text data such as customer responses, conversations, messages, and so on. Lots of other data sources such as audio clips, numerous videos, and images are also included. Their databases are hundreds of petabytes. This data, although difficult to analyze, presents immense opportunities to generate useful insights and information such as product promotion, trend and sentiment analysis, brand management, online reputation management for political outfits and individuals, to name a few. Social media analytics is a rapidly growing field, and several startups and established companies are devoting considerable time and energies to this practice. Table 1-1 compares big data to conventional data.

Table 1-1. *Big Data vs. Conventional Data*

Big Data	Normal or Conventional Data
Huge data sets.	Data set size in control.
Unstructured data such as text, video, and audio.	Normally structured data such as numbers and categories, but it can take other forms as well.
Hard-to-perform queries and analysis.	Relatively easy-to-perform queries and analysis.
Needs a new methodology for analysis.	Data analysis can be achieved by using conventional methods.
Need tools such as Hadoop, Hive, Hbase, Pig, Sqoop, and so on.	Tools such as SQL, SAS, R, and Excel alone may be sufficient.
Raw transactional data.	The aggregated or sampled or filtered data.
Used for reporting, basic analysis, and text mining. Advanced analytics is only in a starting stage in big data.	Used for reporting, advanced analysis, and predictive modeling .
Big data analysis needs both programming skills (such as Java) and analytical skills to perform analysis.	Analytical skills are sufficient for conventional data; advanced analysis tools don't require expert programing skills.
Petabytes/exabytes of data.	Megabytes/gigabytes of data.
Millions/billions of accounts.	Thousands/millions of accounts.
Billions/trillions of transactions.	Millions of transactions.
Generated by big financial institutions, Facebook, Google, Amazon, eBay, Walmart, and so on.	Generated by small enterprises and small banks.

Big Data Use Cases

Big Data is a buzzword today. Every organization wants to use it in the decision-making process. Provided next are just a few examples to give you a feel for its application.

Big Data in Retail

This case is centered on an imaginary large retail chain The retail chain decides to use big data to its advantage to predict trends and to prepare for future demands. The work begins several months before the holiday shopping season begins. The need is to predict the thrust segments and determine the hot items for the season. For analysis, the enterprise data is combined with other relevant information such as industry advertising, web browsing patterns, and social media sentiments. Big data analytics makes use of predictive modeling and generates a future demand pattern for various consumer items. Location-wise demand prediction is also done. Delivery channels such as brick-and-mortar shops and online web sites are also determined, and the delivery of various items is done as per the predicted values. Big data price optimization models are used to determine the right price levels. This may vary from one location to the other.

Big data analytics on the customer demand data combined with current price levels, inventories, and competition data helps the retailer to adjust the prices in real time as per demand and other market dynamics. The expected buying pattern predictions are used to pinpoint customers who are likely to buy a hot-pick item this season. When the customer is at the retail store, they can be contacted using text messaging or e-mail and can be presented with a real-time promotional offer that may prompt a buying decision.

This promotional offer can be in the form of a discount on the purchase value of a product, a gift card, or, for instance, a $10 discount on future purchases. Finally, the retailer might be interested in knowing what other items can be bundled with the item of interest or what other items the client might be interested in purchasing with this item. The retailers are using big data analytics to do the following:

- Demand forecasting
- Determining the right products for promotion
- Anticipating the right delivery channels for different products
- Determining the right locations and availability of the right products at these locations
- Managing inventory
- Knowing the customer to target
- Deciding upon real-time promotional offers
- Performing real-time pricing that depends upon the most current market dynamics

■ **Note**   The techniques listed here may be common in big data and conventional analytics. The only difference is the volume of data. For phenomenally large volumes of data, when you apply these analytical techniques, they are sometimes termed *big data analytics*.

Big Data Use Cases in Financial Industry

Consider this interesting case from the banking industry [4]. The case is similar to the retail use case discussed earlier. The bank analyzes data such as customer spending patterns, recent transaction history, credit information, and social media. The recent transaction history indicates that a customer has just bought a new car. Her social media analysis says that she is an adventure lover. The bank can now consider sending a text message to her immediately, while she is still at the automobile shop, extending an offer of free credit of $5,000 for any purchases made in the next 24 hours. Data analysis has already indicated to the bank her income levels, her available credit, and the costs of adventure accessories such as fog lamps. Based on the big data insights, the bank already knows the financial risk factors involved in making this offer. If that is not enough, the bank may also make an offer of 10 percent cash back on a lunch in the Chinese restaurant of her choice in the nearby locality, based on her food and restaurant choices in her social media interactions. Is this not delightful for the customer and a new revenue stream for the bank?

Consider one more simple case of detecting credit card fraud transactions immediately in collaboration with one's mobile phone operator [5]. The operator is always informed about the client's approximate current location, and if that data shows that a client is somewhere in the United States and his credit card registers a transaction somewhere in the United Kingdom, it indicates a high possibility of fraudulent transaction. Given that both the credit card company and the mobile phone operator have millions of customers and the number of real-time transactions may be even higher, this case may be classified under big data analytics.

Introduction to Data Analysis Tools

This section will discuss some of the more commonly used business analytics tools that are used in the industry today. It is not enough for a data analyst to learn about just one tool. They need to apply different tools as per the situation or what the problem at hand demands. A general knowledge of the strengths and weaknesses of the tools will definitely add value to a data analytics career.

This section will discuss the features of and give resources for further information for three industry-leading tools: SAS, R, and SPSS.

Business analytics aims to model the data in order to discover useful information or business insights and to support decision making. It requires various operations on the data, such as reviewing, cleansing, transforming and modeling, validations, and interpretations to gain useful business insights. Sometimes the sets of data may have a million records or more. Handling and operating such complex data requires automated analysis tools. Fortunately, many such good tools are available today. A simple Google search for *data analysis tools* will give you a list of a number of such tools. Many of them are open source and free for use. SAS, SPSS, and R are the most widely used software packages today, at least for business analytics applications. R is the most popular and widely used statistical analysis package in the open source category, and SAS and SPSS are the two most widely used data analysis packages that are commercially available.

The SAS tool has been around since the 1970s. There are so many legacies built using this tool that most of the companies in the corporate world that are into business analytics at any level continue to use SAS. R was introduced in 1996. Over the years, a lot of new capabilities and features have been built around R. It is one of the most powerful open source data analysis tools available today. This makes it popular in the research and academic community. Many companies from the corporate world have also started using R. SPSS has also existed for more than 20 years now. It has a strong user base in the social sciences and many other communities.

Commonly Used Data Analysis Software

In the following sections, we talk about some commonly used data analysis software and how to make a choice. SAS, SPSS, and R may be termed as the most commonly used software in the industry.

SAS

- Most widely used commercial advanced analytics tool
- Has lot of predictive modeling algorithms

SPSS

- Has good text mining and data mining algorithms

R

- Most widely used open source analytics tool
- Has several packages for data analysis

MATLAB

- Widely used for numerical analysis and computing

RapidMiner

- Good GUI-based tool for segmentation and clustering; can also be used for conventional modeling
- Open source

Weka

- Open source
- Machine learning tool

SAP

- Tool for managing business operations and customer relations
- Most widely used operations tracking tool

Minitab

- A light version analytics tool

Apache Mahout

- Open source
- Advanced analytics tool for big data

Other Tools

- Statistica
- KXEN Modeler
- GNU Octave
- Knime

Choosing a Tool

The final choice of data analysis tool to be used depends upon many considerations.

- The business application, its complexity, and the level of expertise available in the organization.
- The long-term business, information, and analytics strategy of the organization.
- Existing organizational processes.
- Budgetary constraints.
- The investments proposed or already done in the processing hardware systems, which in turn might decide on factors such as processing power and memory available for the software to run.

- Overall organization structure and where the analytics work is being carried out.

- Project environment and governance structure.

- Comfort level of the company in using open source software and warranties and other legal considerations.

- The size of data to be handled.

- The sophistication of graphics and presentation required in the project.

- What analytics techniques to be used and how frequently they will be used.

- How the current data is organized and how comfortable the team is in handling data.

- Whether a data warehouse is in place and how adequately it covers business and customer information that may be required for the analysis.

- Legacy considerations. Is any other similar tool in use already? If yes, how much time and resources are required for any planned switch-over?

- Return-on-investment expectations.

Many more considerations specific to an organization or a project can be added to this list. The order of importance of these considerations may vary from person to person, from project to project, and from one organization to another. The final decision, however, is not an easy one. The later sections of this chapter will list a few comparative features of SAS, SPSS, and R, which might help the decision-making process on the choice of tool that will best suit your needs. Finally, instead of zeroing in on a single tool, deciding to use multiple tools for different business analytics needs is also possible.

In some cases, it might be concluded that a simple spreadsheet application tool, such as Microsoft Excel, is the most convenient and effective and yet gives sufficient insights required to solve the business problem in hand.

Sometimes a single analytics project might require the use of more than one tool. A data analyst will be expected to apply different software tools depending on the problem at hand.

Main Parts of SAS, SPSS, and R

SAS and SPSS have hundreds of functions and procedures and can be broadly divided into five parts.

- Data management and input functions, which help to read, transform, and organize the data prior to the analysis

- Data analysis procedures, which help in the actual statistical analysis

- SAS's output delivery system (ODS) and SPSS's output management system (OMS), which help to extract the output data for final representation or to be used by another procedures as inputs

- Macro languages, which can be used to give sets of commands repeatedly and to conduct programming-related tasks

- Matrix languages (SAS IML and SPSS Matrix), which can be used to add new algorithms

R has all these five areas integrated into one. Most of the R procedures are written using the R language, while SAS and SPSS do not use their native languages to write their procedures. Being open source, R's procedures are available for the users to see and edit to their own advantage.

SAS

As per the SAS web site, the SAS suite of business analytics software has 35+ years of experience and 60,000+ customer sites worldwide. It has the largest market share globally with regard to advanced analytics. It can do practically everything related to advanced analytics, business intelligence, data management, and predictive analytics. It is therefore not strange that the entire book is dedicated to explaining the applications of SAS in advanced business analytics.

SAS development originally started at North Carolina State University, where it was developed from 1966 to 1976. The SAS Institute, founded in 1976, owns this software worldwide. Since 1976, new modules and functionalities have been being added in the core software. The social media analytics module was added in 2010.

The SAS software is overall huge and has more than 200 components. Some of the interesting components are the following:

- *Base SAS*: Basic procedures and data management

- *SAS/STAT*: Statistical analysis

- *SAS/GRAPH*: Graphics and presentation

- *SAS/ETS*: Econometrics and Time Series Analysis

- *SAS/IML*: Interactive matrix language

- *SAS/INSIGHT*: Data mining

- *Enterprise Miner*: Data mining

Analysis Using SAS: The Basic Elements

This section will concentrate on Base SAS procedures. Base SAS helps to read, extract, transform, manage, and do statistical analysis on almost all forms of data. This data can be from a variety of sources such as Excel, flat files, relational databases, and the Internet. SAS provides a point-and-click graphical user interface to perform statistical analysis of data. This option is easy to use and may be useful to nontechnical users or as a means to do a quick analysis. SAS also provides its own programming language, called the SAS programming language. This option provides everything that the GUI has, in addition to several advanced operations and analysis. Many professional SAS users prefer using only the programming option because it gives almost unlimited control to the user on data manipulation, analysis, and presentation.

Most SAS programs have a DATA step and a PROC step. The DATA step is used for retrieval and manipulation of data, while the PROC step contains code for data analysis and reporting. There are approximately 300 PROC procedures. SAS also provides a macro language that can be used to perform routine programming tasks, including repetitive calls to SAS procedures. In the earlier system, SAS provided an ODS, and by using it, SAS data could be published in many commonly used file formats such as Excel, PDF, and HTML. Many of the SAS procedures have the advantage of long history, a wide user base, and excellent documentation.

The Main Advantage Offered by SAS

The SAS programming language is a high-level procedural programming language that offers a plethora of built-in statistical and mathematical functions. It also offers both linear and nonlinear graphics capabilities with advanced reporting features. It is possible to manipulate and conveniently handle the data using SAS programming language, prior to applying statistical techniques. The data manipulation capabilities offered by SAS become even more important because up to three-fourths of the time spent in most analytics project is on data extraction, transformation, and cleaning. This capability is nonexistent in some other popular data analysis packages, which may require data to be manipulated or transformed using several other programs before it can be submitted to the actual statistical analysis procedures. Some statistical techniques such as analysis of variance (ANOVA) procedures are especially strong in the SAS environment.

Listing 1-1 and Listing 1-2 are samples of SAS code. They are just to give you a feel of how SAS code generally looks. More detailed treatments of writing SAS code will be covered later in this book.

Listing 1-1. Regression SAS Code

```
Proc reg data=sales;
Model bill_amount=income  Average_spending  family_members Age;
Run;
```

Listing 1-2. Cluster Analysis Code

```
Proc fastclus  data= sup_market radius=0 replace=full maxclusters = 5 ;
id cust_id;
Var visitsincome age spends;
run;
```

The R Tool

Discussed in the earlier sections, R is an integrated tool for data manipulation, data management, data analysis, and graphics and reporting capabilities. It can do all of the following in an integrated environment:

- Data management functions such as extraction, handling, manipulation and transformation, storage

- The full function and object-oriented R programming language

- Statistical analysis procedures

- Graphics and advanced reporting capabilities

R is open source software maintained by the R Development Core Team and a vast R community (www.r-project.org). It is supported by a large number of packages, which makes it feature rich for the analytics community. About 25 statistical packages are supplied with the core R software as standard and recommended packages. Many more are made available from the CRAN web site at http://CRAN.R-project.org and from other sources. The CRAN site at http://cran.r-project.org/doc/manuals/R-intro.html#Top offers a good resource for an R introduction, including documentation resources.

R's extensibility is one of its biggest advantages. Thousands of packages are available as extensions to the core R software. Developers can see the code behind R procedures and modify it to write their own packages. Most popular programming languages such as C++, Java, and Python can be connected to the R environment. SPSS has a link to R for users who are primarily using the SPSS environment for data analysis. SAS also offers some means to move the data and graphics between the two packages. Table 1-2 lists the most widely used R packages (see http://piccolboni.info/2012/05/essential-r-packages.html).

Table 1-2. *Most Widely Used R Packages*

Rank	Package	Description
1	Stats	Distributions and other basic statistical stuff
2	Methods	Object-oriented programming
3	graphics	Of course, graphics
4	MASS	Supporting material for *Modern Applied Statistics with S*
5	grDevices	Graphical devices
6	utils	In a snub to modularity, a little bit of everything, but very useful
7	lattice	Graphics
8	grid	More graphics
9	Matrix	Matrices
10	mvtnorm	Multivariate normal and t distributions
11	sp	Spatial data
12	tcltk	GUI development
13	splines	Needless to say, splines
14	nlme	Mixed-effects models
15	survival	Survival analysis
16	cluster	Clustering
17	R.methodsS3	Object-oriented programming
18	coda	MCMC
19	igraph	Graphs (the combinatorial objects)
20	akima	Interpolation of irregularly spaced data
21	rgl	3D graphics (openGL)
22	rJava	Interface with Java
23	RColorBrewer	Palette generations
24	ape	Phylogenetics
25	gtools	Functions that didn't fit anywhere else, including macros
26	nnet	Neural networks
27	quadprog	Quadratic programming
28	boot	Bootstrap
29	Hmisc	Yet another miscellaneous package
30	car	Companion to the Applied Regression book
31	lme4	Linear mixed-effects models
32	foreign	Data compatibility
33	Rcpp	R C++ integration

Here are a few R code snippets. It is not necessary to understand them at this stage. The next three chapters are devoted to SAS programming.

```
Input_data=read.csv("Datasets/Insurance_data.csv")
#reads an external CSV file

input_data_final=Input_data[,-c(1)]
#stores the variables of the dataset separately

input_data_final=scale(input_data_final)
#normalizes the data

clusters_five<-kmeans(input_data_final,5)
#creates 5 clusters from the given data

cluser_summary_mean_five=aggregate(input_data_final,by=list(clusters_five$cluster),FUN=mean)
#summarizes clusters by mean

View(cluser_summary_count_five)
#returns the results summarized by size
```

IBM SPSS Analytics Tool

SPSS originally stood for Statistical Package for the Social Sciences. It is a software package used for statistical analysis, originally developed by SPSS Inc. It was acquired by IBM in 2009, and IBM renamed it as SPSS Statistics, with a latest version of SPSS Statistics 22.

Many SPSS users think it has a stronger command menu option compared to R and SAS; its learning curve is also shorter.

The web site at http://fmwww.bc.edu/GStat/docs/StataVSPSS.html has the following opinion about SPSS:

> *SPSS has its roots in the social sciences and the analysis of questionnaires and surveys is where many of its core strengths lie.*

SPSS has been in existence for a long time and hence has a strong user base. Like with any other software, you always have to do a cost-to-benefit analysis while making a buying decision.

Users may find SAS and SPSS similar to each other, and switching from one to the other may be fairly easy. R may look somewhat different for first-time users.

Features of SPSS Statistics 22

SPSS Statistics 22 is built on the philosophy of data-driven decision making anytime, anywhere. It has many new features, such as interaction with mobile devices. It works on Windows, Mac, and Linux desktops. For mobile devices, it supports Apple, Windows 8, and Android devices. It has support for Automatic Linear Modeling (ALM) and heat maps. It enhances the Monte Carlo simulation to help in improving the accuracy of predictive models for uncertain inputs. SPSS Statistics Server is good as far as scalability and performance is concerned. Custom programming is also made easier than before. Python plug-ins can be added as a part of the main installation.

Monte Carlo simulation is a problem-solving technique that is used for approximating certain results by doing multiple trials or simulations that use random variables.

Selection of Analytics Tools

The web site at http://stanfordphd.com/Statistical_Software.html contains a statistical feature comparison for R, Minitab, SAS, STATA, and SPSS. R looks feature-rich given its supporting packages, which are written by the R core development team and many other R enthusiasts. SAS has been around since the 1970s and has a large user base. It has great data management capabilities, which make it a one-stop shop for most of the analytics exercises. The user does not need to go to any other program whether for reading the data files or for the final presentation.

SPSS has great menu-driven features and does not need any training in programming in most cases. The SAS market share and its wide appeal make it the main topic of this book. Several companies in the corporate world find themselves comfortable with SAS, mainly because of the large legacy built around it over the years, its industrial quality code, its good quality of documentation, and the skill availability in the market.

As discussed in this chapter, the fact that all available applications have their own strengths and weaknesses needs to be accepted. No software is fit for all occasions. A data analyst must learn multiple software products and use them as the situation demands to be successful.

The Background Required for a Successful Career in Business Analytics

In this section, we'll talk about what is required if you want to be a data scientist. The main requirement is a love of numbers. A mathematics background at the graduate level usually helps. A good statistics or quantitative techniques course at the graduate level may also be a good start. Candidates with a degree in statistics usually find business analytics to be a natural career choice. An experience working with basic computer programming, databases, and SQL will serve as a solid foundation. You can also start with business intelligence or reporting job and gradually migrate into full-fledged analytics. Knowledge of spreadsheet packages, such as Microsoft Excel, and basic databases, such as Microsoft Access, usually help. Some formal courses in advanced statistics help. With the advent of big data, frameworks such as Hadoop are also gaining importance.

Skills Required for a Business Analytics Professional

The following are the skills required to be successful in the field of business analytics. As you rise in your company's hierarchy, the soft skills take prominence over hardcore technical skills. (See www.villanovau.com/business-analyst-skills for more information.)

Communication Skills

A business analyst needs to interact with users, senior management, technical staff, and clients in order to understand project requirements, business objectives, methodologies to be followed, testing, and other project-related activities. They need to thoroughly understand the business and the underlying problems to be solved. The analysis results are often presented to senior management and other project stakeholders using sophisticated Microsoft PowerPoint presentations and other reporting tools. A high level of verbal and written language skills is necessary to succeed in this career.

Technical Skills

Data analyst use sophisticated statistical packages, reporting tools, presentation, and testing software as part of their day-to-day work life. Mastering a single tool or software is usually not enough; a combination of software may be required depending upon the nature of the problem. Sometimes a plain programming language such as C is used to do the complex operations on the data, prior to building the final model. Multiple databases may be involved in the project, and the analyst may need to extract and process the data from these multiple databases. A variety of reporting and presentation tools may be used to effectively present the project results. Hence, a strong technical aptitude and IT skills are required to succeed.

Analytical Skills

The job here is to analyze large and complex data sets, business documents, surveys, business, and operational requirements. A business analyst needs to develop a high level of understanding of the business, the problem to be solved, analysis methodologies, and presentation requirements. Analysts may need to suggest the prescriptive measures to correct a business problem. Strong analytical aptitude is vital to success in this job.

Problem-Solving Skills

It may be easy to conclude how important the problem-solving skills are for a business analyst, based on the chapter so far. During the course of the project, and even otherwise, an analyst faces a variety of challenges almost every day.

Decision-Making Skills

Business analysts need to take inputs from various sources and stakeholders on complex business matters and then use sound judgment to decide the right course of action. An analyst may need to advise management, developers, and clients on the business, and some of those decisions may have a long-lasting effect on the business, affecting billions of dollars of business. An ability to make right decisions based on correct inputs and methodologies thus is an important skill for a business analyst.

Managerial Skills

Analytics problem solving is often done in the form of projects. An analyst needs to manage the project and direct the staff members in day-to-day activities. All the projects need to be managed within the constraints of time, cost, and quality. It takes high degrees of project management and managerial skills. It involves being skilled in negotiation and persuasion as well.

Conclusion

In this chapter we were introduced to analytics and applications of business analytics. We also discussed the most widely used analytical tools. In future chapters we will get into SAS programming and analytical concepts. Upcoming chapters will have more hands-on exercises and case studies.

CHAPTER 2

■ ■ ■

SAS Introduction

Chapter 1 introduced SAS, the most widely used tool in the world of analytics. SAS is a software suite that can retrieve data from a variety of data sources. It can help you clean the data and perform statistical operations on it. For nontechnical users, it also provides a graphical user interface (GUI) to perform various analytics operations. The soul of SAS is its programming language, which is used by most analysts. It provides more advanced data handling and analytical capabilities than the GUI. The SAS programming language, also known as the SAS scripting language, is much easier to learn than most other programming languages such as FORTRAN, C, and Java.

In this book, you will work with the SAS programming language, not the GUI, but you are encouraged to explore the GUI on your own. This book is not intended to provide in-depth coverage of the SAS programming language; only Base SAS and the procedures required for analytics will be covered here.

SAS was originally used in statistics applications for agriculture projects. Now it's used in the following industries: casinos, communications, education, financial services, government, health insurance, healthcare providers, hotels, insurance, life sciences, manufacturing, media, oil and gas, retail, travel, transportation, utilities, and many more.

Next you will learn some simple SAS programming steps.

Starting SAS in Windows

In Windows, you can start SAS by selecting Start ➤ Programs like you would with any other software. The SAS folder in the Start menu programs list might show all the SAS-related software in SAS. You need to click Base SAS, that is, SAS 9.1 or SAS 9.2 or SAS 9.3, depending on the version of the SAS you have installed. Figure 2-1 shows how to start SAS in Windows.

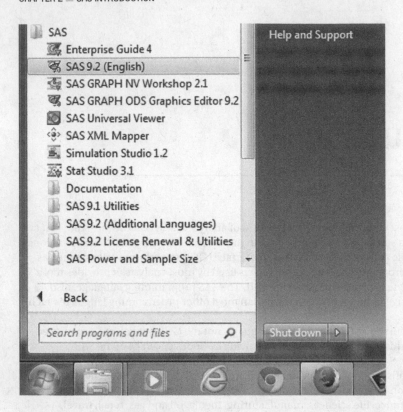

Figure 2-1. *Starting SAS in Windows*

Sometimes you may get the error shown in Figure 2-2. This error pops up when your SAS license is expired, and you may have to renew the license in that case. You can renew SAS by buying a renewal license from SAS. A new SAS installation data (SID) file will be provided by SAS to make it function normally.

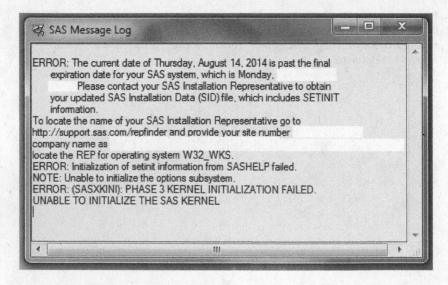

Figure 2-2. *License expiration error in SAS*

The SAS Opening Screen

When you open the SAS program in Windows, your screen, in most cases, should look like the SAS Windows environment shown in Figure 2-3. This may depend upon the operating system of the machine you are using; Figure 2-3 shows a Microsoft Windows SAS session. This is the most usual way your SAS screen will appear. But some windows or icons might be hidden on some systems depending upon the installation procedure and settings you have.

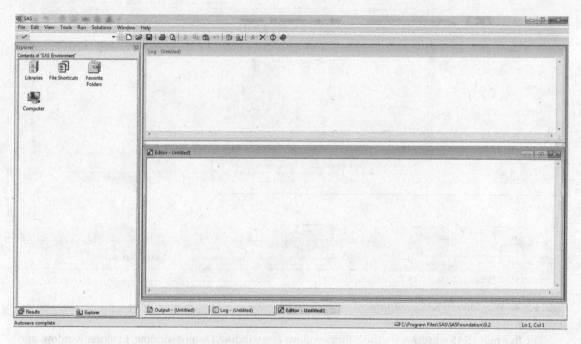

Figure 2-3. *A typical SAS Windows environment*

The Five Main Windows

After opening the SAS screen, you will see many icons and windows. There are five main windows in SAS environment, as shown in Figure 2-4.

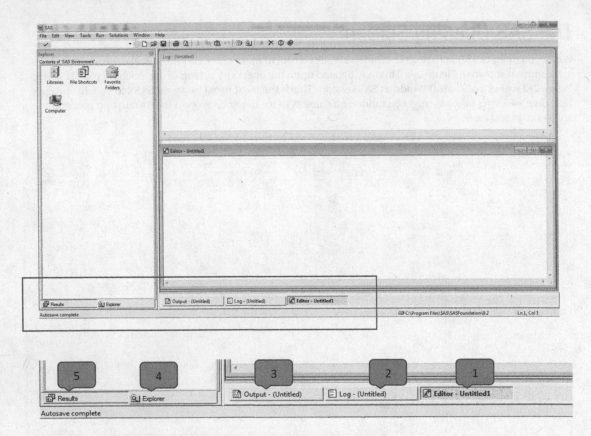

Figure 2-4. *Windows in SAS environment*

The five main SAS windows are the Editor window, Log window, Output window, Explorer window, and Results window. If you can't find any of these windows, you can make them visible by using the View option from the top menu bar.

Editor Window

The Editor window is used for writing SAS scripts that will be used in data modeling and analysis. It's like any other programming text editor. The editor is syntax sensitive and color codes your SAS scripts so that reading the program or identifying errors in the program is easy.

To execute the code, you use either the Submit icon on the top or select Run ➤ Submit (Figure 2-5) from the top menu bar.

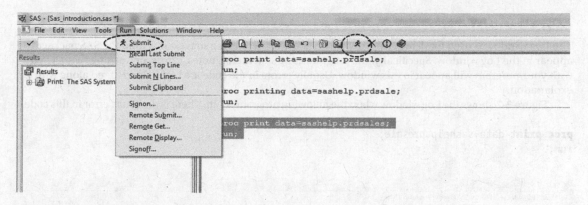

Figure 2-5. *Submitting a SAS program*

Figure 2-6 shows the Editor window with some sample code. This code prints the prdsale table from the SAS help library. (This will be explained in detail later in this book).

```
proc print data=sashelp.prdsale;
run;
```

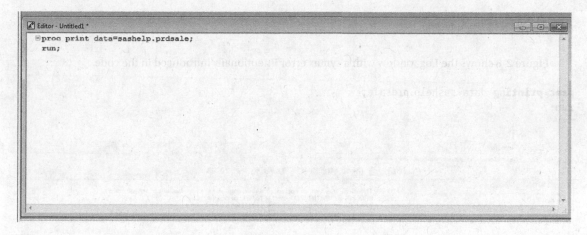

Figure 2-6. *The SAS Editor window*

This SAS code is saved in `.sas` format, which is the usual extension for all the SAS programming script files. You can open these code files with SAS or any other text editor.

Log Window

The Log window is used for debugging. Any observations or debugging suggestions from the SAS package appear in the Log window. Specifically, the programming statements, notes, errors, or warnings associated with your program will appear in this window. Usually errors in the code are highlighted in red along with an explanation.

Figure 2-7 shows the Log window when the following program is run. There is no syntax error in this code.

```
proc print data=sashelp.prdsale;
run;
```

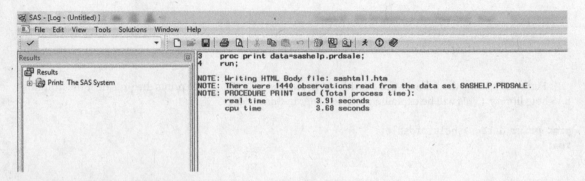

Figure 2-7. *SAS Log window*

Figure 2-8 shows the Log window with a syntax error intentionally introduced in the code.

```
proc printing data=sashelp.prdsale;
run;
```

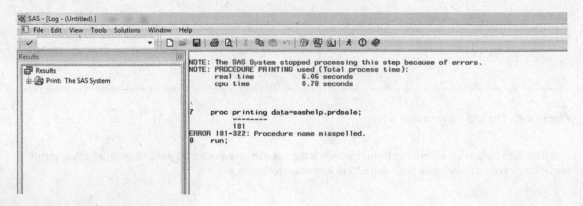

Figure 2-8. *Syntax error in SAS code*

Note that the error correctly indicates that the procedure name is misspelled.

Figure 2-9 shows the Log window with a non-syntax-related error in the code.

```
proc print data=sashelp.prdsales;
run;
```

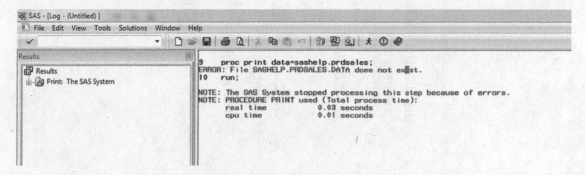

Figure 2-9. *A typical non-syntax-related error*

The error here correctly identifies that the data file name is misspelled and it does not exist in the SAS help library.

If you try to save the log code file, then it will get saved in .log format. Generally, log files are appended. So, you might see all the previous log information also in your current log file. You can press Ctrl+E to erase or clean the log file.

Output Window

The actual program output, like print data or output data from any SAS procedure, is shown in the Output window. Only the printable output from your program will appear in this window. If the program doesn't generate any output or if there is any syntax error in the code that caused the SAS system to stop abruptly, then the Output window might be blank.

Figure 2-10 shows the Output window when you run the following code:

```
proc print data=sashelp.prdsale;
run;
```

35

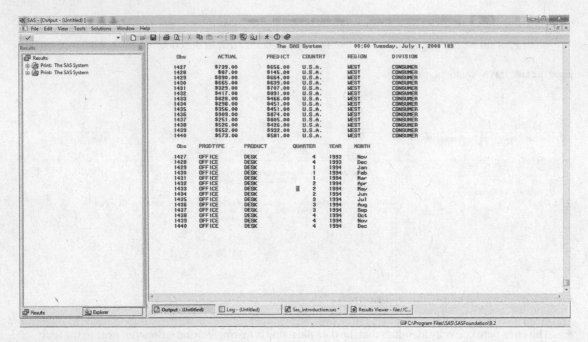

Figure 2-10. *SAS Output window*

I wrote some code for printing the prdsale data set, and the Output window shows the data records of the table. The output shows that the table contains some data for product sales with fields such as actual, predict, country, and division.

The default output is in a list or file listing format. If you try to save this output file, then it will get saved in .lst format. However, for SAS 9.3 and newer, HTML is the default option.

HTML Output

By default SAS generates listing output, but HTML output is a good option to see the output files in a more readable and formatted way. Broadly speaking, HTML files are easy to navigate and understand when compared to default listing files. HTML files can easily be transferred into other formats such as Excel and PowerPoint by using a simple copy and paste command. You can use HTML options in the code to create HTML output, or you can directly set the SAS options to create the HTML output for every program execution.

Here are the steps to set HTML output creation from the SAS menu environment:

1. Select: Tools ➤ Options ➤ Preferences (Figure 2-11).

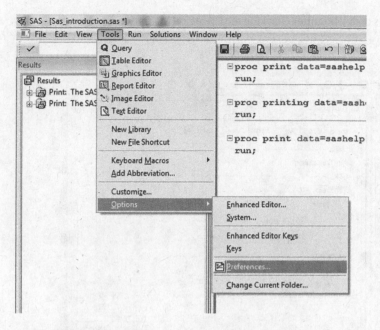

Figure 2-11. *Selecting Preferences*

2. In the Preferences window, select the Results tab and check Create HTML (Figure 2-12).

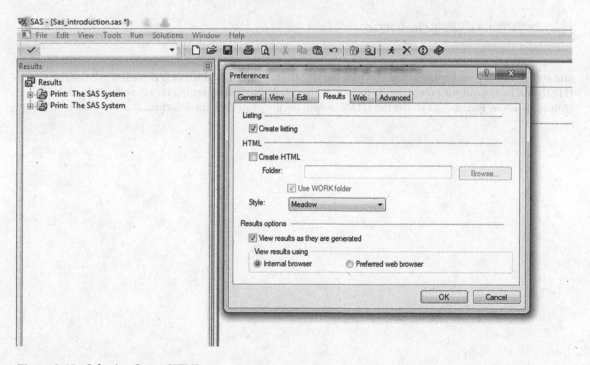

Figure 2-12. *Selecting Create HTML*

Figure 2-13 shows the Preferences dialog box after checking the Create HTML option.

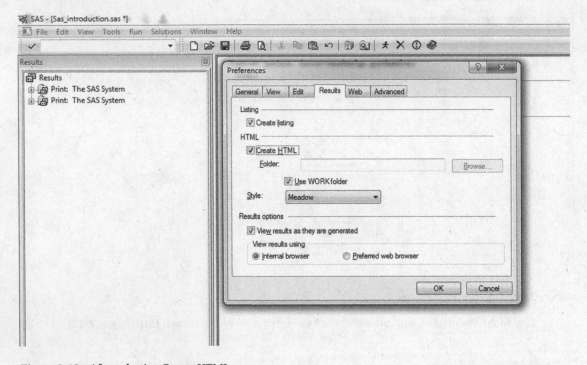

Figure 2-13. *After selecting Create HTML*

The HTML output type gives you several themes to choose from. Depending on the style of your report and the theme of your business, you can pick the HTML style. There is absolutely no difference between HTML and listing output except for the format. In this book, you will be setting HTML code as the default Output window. Figure 2-14 shows the HTML output for the print code from earlier.

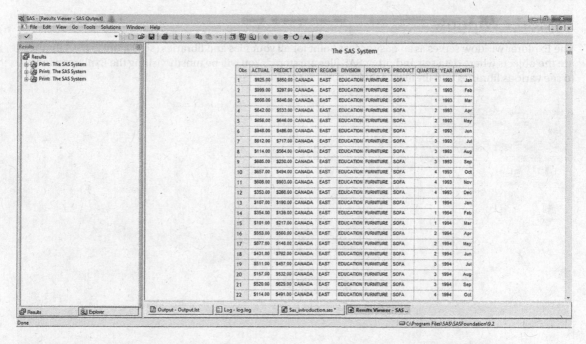

Figure 2-14. *Typical HTML output*

Explorer Window

The Explorer window serves as an easy access point for all your files and libraries (see Figure 2-15). Libraries are the objects where data sets and other SAS files are stored. You will be mostly visiting the Explorer window to see various libraries and the files inside.

Figure 2-15. *SAS Explorer window*

Results Window

The Results window serves as a table of contents (TOC) for the Output window, listing each part of your results in an outline form. The Results window shows both listing and HTML output files as well as the cumulative tree of results that are run in the current session (see Figure 2-16).

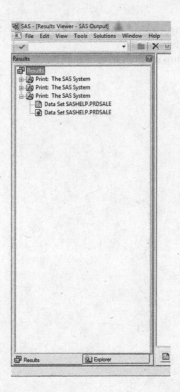

Figure 2-16. *SAS Results window*

The Explorer and Results windows are shown on the left side of the GUI, one below the other. You can toggle between the Explorer and Results windows by clicking Explorer or Results on the bottom taskbar.

Important Menu Options and Icons

In this section we discuss some important SAS menu options, such as creating, closing, and savings SAS program files, which will be used in your day-to-day working with the SAS environment.

To create a new SAS program file, Select: File ➤ New Program from the top menu bar, as shown in Figure 2-17.

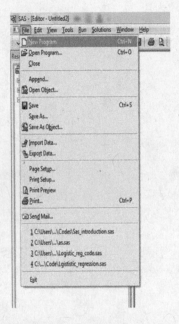

Figure 2-17. *Menu option to create a new program file*

To open an old SAS program file, Select: File ➤ Open Program from the top menu bar, as shown in Figure 2-18.

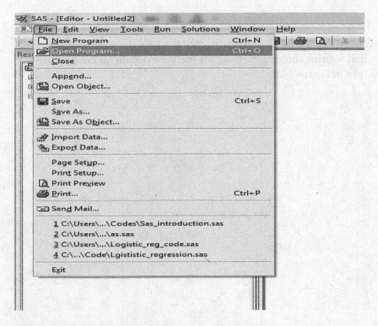

Figure 2-18. *Opening a program file*

To save a SAS program file, Select: File ➤ Save or ➤ Save as from the top menu bar, as shown in Figure 2-19.

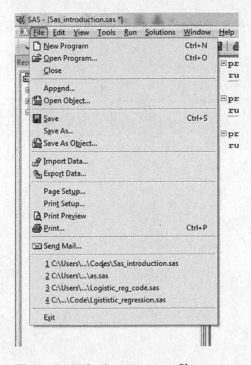

Figure 2-19. *Saving a program file*

View Options

Sometimes you can't find the window you are looking for or maybe you closed one of the windows by mistake. You can use the View menu option to bring them back. Just click the View option on the top menu bar and open the desired window, as shown in Figure 2-20.

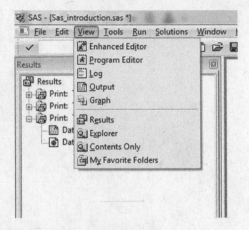

Figure 2-20. *View options*

Run Menu

The Run menu is used for submitting the whole or a selected portion of the SAS program, as shown in Figure 2-21.

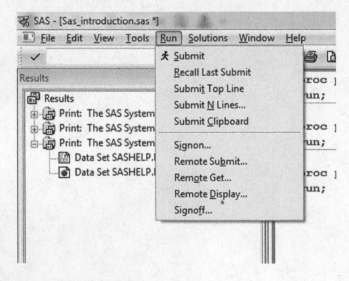

Figure 2-21. *Submitting a SAS program*

Solutions Menu

The Solutions menu gives you access to various customized SAS solutions depending on the software options you have installed, as shown in Figure 2-22. In this book, you will be using only the simple Base SAS scripts; no customized solutions are used.

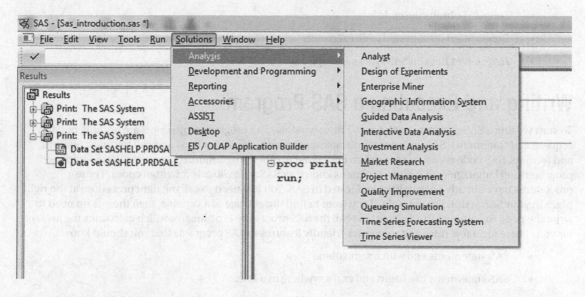

Figure 2-22. *SAS Solutions options*

You have now seen a small SAS print program and all the important windows that will matter to you going forward with analytics using SAS.

Shortcut Icons

Figure 2-23 shows some useful shortcut icons.

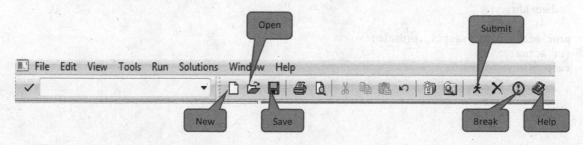

Figure 2-23. *Convenient shortcut icons*

Here are the most useful shortcuts:

- *New*: Starts a new program

- *Open*: Opens an existing code file

- *Save*: Saves the current code file

- *Submit*: Submits the code

- *Break*: Stops a submitted code execution

- *Help*: Gives help on SAS syntax and options

Writing and Executing a SAS Program

To start writing a SAS program, open a new Editor window. SAS programming scripts are nothing but a sequence of statements. SAS code contains statements, expressions, functions and call routines, options, and formats. SAS code is easy to write when compared to other programming languages. It is a simplified programming language with built-in programs known as *SAS procedures* (prewritten code). These procedures have already been written and tested in SAS. You just need to call the right procedure at the right place in your SAS script. For example, if you want to find the average of a variable, then there is no need to write the code to compute the average. Just call the SAS procedure Proc Means, which calculates the average for you. There are a few rules and some user-friendly features in SAS programs that you should know.

- SAS statements end with a semicolon.

- SAS statements can begin and end anywhere in a line.

- One statement can continue over several lines.

- Several statements can be on a line.

- SAS statements are not case sensitive.

- Blanks or special characters separate the "words" in a SAS statement.

- SAS programs end with run or quit, which prompts SAS to start execution of the code.

Here is the code for finding the average of the actual (actual sales) variable in the prdsale data set of the sashelp library:

```
proc means data =sashelp.prdsale;
var actual;
run;
```

This same code can be written as shown here (all these code samples will execute without any error):

```
proc means data =sashelp.prdsale;
var actual;
run;

proc
means data =sashelp.prdsale;
var actual;
run;

proc means data =sashelp.prdsale;var actual;run;

PROC MEANS Data =sashelp.prdsale;
var actual;
Run;
```

In this book, you will be writing simple SAS code scripts to do all the important tasks in an analytics project, such as import the data, clean the data, analyze the data, and report the results.

Comments in the Code

If you have done even a bit of programming, you are aware that you write comments in the code for documentation to explain the logic or flow involved in the code or just to remind you of something later about the code. Similarly, while writing SAS scripts, you can write comments. Most SAS scripts are small compared to conventional COBOL or Java code, which can be pages long.

There are two styles of comments you can use.

- One starts with an asterisk (*) and ends with a semicolon (;). Here's an example:

```
* This script does logistic regression for the price table data;
```

- The other style starts with a slash asterisk (/*) and ends with an asterisk slash (*/). Here is an example:

```
/* This script does logistic regression for the price table data */
```

The following is some sample code with comments.

```
*program to find the mean of actual sales;
PROC MEANS Data =sashelp.prdsale;
var actual;
Run;
```

The previous is the same as the following:

```
/*below program illustrates how to write a SAS code that finds the average actual sales
The dataset used here is prdsale, it is in sashelp library
MEANS is the produce that finds the averages
Var statement is used to specify the variables*/

PROC MEANS Data =sashelp.prdsale;
var actual;
Run;
```

Your First SAS Program

Type the code shown next in your Editor window. This code prints the prdsale data set from the sashelp library. (Chapter 3 discusses libraries.) This code (`sashelp.prdsale`) can be viewed as a table in the database, where sashelp is the database and prdsale is a table in it.

```
proc print data=sashelp.prdsale;
run;
```

The run statement at the end tells SAS to start processing the previous statements. You can use either the submit icon or the Run ➤ Submit option menu option. It is a good habit to first select the code and then execute. If you directly submit, the SAS system will execute all the code present in the program file or the Editor window. When you execute this code, you will see some information in the output file and also in the log file. In this case, the log file shows no error, and the output screen shows the data values in the prdsale table (data set), as shown in Table 2-1.

Table 2-1. Data Values in the prdsale Table

Obs	ACTUAL	PREDICT	COUNTRY	REGION	DIVISION	PRODTYPE	PRODUCT	QUARTER	YEAR	MONTH
1	$925.00	$850.00	CANADA	EAST	EDUCATION	FURNITURE	SOFA	1	1993	Jan
2	$999.00	$297.00	CANADA	EAST	EDUCATION	FURNITURE	SOFA	1	1993	Feb
3	$608.00	$846.00	CANADA	EAST	EDUCATION	FURNITURE	SOFA	1	1993	Mar
4	$642.00	$533.00	CANADA	EAST	EDUCATION	FURNITURE	SOFA	2	1993	Apr
5	$656.00	$646.00	CANADA	EAST	EDUCATION	FURNITURE	SOFA	2	1993	May
6	$948.00	$486.00	CANADA	EAST	EDUCATION	FURNITURE	SOFA	2	1993	Jun
7	$612.00	$717.00	CANADA	EAST	EDUCATION	FURNITURE	SOFA	3	1993	Jul
8	$114.00	$564.00	CANADA	EAST	EDUCATION	FURNITURE	SOFA	3	1993	Aug
9	$685.00	$230.00	CANADA	EAST	EDUCATION	FURNITURE	SOFA	3	1993	Sep
10	$657.00	$494.00	CANADA	EAST	EDUCATION	FURNITURE	SOFA	4	1993	Oct
11	$608.00	$903.00	CANADA	EAST	EDUCATION	FURNITURE	SOFA	4	1993	Nov
12	$353.00	$266.00	CANADA	EAST	EDUCATION	FURNITURE	SOFA	4	1993	Dec
13	$107.00	$190.00	CANADA	EAST	EDUCATION	FURNITURE	SOFA	1	1994	Jan
14	$354.00	$139.00	CANADA	EAST	EDUCATION	FURNITURE	SOFA	1	1994	Feb
15	$101.00	$217.00	CANADA	EAST	EDUCATION	FURNITURE	SOFA	1	1994	Mar
16	$553.00	$560.00	CANADA	EAST	EDUCATION	FURNITURE	SOFA	2	1994	Apr
17	$877.00	$148.00	CANADA	EAST	EDUCATION	FURNITURE	SOFA	2	1994	May
18	$431.00	$762.00	CANADA	EAST	EDUCATION	FURNITURE	SOFA	2	1994	Jun
19	$511.00	$457.00	CANADA	EAST	EDUCATION	FURNITURE	SOFA	3	1994	Jul
20	$157.00	$532.00	CANADA	EAST	EDUCATION	FURNITURE	SOFA	3	1994	Aug

Similarly, you can use the following code for finding the average of an actual variable:

```
proc means data =sashelp.prdsale;
var actual;
run;
```

When you execute the previous code, the log file shows no error, and you get the output shown in Table 2-2.

Table 2-2. *Output of SAS Means Procedure for prdsale*

Analysis Variable : ACTUAL Actual Sales

N	Mean	Std Dev	Minimum	Maximum
1440	507.1784722	287.0313065	3.0000000	1000.00

Here is one more code example:

```
data income_data;
Input income expenses;
Cards;
1200 1000
9000 600
;
run;
Proc print data=income_data;
Run;
```

This generates the output shown in Table 2-3.

Tables 2-3. *Print of income_data*

Obs	income	expenses
1	1200	1000
2	9000	600

Debugging SAS Code Using a Log File

Reading the log file is an important aspect of SAS program execution. Generally, new users write the code and tend to look at the output directly, but looking at the log file is equally important. The log file has mainly three notification types: errors, warnings, and notes.

- An error means that there is a syntax or other error. Sometimes an error can stop a script from executing.

- A warning indicates that the SAS system has automatically corrected your error and executed the code. This is dangerous when SAS misinterprets and executes.

- Notes give a running commentary of important steps while executing programs.

The log file for the SAS code used to print prdsale gives the following information:

```
31    proc print data=sashelp.prdsale;
32    run;
```

```
NOTE: Writing HTML Body file: sashtml8.htm
NOTE: There were 1440 observations read from the data set SASHELP.PRDSALE.
NOTE: PROCEDURE PRINT used (Total process time):
      real time              3.85 seconds
      cpu time               3.79 seconds
```

The previous message from the log file shows that there is no sign of an error in the code and that 1,440 observations were read from the data for printing.

The log file for the SAS code used to find the mean gives the following information. The note in the log file shows 1,440 observations were read from the data set.

```
33    proc means data =sashelp.prdsale;
34    var actual;
35    run;
```

```
NOTE: Writing HTML Body file: sashtml9.htm
NOTE: There were 1440 observations read from the data set SASHELP.PRDSALE.
NOTE: PROCEDURE MEANS used (Total process time):
      real time              0.25 seconds
      cpu time               0.09 seconds
```

Again, there are no errors, and the means procedure ran successfully.

The following is the log file for the third code snippet, which was used to create the print income_data:

```
53    data income_data;
54    Input income expenses;
55    Cards;
```

```
NOTE: The data set WORK.INCOME_DATA has 2 observations and 2 variables.
NOTE: DATA statement used (Total process time):
      real time              0.00 seconds
      cpu time               0.00 seconds
```

```
58    ;
59    run;
60    Proc print data=income_data;
61    Run;
```

```
NOTE: Writing HTML Body file: sashtml11.htm
NOTE: There were 2 observations read from the data set WORK.INCOME_DATA.
NOTE: PROCEDURE PRINT used (Total process time):
      real time              0.29 seconds
      cpu time               0.18 seconds
```

The log file doesn't show any errors. If you deliberately write some incorrect syntax and submit it, you see an error in the log file.

The following code generates the subsequent message in the log file.

```
proc dataprinting data=sashelp.prdsale;
run;
```

```
40   proc dataprinting data=sashelp.prdsale;
ERROR: Procedure DATAPRINTING not found.
41   run;

NOTE: The SAS System stopped processing this step because of errors.
NOTE: PROCEDURE DATAPRINTING used (Total process time):
     real time           0.01 seconds
     cpu time            0.00 seconds
```

The error is clearly mentioned.

The following is one more example of an error:

```
proc print data=sashelp.prdsales;
run;
```

```
42   proc print data=sashelp.prdsales;
ERROR: File SASHELP.PRDSALES.DATA does not exist.
43   run;

NOTE: The SAS System stopped processing this step because of errors.
NOTE: PROCEDURE PRINT used (Total process time):
     real time           0.03 seconds
     cpu time            0.01 seconds
```

Example for Warnings in Log File

The following code is for copying the data from the prdsale data set from the SAS help library into a new data file:

```
data new_data;
set sashelp.prdsale;
where actuals<1000;
run;
```

The following is the log file for the previous code:

```
14   data new_data;
15   set sashelp.prdsale;
16   where actuals<1000;
ERROR: Variable actuals is not on file SASHELP.PRDSALE.
17   run;

NOTE: The SAS System stopped processing this step because of errors.
WARNING: The data set WORK.NEW_DATA may be incomplete.  When this step was stopped there
were
        0 observations and 10 variables.
```

```
WARNING: Data set WORK.NEW_DATA was not replaced because this step was stopped.
NOTE: DATA statement used (Total process time):
      real time             0.07 seconds
      cpu time              0.03 seconds
```

The warnings show that SAS has gone ahead and created a data file with 10 variables and 0 observations.

Tips for Writing, Reading the Log File, and Debugging

Here are a few tips for writing and debugging the SAS programs, meant for beginners:

- Try to keep the log file clean by erasing it after code execution. Press Ctrl+E to clean the log file.

- Do not directly go to output, even when the code executes successfully. Make it a habit to see the log file first and then move to the output file.

- Always start from the first error at the beginning of the log information. Generally SAS shows the last few lines of the log file, but most of the time if you correct the first few errors, the late ones may disappear. However, it depends upon the type of errors that your code contains.

- Understand the notifications; there are three kinds of notifications in log files: errors, warnings, and notes.

- Try to avoid these most common errors.

  - Missing semicolons

  - Missing semicolon at the beginning or earlier in the code, which may create an error in a much later statement

  - Missing run statement

  - Misspelling the key words and data set names

  - Unbalanced quotation marks and parentheses

- Always write the code explanations in the form of comments.

- Instead of executing the whole SAS program in one go, try to execute the code snippets one after the other. That way it will be easier for you to pinpoint the error.

Saving SAS Files

You can use the Save or Save as option in the File menu, which will prompt you to save the SAS program file in the desired location, as shown in Figure 2-24.

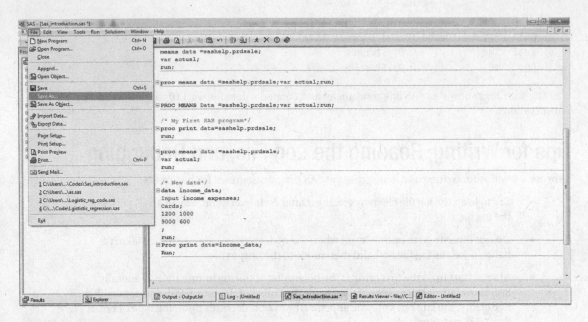

Figure 2-24. *Saving SAS files*

Exercise

Here are some exercises to help you become more familiar with reading log files.

1. Write and execute the following code. Look at the log file and identify the errors, if any.

```
proc print data=sashelp.airr;run;
```

2. Save your SAS file and open it using the File menu.

3. Write and execute the following code. Look at the log file and identify the errors, if any.

```
proc print data=sashelp.buy;run;
```

Conclusion

This chapter introduced you to the SAS programming environment. It discussed navigation in the SAS Windows environment, various menu options, and some shortcut icons. It also discussed writing simple SAS codes and reading log files. In the next few chapters, you will get into more details of SAS programming. Later this programming knowledge will help you in analysis, where you try to interact with SAS by writing SAS programs to analyze the data.

CHAPTER 3

■■■

Data Handling Using SAS

After learning the basics of the SAS tool in the previous chapter, you can now learn about data handling in SAS, the main focus of this chapter. While learning any analytics tool, you should be aware of three main phases: tool basics, basic data handling, and important functions and statistical algorithms (Figure 3-1). Any analyst who wants to work with advanced statistical techniques needs to have a fair understanding of these three areas of a statistical or analytical tool like SAS. There are also more advanced topics such as using macros and other tricks to write efficient code, but you can learn about those topics as your familiarity with the tool increases.

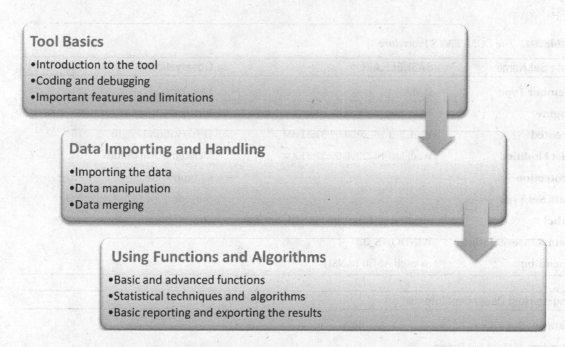

Figure 3-1. *Three phases in learning SAS fundamentals*

This chapter will focus on manipulating data in SAS. You'll learn about SAS data sets, SAS libraries, and SAS statement, import data, and you'll learn how to import data, use SAS constructs such as if-then-else, create new variables, and so on. All of these are fundamental tools for working with SAS.

SAS Data Sets

A SAS data set contains two main portions. First is the descriptive portion of the data, which is similar to the metadata for a typical data set. Another is the data potion, which contains the usual rows and columns containing data typical in a two-dimensional table. In this book, whenever we write data set, we refer to the data portion. SAS data sets have a `.sas7bdat` extension, and you can open them in the SAS environment.

You are already probably familiar with many other formats for storing data. Examples of common storage are text files (`.txt`) and comma-separated files (`.csv`). In addition to working with SAS data, you will also learn how to read data from external files, later in this chapter.

Descriptive Portion of SAS Data Sets

The descriptive portion of a data set contains the metadata, which is data about the data. This is mainly the description of variables and of the data set contained in the SAS data set. You can use the simple `contents` procedure in SAS (`PROC CONTENTS`) to get this metadata. The descriptive portion contains details such as when the data set was created, where it is stored, how many variables and observations it contains, the data type of variables, and so on. Given next is an example that shows the code and output (Table 3-1) associated with `PROC CONTENTS`. It prints the descriptive portion of the `air` data set from the sashelp library.

```
proc contents data=sashelp.air;
run;
```

Table 3-1. *The CONTENTS Procedure*

Data Set Name	SASHELP.AIR	Observations	144
Member Type	DATA	Variables	2
Engine	V9	Indexes	0
Created	Wed, Jan 16, 2008 09:37:31 AM	Observation Length	16
Last Modified	Wed, Jan 16, 2008 09:37:31 AM	Deleted Observations	0
Protection		Compressed	NO
Data Set Type		Sorted	NO
Label	airline data (monthly: JAN49-DEC60)		
Data Representation	WINDOWS_32		
Encoding	us-ascii ASCII (ANSI)		

Engine/Host Dependent Information	
Data Set Page Size	4096
Number of Data Set Pages	1
First Data Page	1
Max Obs per Page	252
Obs in First Data Page	144

(continued)

Table 3-1. (*continued*)

Engine/Host Dependent Information	
Number of Data Set Repairs	0
Filename	C:\Program Files\SAS\SASFoundation\9.2(32-bit)\ets\sashelp\air.sas7bdat
Release Created	9.0201M0
Host Created	XP_PRO

Alphabetic List of Variables and Attributes					
#	**Variable**	**Type**	**Len**	**Format**	**Label**
2	AIR	Num	8		international airline travel (thousands)
1	DATE	Num	8	MONYY.	

Data Portion of Data Set

Dealing with the data potion of data sets is fortunately pretty simple in SAS. As discussed earlier, the data portion in SAS data sets contains the data rows (more appropriately called *observations* or *records*) and datelines along with the columns (called *variables*). You can use the print procedure of SAS (PROC PRINT) to see the data set. The following is the SAS code for printing the first ten observations of the sashelp.air data set:

```
proc print data=sashelp.air(obs=10);
run;
```

Table 3-2 lists the output of this code.

Table 3-2. *Output of PROC PRINT on the sashelp.air Data Set*

Obs	DATE	AIR
1	JAN49	112
2	FEB49	118
3	MAR49	132
4	APR49	129
5	MAY49	121
6	JUN49	135
7	JUL49	148
8	AUG49	148
9	SEP49	136
10	OCT49	119

As a quick recap, Table 3-3 lists the frequently used file icons and extensions in SAS environments.

Table 3-3. *Commonly Used File Icons and Extensions in the SAS Environment*

Icon	Type of File
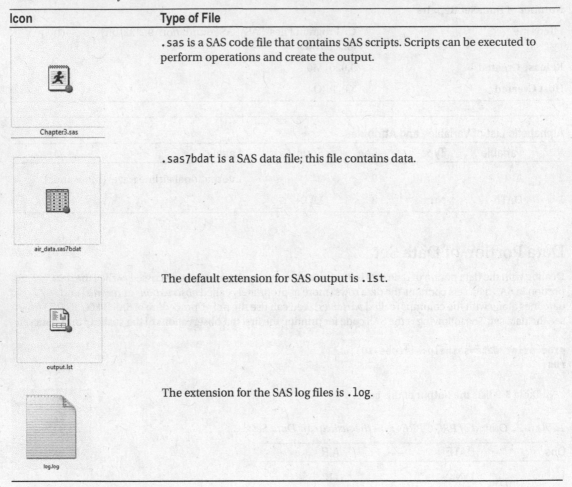	.sas is a SAS code file that contains SAS scripts. Scripts can be executed to perform operations and create the output.
	.sas7bdat is a SAS data file; this file contains data.
	The default extension for SAS output is .lst.
	The extension for the SAS log files is .log.

SAS Libraries

SAS data sets are stored in SAS libraries. The previous example used the air data set from the sashelp library. Every data set is best stored in a library first; otherwise, you wouldn't be able to work with it in the SAS environment. Just to summarize,

- A SAS library is a collection of SAS data sets and other SAS files.

- You can, for example, create a SAS library from an Oracle database containing tables.

You need to define a library to be able to store any data set. A SAS library is not a virtual folder; it is a physical folder in the Windows environment. You just refer to the physical path of a library and the physical folder's name using a *handle* (the library name). So, while creating a library, you need to give two mandatory parameters, the library name and the physical path, where the files will be stored. Described next are the two ways to create a new library: using a graphical user interface (GUI) and using SAS code.

Creating the Library Using the GUI

SAS provides a GUI to create and work with libraries. Here are the steps for creating a library:

1. Open Explorer (Figure 3-2).

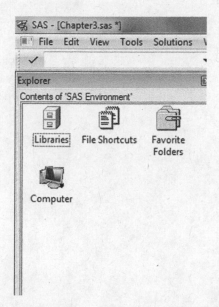

Figure 3-2. *Creating a new library: opening Explorer*

2. Double click the Libraries folder icon. This folder contains all the default libraries in SAS (Figure 3-3).

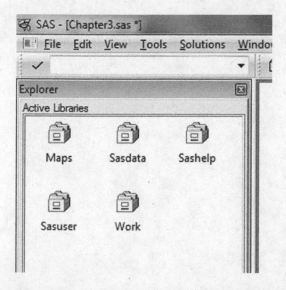

Figure 3-3. *Creating a new library: The active libraries*

3. Right-click within the Explorer window and select New to create a new library (Figure 3-4).

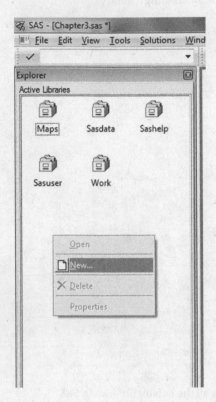

Figure 3-4. *Creating a new library: right-click and select the New menu option*

4. Give the library a name (Figure 3-5).

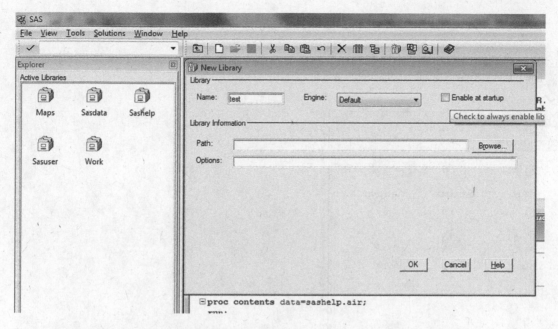

Figure 3-5. *Creating a new library: filling in the library name*

5. Give the physical path of the library (Figure 3-6). All the data sets stored in the library can be seen in that location.

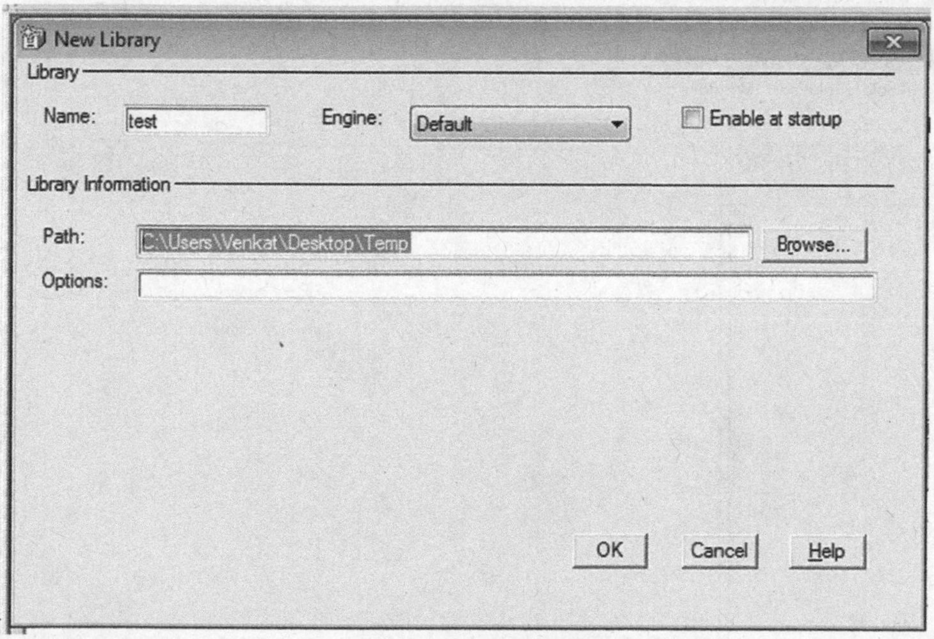

Figure 3-6. *Creating a new library: fill in the physical path of the library*

6.   Click OK, and the new library will be created (Figure 3-7).

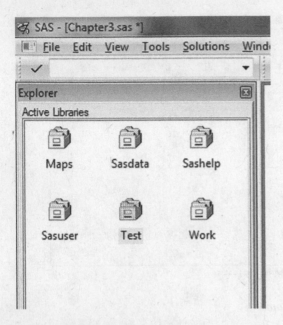

Figure 3-7. *Creating a new library: the new library Test is created*

Obviously, the newly created library is empty as of now. You can verify this by double-clicking the Test icon (Figure 3-8).

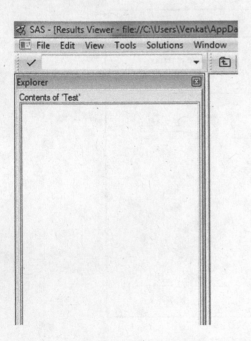

Figure 3-8. *Creating a new library: the new library Test*

What happens if you follow the same steps and create a new library named 1temp (Figure 3-9)?

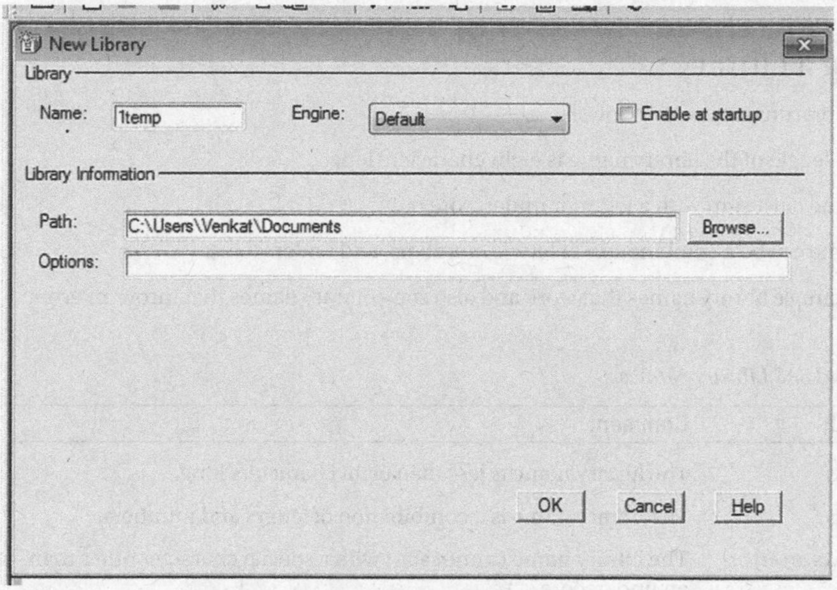

Figure 3-9. *Creating a new library: an invalid name*

The library will not be created because 1temp is an invalid name; you will get the error given in Figure 3-10.

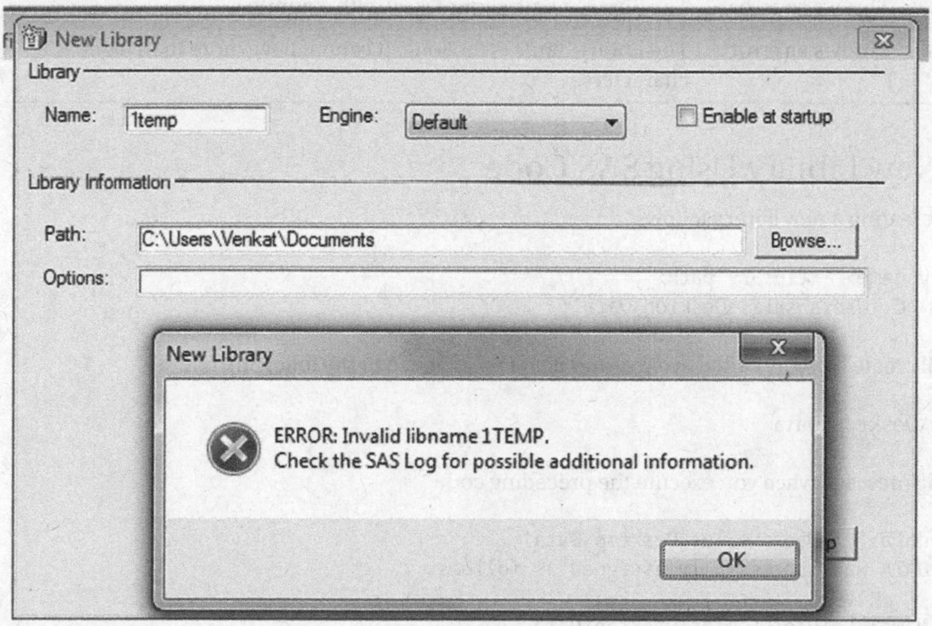

Figure 3-10. *Creating a new library: error, showing an invalid name*

You cannot use a random string as a library name. There are certain naming rules while creating the library, covered next.

Rules of Assigning a Library

The rules for defining the library name are as follows:

- The maximum length of the library name is eight characters long.
- The library name must start with a letter or underscore only.
- The library name can be a combination of letters, numbers, and underscores.

Table 3-4 lists some example library names that work and also some library names that throw an error.

Table 3-4. *Valid and Invalid SAS Library Names*

Library Name	Result	Comment
Mylib	Works	The library name is less than eight characters long.
Mylib1	Works	The library name is a combination of letters and numbers.
$Mylib	Throws an error	The library name cannot start with a special character other than an underscore (_).
_Mylib	Works	The library name can start with an underscore (_).
Mylib$	Throws an error	The library name cannot have any special character other than an underscore (_).
mylib	Works	A library name can include an underscore (_).
2mylib	Throws an error	The library name cannot start with a number.
Datalib2013Sales	Throws an error	The library name is too long. It cannot have more than eight characters.

Creating a New Library Using SAS Code

The SAS code for creating a new library follows:

```
libname <Library name> '<Library path>';
 libname mydata 'C:\Users\sales\Desktop\Data';
```

This code will create a library called mydata, and it will be attached to the following folder:

```
 C:\Users\sales\Desktop\Data
```

Here is the log message when you execute the preceding code:

```
22   libname mydata 'C:\Users\sales\Desktop\Data';
NOTE: Libref MYDATA was successfully assigned as follows:
      Engine:        V9
      Physical Name: C:\Users\sales\Desktop\Data
```

Permanent and Temporary Libraries

There are two types of libraries in SAS: permanent and temporary. The libraries you created by using the libname statement and the GUI are permanent libraries. Once a data set is created in a permanent library, you can use it in the future, and the data set will be available even after a system reboot. By creating or importing a data set into a permanent library, you are physically creating a permanent data set.

Work Library

Work is a temporary library in SAS. The data sets created in the Work library last only for the current SAS session. Once you close and reopen SAS, all the data sets in the Work library will be lost. If you don't specify any library name with a data set, then, by default, SAS stores the data set in Work.

Let's explore these concepts in more depth with an exercise. The following code creates a data set with two columns. You will use this code to create a data set in Work as well as in a permanent library called Mydata. You then close and reopen SAS to see whether the data set is still there.

The following code creates an Income data set in the mydata library with two variables: income and expense. You also put some records the income data set using the datalines keyword.

```
data mydata.income;
input income expense;
datalines;
4500 2000
5000 2300
7890 2810
8900 5400
2300 2000
;
run;
```

The log messages when this code is executed look like this:

```
NOTE: The data set MYDATA.INCOME has 5 observations and 2 variables.
NOTE: DATA statement used (Total process time):
      real time           0.07 seconds
      cpu time            0.01 seconds
```

Figure 3-11 shows the contents of the library Mydata. You can see an Income icon, which is the recently created data set.

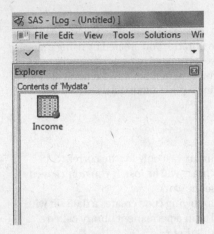

Figure 3-11. *Contents of the Mydata library*

The following code creates the Income data set in the Work library with the same details as you did for the Mydata library:

```
data work.income;
input income expense;
datalines;
4500 2000
5000 2300
7890 2810
8900 5400
2300 2000
;
run;
```

The log messages when you execute this code look like this:

```
NOTE: The data set WORK.INCOME has 5 observations and 2 variables.
NOTE: DATA statement used (Total process time):
      real time           0.01 seconds
      cpu time            0.01 seconds
```

Figure 3-12 shows the contents of the library Work. You can see the Income icon, which is the recently created data set.

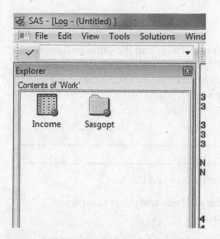

Figure 3-12. *Contents of Work library after creating the Income data set*

The following code also creates Income data set in the Work library. If you don't mention any library name, the data set will be created in Work, by default.

```
Data income;
input income expense;
datalines;
4500 2000
5000 2300
7890 2810
8900 5400
2300 2000
;
run;
```

If you close SAS and reopen it, then you will not be able to see the Income data set in the Work library. By contrast, in Mydata, you will observe that the Income data set is still available.

■ **Note**   If the Mydata library vanishes or gets deleted from the library list, it may not be a reason for worry. All you need to do is rerun the following code to re-create just the library name and assign it the directory path. After doing so, you will again see the Income data set in Mydata. You need not create the data set again. The data set, Income, is permanently created in the given library folder path with a file extension of .sas7bdat.

```
libname mydata 'C:\Users\sales\Desktop\Data';
```

Now that you've learned how to work with SAS libraries, it's time to learn about the basics of SAS coding, which will allow you to work with data and perform some meaningful analysis. The following section introduces you to SAS PROC and DATA steps, the most basic building blocks of writing SAS code.

Two Main Types of SAS Statements

There are two main types of SAS statements (see Table 3-5). One starts with DATA, and the other one starts with PROC. The data step is mainly used for variable creation, data manipulation, and other data-related operations. The PROC step is used for calling (in the SAS code) various algorithms, special functions, and so on. PROC stands for the word *procedure*, which means prewritten code in SAS.

Table 3-5. *DATA Step and PROC Step*

DATA Step	PROC Step
• Starts with a keyword DATA.	• Starts with keyword PROC.
• Used for data manipulations.	• Used for analysis and other complex operations.
• Creating calculated fields.	• You can create the data sets by taking the output of PROC step.
• Mainly used in reporting.	
• Data merging, joining, data cleaning operations.	• **Example:** Predicting the sales by using regression algorithm needs PROC REG.
• **Example:** Preparing the data for analysis, from the orginal raw data, requires lot of data operations.	

Importing Data into SAS

Before you can start working on data, you need to import it into a SAS library in the form of SAS data sets. The data can be imported into SAS from many sources such as CSV files, Excel files, TXT files, database files, and so on. The following sections explain some ways of importing into and creating the data in SAS.

Data Set Creation Using the SAS Program

If you have to create a small data set based on some data provided to you, you create it by writing a small SAS program.

Here is a sample program to create a data set:

```
data mydata.income;
input income expense;
datalines;
4500 2000
5000 2300
7890 2810
8900 5400
2300 2000
;
run;
```

Given in Table 3-6 is the explanation of the preceding syntax.

Table 3-6. *Program Syntax Explanation*

SAS Statement	Explanation
data	This is a keyword to create a new data set in SAS.
mydata.income;	Mydata is the library name, and income is the new data set name.
input	This is a keyword for declaring the variables in the data.
income expense;	These are two variables that will be created in the new data set.
datalines;	This is a keyword for entering the data into the data set.
4500 2000	This is the actual data.
5000 2300	
7890 2810	
8900 5400	
2300 2000	
;	

Here is the log file when you execute this program:

```
NOTE: The data set MYDATA.INCOME has 5 observations and 2 variables.
NOTE: DATA statement used (Total process time):
      real time           0.03 seconds
      cpu time            0.03 seconds
```

Figure 3-13 shows the Income data set in the Mydata library.

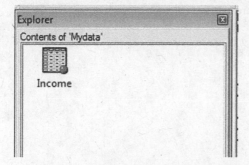

Figure 3-13. *Contents of the Mydata library after creating the Income data set*

Generally, the data set creation step is useful when you quickly want to create a small data set to use in the analysis. If you are creating really large data sets, then manually entering the values is obviously not the desired option. You need to use SAS import features for bringing large data files into the SAS environment.

Using the Import Wizard

You can import the data from external files into SAS by writing SAS code, or you can use the Import Wizard. Let's first look at how to import the external flat files into SAS using the Import Wizard. Here are the steps to import the data:

1. Select: File ➤ Import Data (see Figure 3-14).

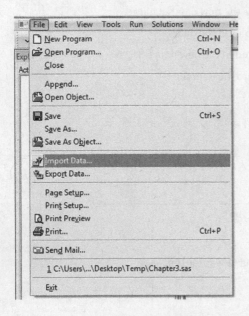

Figure 3-14. *Importing data: select the Import Data option*

2. Select the type of file. In this example, you are going to import a CSV file; hence, you select the CSV format (see Figure 3-15 and Figure 3-16).

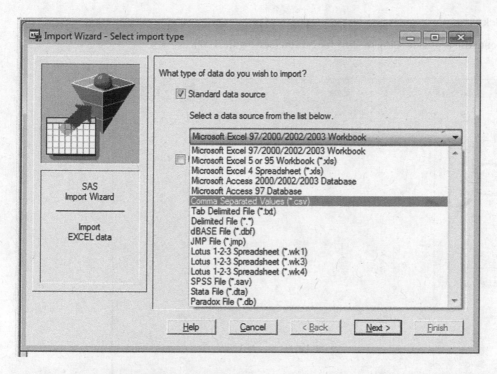

Figure 3-15. *Importing data: select the file type*

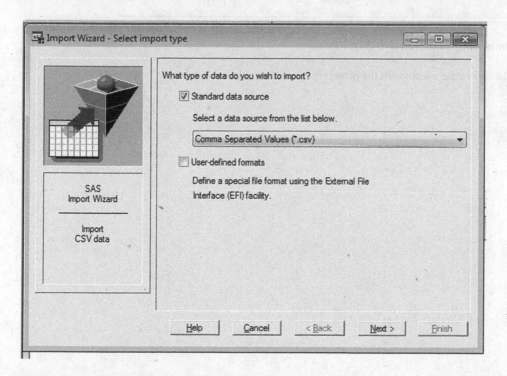

Figure 3-16. *Importing data: the final selection screen*

3.   Click Next.

4.   Locate the input file using the Browse button (see Figure 3-17).

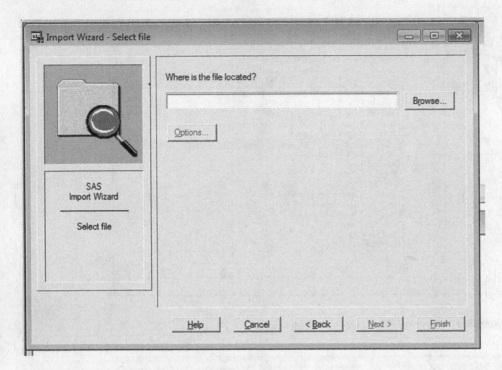

Figure 3-17. *Importing data: browse to the file location*

Figure 3-18 shows the wizard with the proper file path entered.

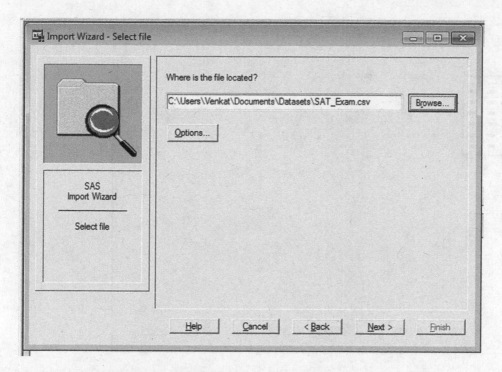

Figure 3-18. *Importing data: the file folder path being selected*

5. Select the file and click Next.

6. Give the details of the destination SAS data set (see Figure 3-19).

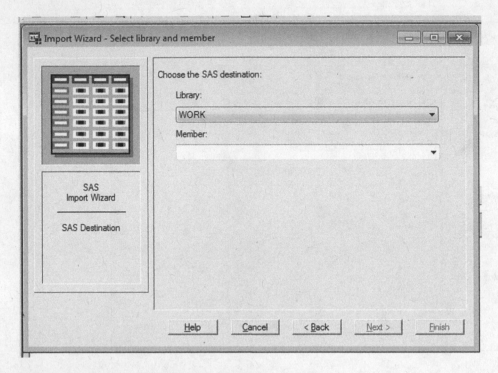

Figure 3-19. *Importing data: SAS data set destination details*

    7.   Choose the library (see Figure 3-20).

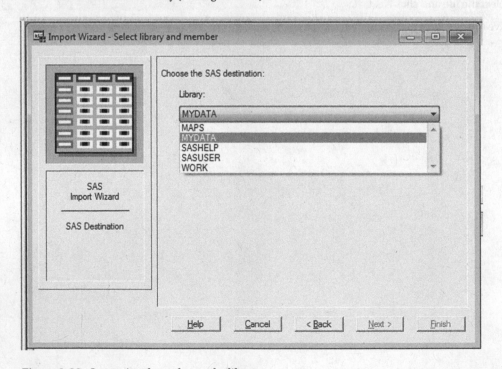

Figure 3-20. *Importing data: choose the library*

8. Name the target data set (see Figure 3-21).

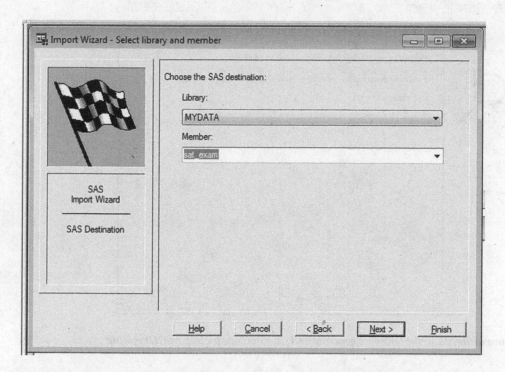

Figure 3-21. *Importing data: name the target data set*

9. If you think you will want to import this data again, you can save the importing steps to a SAS file (see Figure 3-22). This step is optional. Browse to a location and save the .sas file (see Figure 3-23).

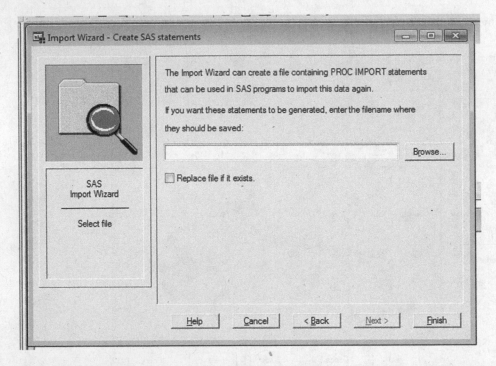

Figure 3-22. *Importing data: the option to generate and save the code for import GUI options*

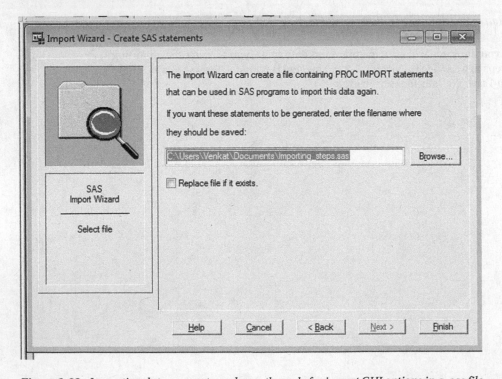

Figure 3-23. *Importing data: generate and save the code for import GUI options in a .sas file*

10.   Click Finish to end the process.

The data set will be in the library, and the SAS code will be in the `.sas` file. Figure 3-24 shows the `sat_exam` data set stored in the Mydata library.

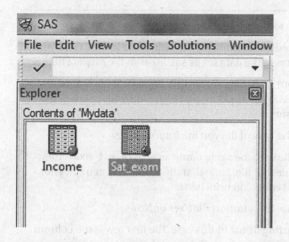

Figure 3-24. *Importing data: contents of the Mydata library*

The SAS code from the file you saved follows:

```
PROC IMPORT OUT= MYDATA.sat_exam
           DATAFILE= "C:\Users\Documents\Datasets\SAT_Exam.csv"
           DBMS=CSV REPLACE;
    GETNAMES=YES;
    DATAROW=2;
RUN;
```

In the same fashion, you can import different types of files such as TXT files, Excel files, and so on, using the Import Wizard. The next section explains the SAS code to import the data.

Import Using the Code

In the previous section you imported data using the GUI. You can also import data using SAS code. While analyzing data, usually some data manipulation code is also need after the data import, so it's convenient that you write the data import code in the same file.

Importing a CSV File Using Code

The following code imports a simple CSV file into SAS:

```
PROC IMPORT OUT= MYDATA.sat_exam
           DATAFILE= "C:\Users\Documents\Datasets\SAT_Exam.csv"
           DBMS=CSV REPLACE;
    GETNAMES=YES;
    DATAROW=2;
RUN;
```

Table 3-7 explains this code.

Table 3-7. *Code to Import a .csv File*

SAS Statement	Explanation
PROC IMPORT	This is the procedure for importing data into SAS.
OUT= MYDATA.sat_exam	This is the OUT keyword for mentioning the output file in SAS. Here you are importing and storing the data set as sat_exam in the Mydata library.
DATAFILE= "C:\Users\Documents\Datasets\SAT_Exam.csv"	The DATAFILE keyword is for mentioning the input data file that needs to be imported into SAS. You need to specify the full folder path of the file along with extension.
DBMS=CSV	DBMS=CSV indicates the type of file you are importing.
REPLACE;	If there is already a file with the same name as MYDATA.sat_exam, then this option will replace that file. Note that the first SAS statement ends here, and there is no semicolon until here.
GETNAMES=YES;	Get the column names from import file? Yes or No?
DATAROW=2;	Where is the data starting from? In this case, the first row is the column names, and the data is starting from the second row.

Here is the code for importing a CSV file containing burger sales:

```
PROC IMPORT OUT= MYDATA.burger
           DATAFILE= "C:\Users\Documents\Datasets\Burger_sales.csv"
          DBMS=CSV REPLACE;
    GETNAMES=YES;
    DATAROW=2;
RUN;
```

The important notes from the log file when you execute this code follows:

```
NOTE: The infile 'C:\Users\Documents\Datasets\Burger_sales.csv' is:
      Filename=C:\Users\Documents\Datasets\Burger_sales.csv,
      RECFM=V,LRECL=32767,File Size (bytes)=375,
      Last Modified=10Jun2014:15:33:18,
      Create Time=05Oct2014:17:02:57

NOTE: 35 records were read from the infile 'C:\Users\Documents\Datasets\
Burger_sales.csv'.
      The minimum record length was 8.
      The maximum record length was 9.
NOTE: The data set MYDATA.BURGER has 35 observations and 2 variables.
NOTE: DATA statement used (Total process time):
      real time            0.03 seconds
      cpu time             0.03 seconds

35 rows created in MYDATA.BURGER                              from
C:\Users\Documents\Datasets\Burger_sales.csv.
```

```
NOTE: MYDATA.BURGER data set was successfully created.
NOTE: PROCEDURE IMPORT used (Total process time):
      real time           0.10 seconds
      cpu time            0.10 seconds
```

Let's look at how to import other types of files.

Importing an Excel File Using SAS Code

There is an additional parameter RANGE in the following Excel import code. This refers to a worksheet name in an Excel sheet. Generally an Excel file has multiple worksheets, so you need to mention the sheet that needs to be imported.

```
PROC IMPORT OUT= MYDATA.BURGER_sales_from_excel
           DATAFILE= "C:\Users\Documents\Datasets\Burger_sales.xls"
           DBMS=EXCEL REPLACE;
    RANGE="Burger_sales$";
    GETNAMES=YES;
  RUN;
```

Here are the log messages when you execute this code:

```
153  PROC IMPORT OUT= MYDATA.BURGER_sales_from_excel
154              DATAFILE= "C:\Users\Documents\Datasets\Burger_sales.xls"
155              DBMS=EXCEL REPLACE;
156        RANGE="Burger_sales$";
157        GETNAMES=YES;
158      RUN;
```

```
NOTE: MYDATA.BURGER_SALES_FROM_EXCEL data set was successfully created.
NOTE: PROCEDURE IMPORT used (Total process time):
      real time           0.13 seconds
      cpu time            0.06 seconds
```

Similarly, the following code imports a text file:

```
PROC IMPORT OUT= MYDATA.SAT_EXAM_data_from_text_file
           DATAFILE= "C:\Users\Documents\Datasets\SAT_Exam.txt"
           DBMS=TAB REPLACE;
    GETNAMES=YES;
    DATAROW=2;
RUN;
```

You can import various types of files using PROC IMPORT; the only change is in some of the parameters in the import code.

Data Manipulations

You import raw data using the import code. Most of the time you need to manipulate the raw data to prepare it for analysis. In the process, you may have to subset the data, merge it with a different data set, create new calculated variables, derive variables from existing variables, and so on. The following sections take you through some important steps, which you will need while manipulating data for almost any analysis.

Making a Copy of a SAS Data Set

Imagine that you have imported some data and you further want to prepare the data for analysis. The following code creates a copy of the data set in SAS. For this, you use a SET statement in the data step:

```
data MYDATA.sat_exam_copy;
set  MYDATA.sat_exam;
run;
```

The preceding code simply copies the data from sat_exam and stores it in sat_exam_copy in the MYDATA library. Here is the log file for this code:

```
NOTE: MYDATA.SAT_EXAM_DATA_FROM_TEXT_FILE data set was successfully created.
NOTE: PROCEDURE IMPORT used (Total process time):
      real time            0.60 seconds
      cpu time             0.18 seconds

1761  data MYDATA.sat_exam_copy;
1762  set  MYDATA.sat_exam;
1763  run;

NOTE: There were 96 observations read from the data set MYDATA.SAT_EXAM.
NOTE: The data set MYDATA.SAT_EXAM_COPY has 96 observations and 5 variables.
NOTE: DATA statement used (Total process time):
      real time            0.03 seconds
      cpu time             0.01 seconds
```

Similarly, if you want to copy the data set into a different library, say Work, then you can use the following code:

```
data sat_exam_copy;
set  MYDATA.sat_exam;
run;
```

Here is the log file for this code:

```
NOTE: There were 96 observations read from the data set MYDATA.SAT_EXAM.
NOTE: The data set MYDATA.SAT_EXAM_COPY has 96 observations and 5 variables.
NOTE: DATA statement used (Total process time):
      real time            0.03 seconds
      cpu time             0.03 seconds
```

Let's use an example of a market campaign's data to understand this topic better. The following code imports the market data file:

```
PROC IMPORT OUT= MYDATA.market
            DATAFILE= "C:\Users\Documents\Datasets\Market_campaign.csv"
            DBMS=CSV REPLACE;
       GETNAMES=YES;
       DATAROW=2;
RUN;
```

Table 3-8 lists the snapshot of the market campaign data.

Table 3-8. *A Snapshot of the Market Campaign Data*

Obs	Camp_ID	Name	start_date	end_date	Category	reach	budget
1	1	Citrix WP_Q1_13 CPG	10/25/2008	01/20/2010	Technology	36.6	15396
2	2	Evergreen WP27 CPG	11/01/2008	09/03/2009	Technology	36.6	9603
3	3	Stratus WP CPG	11/06/2008	02/23/2009	Technology	36.6	19173
4	4	Oco WP_PAN CPG	11/06/2008	02/23/2009	Technology	36.6	5305
5	5	DTI wp CPG	11/08/2008	11/16/2012	Business/Finance	100	2735
6	6	APSFC yourlessthan$2, Unlimiteddomains CPG	11/09/2008	11/09/2008	Technology	36.6	1937
7	7	Hubspot WP CPG	11/10/2008	01/11/2009	Technology	36.6	2144
8	8	Info WP CPG	11/10/2008	11/14/2010	Marketing	69.2	6657
9	9	Cognos WP CPG	11/10/2008	11/12/2012	Technology	38.5	8783
10	10	HP WP CPG	11/11/2008	11/06/2017	Technology	29.4	1414

Table 3-9 lists the variables and their descriptions.

Table 3-9. *The Variables and Their Descriptions for Market Data Examples*

Variable	Description
Camp_id	Market campaign ID
name	Name of the market campaign
start_date	Campaign start date
end_date	Campaign end date
Category	Category or type of campaign
reach	Campaign reach (100 percent reach is the target)
budget	Market campaign budget

In the next section, you will use this market campaign case study to try various data manipulation operations using SAS.

Creating New Variables

To create a new variable, you, once again, use a set statement along with new variable statement. Here is a generic sample:

```
data new_data;
set old_data;
<new var statements>;
run;
```

Let's try to write some actual SAS code snippets to elaborate on the previous generic code to create new variables.

Create a new variable, budget_new, by adding 12 percent tax.

```
data MYDATA.market_v1;
set MYDATA.market;
budget_new=budget*1.12;
run;
```

The log file for the preceding code follows:

```
NOTE: There were 7843 observations read from the data set MYDATA.MARKET.
NOTE: The data set MYDATA.MARKET_V1 has 7843 observations and 8 variables.
NOTE: DATA statement used (Total process time):
      real time            0.01 seconds
      cpu time             0.01 seconds
```

Note the first line of the SAS log file. It translates as follows. Whenever a record has a missing value of budget, the budget_new variable will also have a missing value. This is an important point to be remembered while dealing with data in SAS.

Table 3-10 lists the snapshot of the new data set, market_v1.

Table 3-10. *A Snapshot of the New Data Set: market_v1*

Camp_ID	Name	start_date	end_date	Category	reach	budget	budget_new
1	Citrix WP_Q1_13 CPG	10/25/2008	01/20/2010	Technology	36.6	15396	17243.52
2	Evergreen WP27 CPG	11/01/2008	09/03/2009	Technology	36.6	9603	10755.36
3	Stratus WP CPG	11/06/2008	02/23/2009	Technology	36.6	19173	21473.76
4	Oco WP_PAN CPG	11/06/2008	02/23/2009	Technology	36.6	5305	5941.60
5	DTI wp CPG	11/08/2008	11/16/2012	Business/ Finance	100	2735	3063.20
6	APSFC yourlessthan$2, Unlimiteddomains CPG	11/09/2008	11/09/2008	Technology	36.6	1937	2169.44
7	Hubspot WP CPG	11/10/2008	01/11/2009	Technology	36.6	2144	2401.28
8	Info WP CPG	11/10/2008	11/14/2010	Marketing	69.2	6657	7455.84
9	Cognos WP CPG	11/10/2008	11/12/2012	Technology	38.5	8783	9836.96
10	HP WP CPG	11/11/2008	11/06/2017	Technology	29.4	1414	1583.68

Let's suppose that the reach values are inflated. Take only 80 percent of the reach and name the new variable net_reach, as shown here:

```
data MYDATA.market_v2;
set MYDATA.market_v1;
net_reach=reach*0.8;
run;
```

Here is the log file for this code:

```
NOTE: There were 7843 observations read from the data set MYDATA.MARKET_V1.
NOTE: The data set MYDATA.MARKET_V2 has 7843 observations and 9 variables.
NOTE: DATA statement used (Total process time):
      real time           0.01 seconds
      cpu time            0.01 seconds
```

Table 3-11 lists the snapshot of the new data set, market_v2.

Table 3-11. *A Snapshot of the New Data Set, market_v2*

Camp_ID	Name	start_date	end_date	Category	reach	budget	budget_new	net_reach
1	Citrix WP_Q1_13 CPG	10/25/2008	01/20/2010	Technology	36.6	15396	17243.52	29.28
2	Evergreen WP27 CPG	11/01/2008	09/03/2009	Technology	36.6	9603	10755.36	29.28
3	Stratus WP CPG	11/06/2008	02/23/2009	Technology	36.6	19173	21473.76	29.28
4	Oco WP_PAN CPG	11/06/2008	02/23/2009	Technology	36.6	5305	5941.60	29.28
5	DTI wp CPG	11/08/2008	11/16/2012	Business/ Finance	100	2735	3063.20	80.00
6	APSFC yourlessthan$2, Unlimiteddomains CPG	11/09/2008	11/09/2008	Technology	36.6	1937	2169.44	29.28
7	Hubspot WP CPG	11/10/2008	01/11/2009	Technology	36.6	2144	2401.28	29.28
8	Info WP CPG	11/10/2008	11/14/2010	Marketing	69.2	6657	7455.84	55.36
9	Cognos WP CPG	11/10/2008	11/12/2012	Technology	38.5	8783	9836.96	30.80
10	HP WP CPG	11/11/2008	11/06/2017	Technology	29.4	1414	1583.68	23.52

In fact, you can execute both operations in the same data step as follows:

```
Data MYDATA.market_v3;
set MYDATA.market;
budget_new=budget*1.12;
net_reach=reach*0.8;
run;
```

Here is the log file for this code:

```
NOTE: There were 7843 observations read from the data set MYDATA.MARKET.
NOTE: The data set MYDATA.MARKET_V3 has 7843 observations and 9 variables.
NOTE: DATA statement used (Total process time):
      real time            0.01 seconds
      cpu time             0.01 seconds
```

Table 3-12 lists the snapshot of the new data set, market_v3.

Table 3-12. *A Snapshot of the New Data Set, market_v3*

Camp_ID	Name	start_date	end_date	Category	reach	budget	budget _new	net_ reach
1	Citrix WP_Q1_13 CPG	10/25/2008	01/20/2010	Technology	36.6	15396	17243.52	29.28
2	Evergreen WP27 CPG	11/01/2008	09/03/2009	Technology	36.6	9603	10755.36	29.28
3	Stratus WP CPG	11/06/2008	02/23/2009	Technology	36.6	19173	21473.76	29.28
4	Oco WP_PAN CPG	11/06/2008	02/23/2009	Technology	36.6	5305	5941.60	29.28
5	DTI wp CPG	11/08/2008	11/16/2012	Business/ Finance	100	2735	3063.20	80.00
6	APSFC yourlessthan$2, Unlimiteddomains CPG	11/09/2008	11/09/2008	Technology	36.6	1937	2169.44	29.28
7	Hubspot WP CPG	11/10/2008	01/11/2009	Technology	36.6	2144	2401.28	29.28
8	Info WP CPG	11/10/2008	11/14/2010	Marketing	69.2	6657	7455.84	55.36
9	Cognos WP CPG	11/10/2008	11/12/2012	Technology	38.5	8783	9836.96	30.80
10	HP WP CPG	11/11/2008	11/06/2017	Technology	29.4	1414	1583.68	23.52

Creating New Variables Using Multiple Variables

The syntax for creating a new field in a table using multiple variables is similar to what you have already seen for creating a single variable. The following is the generic code:

```
data new_data;
set old_data;
<new var statements>;
run;
```

Let's look at an example: A market campaign's effectiveness is measured by its reach with respect to the budget. Reach per dollar is the new variable that you want to create to determine the effectiveness of each campaign. Here is the code:

```
Data MYDATA.market_v4;
set  MYDATA.market;
reach_per_dollar=reach/budget;
run;
```

Table 3-13 lists the snapshot of the new data set, market_v4.

Table 3-13. *A Snapshot of the New Data Set, market_v4*

Camp_ID	Name	start_date	end_date	Category	reach	budget	reach_per_dollar
1	Citrix WP_Q1_13 CPG	10/25/2008	01/20/2010	Technology	36.6	15396	0.002377
2	Evergreen WP27 CPG	11/01/2008	09/03/2009	Technology	36.6	9603	0.003811
3	Stratus WP CPG	11/06/2008	02/23/2009	Technology	36.6	19173	0.001909
4	Oco WP_PAN CPG	11/06/2008	02/23/2009	Technology	36.6	5305	0.006899
5	DTI wp CPG	11/08/2008	11/16/2012	Business/Finance	100	2735	0.036563
6	APSFC yourlessthan$2, Unlimiteddomains CPG	11/09/2008	11/09/2008	Technology	36.6	1937	0.018895
7	Hubspot WP CPG	11/10/2008	01/11/2009	Technology	36.6	2144	0.017071
8	Info WP CPG	11/10/2008	11/14/2010	Marketing	69.2	6657	0.010395
9	Cognos WP CPG	11/10/2008	11/12/2012	Technology	38.5	8783	0.004383
10	HP WP CPG	11/11/2008	11/06/2017	Technology	29.4	1414	0.020792

Creating New Variables Using if-then-else

You might have to use an if-then-else structure to create new categorical variables that contain values such as High, Medium, Low, Yes or No, and so on. The following is the generic syntax for creating variables using an if-then-else structure:

```
data new_data;
set old_data;
<if then else statements>;
run;
```

Again, let's look at an example: Create a new variable called reach_ind (reach indicator), which takes a value of High when the reach is more than 66 percent, Medium when the reach is in between 33 percent and 66 percent, and Low when the reach is less than 33 percent.

```
Data MYDATA.market_v5;
set  MYDATA.market;
if reach<33 then reach_ind='Low';
else if reach >= 33 and reach <=66 then reach_ind='Med';
else reach_ind='High';
run;
```

Table 3-14 lists the snapshot of the new data set, market_v5.

Table 3-14. *A Snapshot of the New Data Set, market_v5*

Camp_ID	Name	start_date	end_date	Category	reach	budget	reach_ind
1	Citrix WP_Q1_13 CPG	10/25/2008	01/20/2010	Technology	36.6	15396	Med
2	Evergreen WP27 CPG	11/01/2008	09/03/2009	Technology	36.6	9603	Med
3	Stratus WP CPG	11/06/2008	02/23/2009	Technology	36.6	19173	Med
4	Oco WP_PAN CPG	11/06/2008	02/23/2009	Technology	36.6	5305	Med
5	DTI wp CPG	11/08/2008	11/16/2012	Business/ Finance	100	2735	Hig
6	APSFC yourlessthan$2, Unlimiteddomains CPG	11/09/2008	11/09/2008	Technology	36.6	1937	Med
7	Hubspot WP CPG	11/10/2008	01/11/2009	Technology	36.6	2144	Med
8	Info WP CPG	11/10/2008	11/14/2010	Marketing	69.2	6657	Hig
9	Cognos WP CPG	11/10/2008	11/12/2012	Technology	38.5	8783	Med
10	HP WP CPG	11/11/2008	11/06/2017	Technology	29.4	1414	Low

Here is another example. Create a new variable called budget indicator, which takes a value of High when the budget is more than 75000, Medium when the reach is in between 20000 and 75000, and Low when the reach is less than 20000.

```
Data MYDATA.market_v6;
set  MYDATA.market;
if budget<20000 then budget_ind='Low';
else if budget >= 20000 and budget <=75000 then budget_ind='Med';
else budget_ind='High';
run;
```

Table 3-15 lists the snapshot of the new data set, market_v6.

Table 3-15. *A Snapshot of the New Data Set, market_v6*

Camp_ID	Name	start_date	end_date	Category	reach	budget	budget_ind
1	Citrix WP_Q1_13 CPG	10/25/2008	01/20/2010	Technology	36.6	15396	Low
2	Evergreen WP27 CPG	11/01/2008	09/03/2009	Technology	36.6	9603	Low
3	Stratus WP CPG	11/06/2008	02/23/2009	Technology	36.6	19173	Low
4	Oco WP_PAN CPG	11/06/2008	02/23/2009	Technology	36.6	5305	Low
5	DTI wp CPG	11/08/2008	11/16/2012	Business/Finance	100	2735	Low
6	APSFC yourlessthan$2, Unlimiteddomains CPG	11/09/2008	11/09/2008	Technology	36.6	1937	Low
7	Hubspot WP CPG	11/10/2008	01/11/2009	Technology	36.6	2144	Low
8	Info WP CPG	11/10/2008	11/14/2010	Marketing	69.2	6657	Low
9	Cognos WP CPG	11/10/2008	11/12/2012	Technology	38.5	8783	Low
10	HP WP CPG	11/11/2008	11/06/2017	Technology	29.4	1414	Low

Updating the Same Data Set

Until now, you have seen how to create a new data set by adding a new variable to the original one. If the situation demands, you can add new variables to the same data set. Here is the generic code:

```
data old_data;
set old_data;
<New variable statements>;
run;
```

The following example updates the market data while creating budget_new and net_reach variables:

```
Data MYDATA.market;
set MYDATA.market;
budget_new=budget*1.12;
net_reach=reach*0.8;
run;
```

Table 3-16 lists the snapshot of the updated data set (market).

Table 3-16. *A Snapshot of the Updated Data Set (market)*

Camp_ID	Name	start_date	end_date	Category	reach	budget	budget _new	net _reach
1	Citrix WP_Q1_13 CPG	10/25/2008	01/20/2010	Technology	36.6	15396	17243.52	29.28
2	Evergreen WP27 CPG	11/01/2008	09/03/2009	Technology	36.6	9603	10755.36	29.28
3	Stratus WP CPG	11/06/2008	02/23/2009	Technology	36.6	19173	21473.76	29.28
4	Oco WP_PAN CPG	11/06/2008	02/23/2009	Technology	36.6	5305	5941.60	29.28
5	DTI wp CPG	11/08/2008	11/16/2012	Business/ Finance	100	2735	3063.20	80.00
6	APSFC yourlessthan$2, Unlimiteddomains CPG	11/09/2008	11/09/2008	Technology	36.6	1937	2169.44	29.28
7	Hubspot WP CPG	11/10/2008	01/11/2009	Technology	36.6	2144	2401.28	29.28
8	Info WP CPG	11/10/2008	11/14/2010	Marketing	69.2	6657	7455.84	55.36
9	Cognos WP CPG	11/10/2008	11/12/2012	Technology	38.5	8783	9836.96	30.80
10	HP WP CPG	11/11/2008	11/06/2017	Technology	29.4	1414	1583.68	23.52

Drop and Keep Variables

Sometimes you might not need all the variables from a raw data set. On a few occasions, you might want to keep some important variables and delete the ones that are not relevant to the analysis. For example, you may want to keep just 10 important variables out of a list of 200 in the raw data set. The syntax of this "keep and drop" SAS construct follows:

Syntax for Keep

```
data new_data;
set old_data(Keep=Var1 Var2 Var3);
<Rest of the statements>
run;
```

Syntax for Drop

```
data new_data;
set old_data(Drop=Var5 Var6 Var7);
<rest of the statements>
run;
```

In this first example, you create a new data set from the market data; keep only the campaign ID, the name, and the budget. The idea is to perform a univariate analysis solely on budget. You will learn about univariate analysis in later chapters.

```
Data MYDATA.market_v7;
set MYDATA.market(keep=camp_id name  budget);
run;
```

Table 3-17 lists the snapshot of the updated data set (market_v7).

Table 3-17. *A Snapshot of the Updated Data Set (market_v7)*

Obs	Camp_ID	Name	budget
1	1	Citrix WP_Q1_13 CPG	15396
2	2	Evergreen WP27 CPG	9603
3	3	Stratus WP CPG	19173
4	4	Oco WP_PAN CPG	5305
5	5	DTI wp CPG	2735
6	6	APSFC yourlessthan$2,Unlimiteddomains CPG	1937
7	7	Hubspot WP CPG	2144
8	8	Info WP CPG	6657
9	9	Cognos WP CPG	8783
10	10	HP WP CPG	1414

In this second example, you create a new data set from the market data, with the drop start date and end date. They will not be used anywhere in the analysis.

```
Data MYDATA.market_v8;
set MYDATA.market(Drop=start_date end_date);
run;
```

Table 3-18 lists the snapshot of the updated data set (market_v8).

Table 3-18. *A Snapshot of the Updated Data Set (market_v8)*

Obs	Camp_ID	Name	Category	reach	budget	budget_new	net_reach
1	1	Citrix WP_Q1_13 CPG	Technology	36.6	15396	17243.52	29.28
2	2	Evergreen WP27 CPG	Technology	36.6	9603	10755.36	29.28
3	3	Stratus WP CPG	Technology	36.6	19173	21473.76	29.28
4	4	Oco WP_PAN CPG	Technology	36.6	5305	5941.60	29.28
5	5	DTI wp CPG	Business/ Finance	100	2735	3063.20	80.00
6	6	APSFC yourlessthan$2, Unlimiteddomains CPG	Technology	36.6	1937	2169.44	29.28
7	7	Hubspot WP CPG	Technology	36.6	2144	2401.28	29.28
8	8	Info WP CPG	Marketing	69.2	6657	7455.84	55.36
9	9	Cognos WP CPG	Technology	38.5	8783	9836.96	30.80
10	10	HP WP CPG	Technology	29.4	1414	1583.68	23.52

These are examples of dropping the fields or columns or variables from the data. Now you will see how to subset or filter the data, that is, using some specific rows based on a condition.

Subsetting the Data

Subsetting or filtering is almost an inevitable operation while manipulating the data. You may not need all the rows of raw data for analysis. You may just want to filter out irrelevant data and consider only the data relevant for the further analysis. You are going to use the same set statement again. Here is the generic code:

```
data new_data;
set old_data;
<Where condition>;
run;
```

Here is another example:

```
data new_data;
set old_data;
<if condition>;
run;
```

In this example, you extract a subset from the market data. The resultant data should contain campaigns from the healthcare category only. The following code fulfills this requirement:

```
Data MYDATA.market_v9;
set MYDATA.market;
where category='Healthcare';
run;
```

or
```
Data MYDATA.market_v10;
set MYDATA.market;
if category='Healthcare';
run;
```

Table 3-19 lists the snapshot of the updated data set (market_v9).

Table 3-19. *A Snapshot of the Updated Data Set (market_v9)*

Camp_ID	Name	start_date	end_date	Category	reach	budget	budget _new	net _reach
4313	test DisasterLeads91508 CPG	12/22/2012	.	Healthcare	39.9	745	834.40	31.92
4316	Test DirectUS20310 CPG	12/23/2012	.	Healthcare	26.2	877	982.24	20.96
4317	accel DirectionHLOTD CPG	12/24/2012	.	Healthcare	100	806	902.72	80.00
4331	N2WDMS,Compliance DirectGovernment CPG	12/27/2012	.	Healthcare	100	641	717.92	80.00
4382	test Dell CPG	01/04/2013	.	Healthcare	100	665	744.80	80.00
4385	Trizetto DellTrialDownload CPG	01/04/2013	02/23/2013	Healthcare	1.5	673	753.76	1.20
4398	Nuance Dec CPG	01/05/2013	01/25/2013	Healthcare	0	792	887.04	0.00
4418	Test DataITwhitepapersApril11 CPG	01/11/2013	.	Healthcare	100	692	775.04	80.00
4446	Trizetto Data CPG	01/14/2013	.	Healthcare	100	187	209.44	80.00
4448	Cisco Data CPG	01/14/2013	01/25/2013	Healthcare	23.2	446	499.52	18.56

Table 3-20 lists the snapshot of the updated data set (market_v10).

Table 3-20. *A Snapshot of the Updated Data Set (market_v10)*

Camp_ID	Name	start_date	end_date	Category	reach	budget	budget _new	net _reach
4313	test DisasterLeads91508 CPG	12/22/2012	.	Healthcare	39.9	745	834.40	31.92
4316	Test DirectUS20310 CPG	12/23/2012	.	Healthcare	26.2	877	982.24	20.96
4317	accel DirectionHLOTD CPG	12/24/2012	.	Healthcare	100	806	902.72	80.00
4331	N2WDMS,Compliance DirectGovernment CPG	12/27/2012	.	Healthcare	100	641	717.92	80.00
4382	test Dell CPG	01/04/2013	.	Healthcare	100	665	744.80	80.00
4385	Trizetto DellTrialDownload CPG	01/04/2013	02/23/2013	Healthcare	1.5	673	753.76	1.20
4398	Nuance Dec CPG	01/05/2013	01/25/2013	Healthcare	0	792	887.04	0.00
4418	Test DataITwhitepapersApril11 CPG	01/11/2013	.	Healthcare	100	692	775.04	80.00
4446	Trizetto Data CPG	01/14/2013	.	Healthcare	100	187	209.44	80.00
4448	Cisco Data CPG	01/14/2013	01/25/2013	Healthcare	23.2	446	499.52	18.56

To subset the data, you can use either a where or an if condition. Yes, there is a difference between using the two conditions, even though the resulting data set appears to be the same.

Differences Between where and if

To understand the difference, you need to take a closer look at the log messages (Table 3-21).

Table 3-21. *Differences Between where and if Clauses*

Where Condition	If Condition

NOTE: There were 112 observations read from the data set MYDATA.MARKET.
 	
   WHERE category ='Healthcare';
NOTE: The data set MYDATA.MARKET_V9 has 112 observations and 7 variables.
NOTE: DATA statement used (Total process time):
 	
  real time   0.03 seconds
  cpu time   0.01 seconds
	NOTE: There were 7843 observations read from the data set MYDATA.MARKET.
	NOTE: The data set MYDATA.MARKET_V10 has 112 observations and 7 variables.
	NOTE: DATA statement used (Total process time):
	
	  real time 0.01 seconds
	  cpu time 0.00 seconds

The where clause reads only 112 observations.	The if clause reads complete data set into SAS.
The where clause first applies the filter and then reads the observations.	The if clause reads all the records and then applies the filters.
Even when there are 100,000,000 observations (for example), the where condition will apply the filter first and then read the remaining records, so it is fast compared to an if condition.	When there are 100,000,000 observations (for example), it will take lot of time to read all of them into SAS and then start applying the filters, so it takes more time compared to the where condition.

The where clause works not only in the data step; it works in the PROCstep also.
`proc print data= MYDATA.market;`
`where category='Healthcare';`
`run;`
The previous code successfully prints healthcare data.
	If works only in the data step; it doesn't works in the PROC step.
	`proc print data= MYDATA.market;`
	`if category='Healthcare';`
	`run;`
	This code generates the error, given below
	`580  proc print data= MYDATA.market;`
	`581  if category='Healthcare';`
	`      --`
	`      180`
	`ERROR 180-322: Statement is not valid or it is used out of proper order.`
	`582  run;`

Conclusion

In this chapter, you learned the basic concepts related to data manipulation in the SAS programming environment. The chapter focused on how to handle data in SAS, how to import data into SAS, how to create derived variables, and how to create a subset of the data. These coding concepts are useful when preparing data for analysis.

In the next chapter, you will see some of the important SAS procedures and start working with them. You will also learn how to join various data sets.

CHAPTER 4

■ ■ ■

Important SAS Functions and Procs

In the previous chapter, you learned about some important data manipulation techniques. You will typically take the raw data and prepare it for analysis. Once the final data is ready, you can go ahead with the analysis. The analysis might involve representing simple aggregated tables in meaningful graphs, applying simple descriptive statistics to advanced analytics, and predictive modeling. You may need advanced algorithms and functions to perform the analysis. While learning an analytics tool, it's vital to know how to use some of the important functions and algorithms.

In this chapter, you will learn about basic functions and important SAS procedures. The chapter discusses numeric, character, and date functions, as well as some important SAS procedures such as CONTENTS and SORT. Finally, you'll learn about some graphic functions such as GCHART and GPLOT.

SAS Functions

SAS has prewritten routines for most of the numerical, string, and date operations. They are called *functions*. Using functions in SAS is not very different from the other programing languages. In fact, it's easier. You need to write the function name and pass some mandatory and optional parameters. As usual, you expect each function to come with parentheses. If there are no parameters expected for a certain function, then the function will have nothing in the parentheses, as in today().

Let's start with an activity to make some data ready for use later. Let's import the Market_data_two data set, which will be used in several examples. This data set contains various types of assets used in typical marketing campaigns. Variables named White_Paper, Webinar, Software_Download, Free_Offer, Live_Event, and Case_Study represent different types of assets used for marketing campaigns.

The following SAS code imports and prints a snapshot of the data:

```
PROC IMPORT OUT= WORK.market_asset
            DATAFILE= "C:\Users\ Documents\Training\Books\Content\4. SAS Programs and
            Analytics\
            Datasets\Market_data_two.csv"
            DBMS=CSV REPLACE;
    GETNAMES=YES;
    DATAROW=2;
RUN;

proc print data=market_asset(obs=10) noobs;
run;
```

Table 4-1 lists the result of this code.

Table 4-1. *A Snapshot of the Market Asset Data*

id	name	num_ assets	White_ Paper	Webinar	Software_ Download	Free_Offer	Live_Event	Case_Study
1	Citrix WP_Q1_13 CPG	8	3	2	0	3	0	0
2	Evergreen WP27 CPG	0	0	0	0	0	0	0
3	Stratus WP CPG	3	0	0	0	0	0	0
4	Oco WP_PAN CPG	0	0	0	0	0	0	0
5	DTI wp CPG	0	0	0	0	0	0	0
6	APSFC yourlessthan$2, Unlimiteddomains CPG	0	0	0	0	0	0	0
7	Hubspot WP CPG	10	9	1	0	0	0	0
8	Info WP CPG	5	3	1	0	0	0	1
9	Cognos WP CPG	3	3	0	0	0	0	0
10	HP WP CPG	2	2	0	0	0	0	0

Now that the data set is ready, you will learn how to use functions.

Numeric Functions

Numeric functions perform arithmetic operations on numerical variables. The following sections provide some examples.

Numerical Function Example: Mean

In this example, the numerical SAS function returns the arithmetic mean of a given number of numeric variables or a given set of numbers. The mean that we're talking about here is row-wise; it's not the mean of a particular column. If the number of assets is the field name, then the mean (of the number of assets) in SAS returns the same result as the numeric value in the row, since it is a row-wise mean. The average or mean is usually calculated for more than one variable, and it appears in each row.

The syntax of a mean function is as follows:

```
MEAN (Var1, Var2, Var3...)
```

In this first example, you'll determine, what is the average of two variables white paper and case study for each campaign. The following code finds the average. Here you will use the data set market_asset, created in the previous section.

```
Data market_asset_v1;
set market_asset;
mean_val=Mean(White_Paper, Case_Study);
run;

proc print data=market_asset_v1(obs=10) noobs;
run;
```

Table 4-2 lists the snapshot of the new data set with the mean value (mean_val) added as the last column.

Table 4-2. *Mean of Variables White_Paper and Case_Study Added in the Last Column*

id	name	num_ assets	White_ Paper	Webinar	Software_ Download	Free_ Offer	Live Event	Case_ Study	mean_val
1	Citrix WP_Q1_13 CPG	8	3	2	0	3	0	0	1.5
2	Evergreen WP27 CPG	0	0	0	0	0	0	0	0.0
3	Stratus WP CPG	3	0	0	0	0	0	0	0.0
4	Oco WP_PAN CPG	0	0	0	0	0	0	0	0.0
5	DTI wp CPG	0	0	0	0	0	0	0	0.0
6	APSFC yourlessthan$2, Unlimiteddomains CPG	0	0	0	0	0	0	0	0.0
7	Hubspot WP CPG	10	9	1	0	0	0	0	4.5
8	Info WP CPG	5	3	1	0	0	0	1	2.0
9	Cognos WP CPG	3	3	0	0	0	0	0	1.5
10	HP WP CPG	2	2	0	0	0	0	0	1.0

Here are a few more functions that are similar to the mean: sum, Min, and Max.

```
data market_asset_v2;
set market_asset;
sum_two=Sum(White_Paper,Case_Study);
min_two=min(White_Paper,Case_Study);
max_two=max(White_Paper,Case_Study);
mean_val=Mean(White_Paper, Case_Study);
run;

proc print data=market_asset_v2(obs=10)noobs;
run;
```

Table 4-3 contains a snapshot of the new data set, where the last four columns list the computed output from the SAS numeric functions.

Table 4-3. Numeric Functions Applied on the Variables White_Paper and Case_Study, Added in the Last Four Columns

id	name	num_assets	White_Paper	Webinar	Software Download	Free_Offer	Live_Event	Case_Study	sum_two	min_two	max_two	mean_val
1	Citrix WP_Q1_13 CPG	8	3	2	0	3	0	0	3	0	3	1.5
2	Evergreen WP27 CPG	0	0	0	0	0	0	0	0	0	0	0.0
3	Stratus WP CPG	3	0	0	0	0	0	0	0	0	0	0.0
4	Oco WP_PAN CPG	0	0	0	0	0	0	0	0	0	0	0.0
5	DTI wp CPG	0	0	0	0	0	0	0	0	0	0	0.0
6	APSFC yourlessthan$2, Unlimiteddomains CPG	0	0	0	0	0	0	0	0	0	0	0.0
7	Hubspot WP CPG	10	9	1	0	0	0	0	9	0	9	4.5
8	Info WP CPG	5	3	1	0	0	0	1	4	1	3	2.0
9	Cognos WP CPG	3	3	0	0	0	0	0	3	0	3	1.5
10	HP WP CPG	2	2	0	0	0	0	0	2	0	2	1.0

In this second example, in the market asset data, the number of assets (num_assets) populated in the data set is supposed to be a simple sum of all assets: White_Paper, Webinar, Software_Download, Free_Offer, Live_Event, and Case_Study. An analyst claims that num_assets is not the exact sum. You need to verify this claim.

The following is the code along with the explanatory comments that the analyst writes to substantiate her claim:

```
/* introducing of a new dataset */

data market_asset_v3;

/* Using market_asset dataset */
set market_asset;

/* Finding sum of the columns in the dataset and storing it in num_asset_new*/

num_asset_new=Sum(White_Paper, Webinar, Software_Download, Free_Offer, Live_Event, Case_Study);

/* Finding the absolute difference between num_asset_new
and the column num_assets and storing it into Difference variable */

Difference=    abs( num_asset_new-num_assets);
run;

/* Printing first 10 observations of the dataset, where Difference is not equal to zero */

proc print data=market_asset_v3(obs=10) noobs;
where  Difference ne  0;
run;
```

For easier reading, the following is the same code without comments. Yes, SAS code files are really smaller than most other programming languages. The SAS code used throughout this book is similar, equal, or even lesser in size than this. Very cool!

```
data market_asset_v3;
set market_asset;
num_asset_new=Sum(White_Paper, Webinar, Software_Download, Free_Offer, Live_Event, Case_Study);
Difference=    abs( num_asset_new-num_assets);
run;

proc print data=market_asset_v3(obs=10) noobs;
where  Difference ne  0;
run;
```

Table 4-4 shows a snapshot of the new data set where the absolute difference is not equal to zero.

Table 4-4. Listing of the Data Set for the Second Example, Verifying the Analyst Claim

id	name	num_assets	White_Paper	Webinar	Software_Download	Free_Offer	Live_Event	Case_Study	num_asset_new	Difference
3	Stratus WP CPG	3	0	0	0	0	0	0	0	3
14	Trade WPITHMay-August'11 CPG	3	2	0	0	0	0	0	2	1
24	Vendor Word CPG	2	7	0	0	0	0	1	8	6
41	VeriSign WindsTrialNTA81310 CPG	5	0	0	0	0	0	0	0	5
42	uptime WindsTrialNPM81310 CPG	8	0	0	0	0	0	0	0	8
54	Fathom WindowsHubonly Nov12 CPG	9	0	0	0	0	0	0	0	9
61	Motorola Win8 CPG	10	0	0	0	0	0	0	0	10
64	IBM WightForm) CPG,	7	3	0	0	0	0	0	3	4
77	NexGen whitepaper CPG	5	0	0	0	0	0	0	0	5
84	Adaptive Whitepaper CPG	2	6	0	0	0	0	4	10	8

Table 4-5 lists some other commonly used numerical functions and their usage.

Table 4-5. *Commonly Used SAS Numeric Functions*

Function	Feature	Example

LOG	Returns the natural (base e) logarithm value	x=log(5);
		y=log(2.2);
		z=log(x+y);

EXP	Returns the value of the exponential function	x=exp(1);
		y=exp(0);
		z=exp(x*y);

SQRT	Returns the square root	x=sqrt(14);
		y=sqrt(2*3*4);
		z=sqrt(abs (x-y));

N	Returns the count of nonmissing values	x=n(2,4,.);
STD	Returns the standard deviation of input variables	y=std(2,4,.);
VAR	Returns the variance of the input variable	z=var(2,4,.);
INT	Returns the integer value of the variable	x1=int(543.210);

Character Functions

Character or string functions are required to handle text variables. The following sections explain some of the commonly used character functions.

Substring

The substring function extracts a portion of a string variable. Imagine you want to use only the first four or five characters of a variable. The substring function will do the job for you.

Here is the syntax of the substring function:

```
New variable= SUBSTR(variable, start character, number of characters)
```

In this example, you take only the first ten characters of market campaign name (the name variable in the market_asset data set) and store them in the new variable name_part1. Once this is done, you take the next ten characters and store them in the variable name_part2.

Here is the code, which is self-explanatory:

```
data market_asset_v4;
set market_asset;
name_part1=substr(name,1,10); /* Substring first 10 characters */
name_part2=substr(name,11,10); /* Substring characters from 11 to 20 */
run;

proc print data=market_asset_v4(obs=10) noobs;
run;
;
```

Table 4-6 lists a snapshot of the new data set with two new string columns added at the end.

Table 4-6. A Snapshot of the Market Asset Data with Two New String Columns Added at the End

id	name	num_assets	White_Paper	Webinar	Software_Download	Free_Offer	Live_Event	Case_Study	name_part1	name_part2
1	Citrix WP_Q1_13 CPG	8	3	2	0	3	0	0	Citrix WP_	Q1_13 CPG
2	Evergreen WP27 CPG	0	0	0	0	0	0	0	Evergreen	WP27 CPG
3	Stratus WP CPG	3	0	0	0	0	0	0	Stratus WP	CPG
4	Oco WP_PAN CPG	0	0	0	0	0	0	0	Oco WP_PAN	CPG
5	DTI wp CPG	0	0	0	0	0	0	0	DTI wp CPG	
6	APSFC yourlessthan$2, Unlimiteddomains CPG	0	0	0	0	0	0	0	APSFC your	lessthan$2
7	Hubspot WP CPG	10	9	1	0	0	0	0	Hubspot WP	CPG
8	Info WP CPG	5	3	1	0	0	0	1	Info WP CP	G
9	Cognos WP CPG	3	3	0	0	0	0	0	Cognos WP	CPG
10	HP WP CPG	2	2	0	0	0	0	0	HP WP CPG	CPG

Here are some more string functions:

```
data market_asset_v5;
set market_asset;

/* LENGTH: returns an integer number, which is the position of the rightmost character in the string
variable */

length_name=length(name);

/* TRIM: it removes trailing and following blanks. This function is useful for string
concatenation */

trim_name=trim(name);

/* UPCASE: converts all lowercase letters in the string to uppercase letters */

Captial_name=Upcase(name);

/* LOWCASE: converts all uppercase letters in the string to lowercase letters */

Small_name=Lowcase(name);
run;

/* This print code lists the selected variables only */

proc print data=market_asset_v5(obs=10) noobs;
var id name length_name trim_name Captial_name    Small_name ;
run;

proc print data=market_asset_v5(obs=10) noobs;
var id name length_name trim_name Captial_name    Small_name ;
run;
```

Table 4-7 lists the output of this code. You need to carefully observe the original string names in the second column and its transformations in the subsequent columns to comprehend what these functions are doing.

Table 4-7. Market Asset Data with Selected Variables, Application of String Functions

id	name	length_name	trim_name	Captial_name	Small_name
1	Citrix WP_Q1_13 CPG	19	Citrix WP_Q1_13 CPG	CITRIX WP_Q1_13 CPG	citrix wp_q1_13 cpg
2	Evergreen WP27 CPG	18	Evergreen WP27 CPG	EVERGREEN WP27 CPG	evergreen wp27 cpg
3	Stratus WP CPG	14	Stratus WP CPG	STRATUS WP CPG	stratus wp cpg
4	Oco WP_PAN CPG	14	Oco WP_PAN CPG	OCO WP_PAN CPG	oco wp_pan cpg
5	DTI wp CPG	10	DTI wp CPG	DTI WP CPG	dti wp cpg
6	APSFC yourlessthan$2, Unlimiteddomains CPG	41	APSFC yourlessthan$2, Unlimiteddomains CPG	APSFC YOURLESSTHAN$2, UNLIMITEDDOMAINS CPG	apsfc yourlessthan$2, unlimiteddomains cpg
7	Hubspot WP CPG	14	Hubspot WP CPG	HUBSPOT WP CPG	hubspot wp cpg
8	Info WP CPG	11	Info WP CPG	INFO WP CPG	info wp cpg
9	Cognos WP CPG	13	Cognos WP CPG	COGNOS WP CPG	cognos wp cpg
10	HP WP CPG	9	HP WP CPG	HP WP CPG	hp wp cpg

Date Functions

This section explains some commonly needed date functions that are used by the data analysts. Using date functions, you can manipulate date variables like formatting, finding durations in weeks and days, and so on.

Date Interval

You'll often need to find the interval between two dates. It is not as simple as finding the difference (in number of days) between the start date and the end date. The difference may be required in days, months, years, and quarters. You can use the INTCK function to find the interval length between two dates.

This example again refers to the market campaign data set. It has a start date and an end date for each campaign. You need to find the duration of each campaign.

Here is the code:

```
data market_campaign_v1;
set market_campaign;
Duration_days=INTCK('day',start_date,end_date); /* Finds the duration in days */
Duration_months=INTCK('month',start_date,end_date); /* Finds the duration in months */
Duration_weeks=INTCK('week',start_date,end_date); /* Finds the duration in weeks */
run;

proc print data=market_campaign_v1(obs=10) noobs;
var camp_id name Duration_days  Duration_months Duration_weeks;
run;
```

Table 4-8 gives the snapshot of the output.

Table 4-8. *Campaign Duration Calculated in Days, Months, and Week Using the INTCK Function*

Camp_ID	Name	Duration_days	Duration_months	Duration_weeks
1	Citrix WP_Q1_13 CPG	452	15	65
2	Evergreen WP27 CPG	306	10	44
3	Stratus WP CPG	109	3	16
4	Oco WP_PAN CPG	109	3	16
5	DTI wp CPG	1469	48	210
6	APSFC yourlessthan$2, Unlimiteddomains CPG	0	0	0
7	Hubspot WP CPG	62	2	9
8	Info WP CPG	734	24	105
9	Cognos WP CPG	1463	48	209
10	HP WP CPG	3282	108	469

The following are some more useful date functions:

```
data market_campaign_v2;
set market_campaign;

/* MONTH(date:returns the value as numeric for the month of the year for a SAS date */

start_month=month(start_date);

/* YEAR(date): returns the year from a valid SAS date */

start_year=year(start_date);
run;

proc print data=market_campaign_v2(obs=10) noobs;
run;
```

Table 4-9 is the snapshot of this code. The newly calculated start_month and start_year are added as the last two columns.

Table 4-9. Market Data Snapshot with Start Month and Year Calculated Using Date Functions

Camp_ID	Name	start_date	end_date	Category	reach	budget	start_month	start_year
1	Citrix WP_Q1_13 CPG	10/25/2008	01/20/2010	Technology	36.6	15396	10	2008
2	Evergreen WP27 CPG	11/01/2008	09/03/2009	Technology	36.6	9603	11	2008
3	Stratus WP CPG	11/06/2008	02/23/2009	Technology	36.6	19173	11	2008
4	Oco WP_PAN CPG	11/06/2008	02/23/2009	Technology	36.6	5305	11	2008
5	DTI wp CPG	11/08/2008	11/16/2012	Business/Finance	100	2735	11	2008
6	APSFC yourlessthan$2, Unlimiteddomains CPG	11/09/2008	11/09/2008	Technology	36.6	1937	11	2008
7	Hubspot WP CPG	11/10/2008	01/11/2009	Technology	36.6	2144	11	2008
8	Info WP CPG	11/10/2008	11/14/2010	Marketing	69.2	6657	11	2008
9	Cognos WP CPG	11/10/2008	11/12/2012	Technology	38.5	8783	11	2008
10	HP WP CPG	11/11/2008	11/06/2017	Technology	29.4	1414	11	2008

There are many more numeric, character, and date functions in SAS. Most of them are straightforward. They are not listed here, but you can easily Google them, as and when you need them.

Important SAS PROCs

In this section, you will see some of the important procedures (*procs*) in SAS. The data step is for data manipulations, and the proc, or procedure, step is for referring to algorithms and prewritten SAS routines. SAS procedures are useful for everything from producing simple data insights to getting inferences based upon advanced analysis.

The Proc Step

The proc step starts with the keyword PROC, and there is always a procedure name followed by the word PROC. You need to clearly understand the difference between functions and procedures in SAS. A function is a routine. In the following section, you start with a procedure by the name of PROC CONTENTS.

PROC CONTENTS

PROC CONTENTS gives you the SAS data set information that is similar to metadata of a database table. Before attempting to proceed with analysis, it is always a good idea to first look at the output of proc contents to get some quick information about data (or a data set to be precise). Contents data provides you with two levels of information.

- *Data set level*: This includes details of the data such as the name, total number of records, date created, number of variables, file size, file location and access permissions, and so on.

- *Variable level*: Details provided are the type of variable, length of variable, format of variable, and label of the variable.

Proc contents can be used for the following:

- Quickly checking the overall details of a data set instead of opening or printing it

- Checking the total number of observations after importing an external file into SAS

- Checking for a variable's presence in a data set

- Verifying the type and format of a variable

PROC CONTENTS Example

The following code displays the contents of the Stocks data set from the sashelp library:

```
proc contents data=sashelp.stocks;
run;
```

Table 4-10 lists the output of this SAS code.

Table 4-10. *SAS Code Output of PROC CONTENTS on sashelp.stocks*

Data Set Name	SASHELP.STOCKS	Observations	699
Member Type	DATA	Variables	8
Engine	V9	Indexes	0
Created	Wed, Jan 16, 2008 09:50:30 AM	Observation Length	72
Last Modified	Wed, Jan 16, 2008 09:50:30 AM	Deleted Observations	0
Protection		Compressed	NO
Data Set Type		Sorted	NO
Label	Performance of Three Stocks from 1996 to 2005		
Data Representation	WINDOWS_32		
Encoding	us-ascii ASCII (ANSI)		

Engine/Host Dependent Information	
Data Set Page Size	8192
Number of Data Set Pages	7
First Data Page	1
Max Obs·per Page	113
Obs in First Data Page	86
Number of Data Set Repairs	0
Filename	C:\Program Files\SAS\SASFoundation\9.2(32-bit)\ graph\sashelp\stocks.sas7bdat
Release Created	9.0201M0
Host Created	XP_PRO

Here is the output of proc contents on the sashelp.stocks data set.

Alphabetic List of Variables and Attributes						
#	Variable	Type	Len	Format	Informat	Label
8	AdjClose	Num	8	DOLLAR8.2	BEST32.	Adjusted Close
6	Close	Num	8	DOLLAR8.2	BEST32.	
2	Date	Num	8	DATE.	DATE.	
4	High	Num	8	DOLLAR8.2	BEST32.	
5	Low	Num	8	DOLLAR8.2	BEST32.	
3	Open	Num	8	DOLLAR8.2	BEST32.	
1	Stock	Char	9			
7	Volume	Num	8	COMMA12.	BEST32.	

There are three tables in the output. The first one is all about the data set properties. The second table may be slightly less important from an analyst point of view. It lists how SAS stored this data. The third table, probably the most important one, is about variables in the data set and their properties.

Two useful options in PROC Contents are VARNUM and Short.

VARNUM Option

The variables in the Proc Contents output are printed in alphabetical order, not in the order in which they appear in the data set. To print them in the original order of data set, you use the varnum option. Here is the code using this option:

```
proc contents data=sashelp.stocks varnum;
run;
```

Table 4-11 lists the output of this code. You can see that the variables in the third table of this output are printed in the original order of the data set. This order is different from that of Table 4-10, which is alphabetical.

Table 4-11. *The Output of proc contents on sashelp.stocks Data Set with varnum Option*

Data Set Name	SASHELP.STOCKS	**Observations**	699
Member Type	DATA	**Variables**	8
Engine	V9	**Indexes**	0
Created	Wed, Jan 16, 2008 09:50:30 AM	**Observation Length**	72
Last Modified	Wed, Jan 16, 2008 09:50:30 AM	**Deleted Observations**	0
Protection		**Compressed**	NO
Data Set Type		**Sorted**	NO
Label	Performance of Three Stocks from 1996 to 2005		
Data Representation	WINDOWS_32		
Encoding	us-ascii ASCII (ANSI)		

Engine/Host Dependent Information

Data Set Page Size	8192
Number of Data Set Pages	7
First Data Page	1
Max Obs per Page	113
Obs in First Data Page	86
Number of Data Set Repairs	0
Filename	C:\Program Files\SAS\SASFoundation\9.2(32-bit)\ graph\sashelp\stocks.sas7bdat
Release Created	9.0201M0
Host Created	XP_PRO

Variables in Creation Order

#	Variable	Type	Len	Format	Informat	Label
1	Stock	Char	9			
2	Date	Num	8	DATE.	DATE.	
3	Open	Num	8	DOLLAR8.2	BEST32.	
4	High	Num	8	DOLLAR8.2	BEST32.	
5	Low	Num	8	DOLLAR8.2	BEST32.	
6	Close	Num	8	DOLLAR8.2	BEST32.	
7	Volume	Num	8	COMMA12.	BEST32.	
8	AdjClose	Num	8	DOLLAR8.2	BEST32.	Adjusted Close

SHORT Option

Sometimes you just want to see the list of variables in a data set and nothing else. The short option in proc contents prints just that.

```
proc contents data=sashelp.stocks short;
run;
```

Table 4-12 lists the output.

Table 4-12. *The Output of proc contents on sashelp.stocks Data Set with short Option*

Alphabetic List of Variables for SASHELP.STOCKS

AdjClose Close Date High Low Open Stock Volume

Here is the code to print the variables in the original order:

```
proc contents data=sashelp.stocks varnum short;
run;
```

Table 4-13 lists the output.

Table 4-13. *The Output of proc contents on sashelp.stocks Data Set with varnum short Option*

Variables in Creation Order

Stock Date Open High Low Close Volume AdjClose

PROC SORT

Data sorting is used to order the data or present the results in ascending or descending order. Table 4-14 is the code and its explanation.

```
proc sort data=<dataset>;
by <variable>;
run;

proc sort data=<dataset> out = <New Data set>;
by <variable>;
run;
```

Table 4-14. *Generic Code for proc sort with Explanation*

Code	Explanation
proc sort	Calling sort procedure
data=<data set>;	Data set name
by <variable>;	Sorting variable

The option out = <New Dataset>; is used when you don't want to overwrite the original data set with the sort. Also, it is important to note that if there is any error while executing the proc sort code and if you don't give an out option, then you may lose the original data set.

PROC SORT Example

Telco bill data contains the customer ID's account start date, bill number, bill amount, and other details. You want to sort the data based on bill amount. You want to see the top 20 bill payers in the beginning of the data set.

The following code sorts the data based on bill amount. The output is saved in a different data set.

```
proc sort data=MYDATA.bill out=mydata.bill_top;
by Bill_Amount;
run;
proc print data=mydata.bill_top (obs=20);
run;
```

Table 4-15 lists the result of this code.

Table 4-15. *Output of proc sort on bill Data Set, Sorted on Bill Amount in Ascending Order*

Obs	Cust_id	Cust_type	ACCOUNT_catid_NO	cust_account_start_date	Bill Id	Bill_Amount
1	2019204440	ATTI -BC	1443853747	05/09/2010	1215319470	-2058248
2	2019204438	ATTI -AB	1443857521	09/11/2010	1215389205	-377764
3	2019125220	ATTI -CC	1443793995	12/29/2009	1215446415	-348840
4	2019136109	ATTI -EE	1443823015	06/25/2009	1215442222	-313563
5	2019125239	ATTI -DC	1443794102	06/10/2009	1229801727	-209281
6	2019125217	ATTI -FC	1443793992	10/11/2009	1229827979	-206416
7	2019135322	ATTI -FA	1443826795	12/18/2009	1215370453	-40606
8	2019134557	ATTI -DF	1443824017	05/21/2009	1215453672	-40316
9	2019134558	ATTI -CD	1443824018	09/11/2009	1215368904	-40316
10	2019134559	ATTI -AD	1443824019	06/13/2010	1215452882	-40316
11	2019136069	ATTI -AD	1443826797	10/17/2009	1215443526	-39431
12	2019136070	ATTI -DE	1443826798	10/11/2009	1215455626	-39431
13	2019136071	ATTI -BA	1443826799	03/29/2010	1215316125	-39431
14	2019139219	ATTI -BD	1443826096	04/05/2010	1215453906	-30168
15	2019139216	ATTI -CE	1443826097	02/07/2010	1215417828	-30168
16	2019133998	ATTI -DA	1443812661	03/28/2009	1215391817	-24298
17	2019133999	ATTI -CC	1443812662	09/04/2009	1215385903	-24298
18	2019134001	ATTI -CB	1443812665	02/25/2010	1215455806	-24298
19	2019135323	ATTI -EB	1443826796	09/26/2009	1215378799	-16778
20	2019136729	ATTI -FC	1443823483	12/12/2009	1215438097	-16646

You got the output, and the new table is sorted on bill amount, but the final output has a small surprise for you. The data is sorted in ascending order. Yes! By default SAS sorts data in ascending order. If you want to sort the data, you need to give the descending option. In this example, you want to see the top ten bill payers. Also note that the negative numbers might indicate that the customers have paid more than their billed amount in previous cycles. The following code sorts the table in descending order of bill amount:

```
proc sort data=MYDATA.bill out=mydata.bill_top;
by descending Bill_Amount ;
run;
proc print data=mydata.bill_top (obs=20);
run;
```

Table 4-16 lists the output.

Table 4-16. *Output of proc sort on bill Data Set, Sorted on Bill Amount Descending Order*

Obs	Cust_id	Cust_type	ACCOUNT_catid_NO	cust_account_start_date	Bill Id	Bill_Amount
1	2019163645	ATTI -ED	1443831475	12/12/2009	1208136629	2498886
2	6019124803	ATTI -FA	1443792516	10/17/2008	1229767424	1264144
3	9077649354	ATTI -FF	1442268246	02/16/2007	1208060015	842793
4	2019163654	ATTI -AF	1443831587	02/21/2010	1208133216	637458
5	9077692456	ATTI -AA	1439417474	11/22/2005	1227041576	611285
6	10254469134	ATTI -FC	2146374084	02/14/2012	1234675885	582857
7	10254646912	ATTI -AB	2146412749	08/18/2012	1234570950	506827
8	2019163646	ATTI -EE	1443831477	12/28/2009	1208069539	493230
9	10254658023	ATTI -CE	2146412750	01/15/2012	1234688311	490434
10	2019165914	ATTI -EC	1443836616	01/26/2010	1208155818	449101
11	9077692456	ATTI -CB	1439417474	10/31/2006	1239190652	435840
12	2019145910	ATTI -FC	1443832534	01/07/2010	1208101338	354218
13	10254315887	ATTI -DF	2146895608	11/08/2012	1234667240	331287
14	9077583456	ATTI -CC	1441012525	06/21/2007	1229999711	329104
15	2019145904	ATTI -EC	1443832496	07/18/2009	1208054195	324730
16	9078289456	ATTI -AC	2142893560	10/30/2011	1239181341	275988
17	6019123940	ATTI -FF	1443795885	01/16/2009	1208032704	252900
18	9077651540	ATTI -FE	1442286993	12/19/2007	1208104760	236049
19	2019134657	ATTI -BE	1443814020	11/15/2009	1238345134	233339
20	2019134656	ATTI -FF	1443814018	12/20/2009	1238346617	233339

Now Table 4-16 clearly shows the top ten bill payers. Let's proceed from here.

Using the same example, imagine that you want to first sort the customers based on their account start date old to new, and if there are any ties, then you want to sort them based on bill amount. Here is the SAS code to do that:

```
proc sort data=MYDATA.bill out=mydata.StartDate_bill_top;
by cust_account_start_date descending  Bill_Amount ;
run;
proc print data=mydata.StartDate_bill_top(obs=20);
run;
```

Table 4-17 lists the output.

The data in Table 4-17 is sorted based on account start date in ascending order and bill amount in descending order.

Table 4-17. *Output of proc sort on Bill Data Set, Sorted on Account Start Date and Bill Amount*

Obs	Cust_id	Cust_type	ACCOUNT_catid_NO	cust_account_start_date	Bill Id	Bill_Amount
1	9077106503	ATTI -CA	1434718728	03/14/2003	1210568508	295
2	9077104492	ATTI -AA	1434730232	04/28/2003	1211768733	737
3	9076721940	ATTI -DE	1434576799	05/03/2003	1231816553	624
4	9077106815	ATTI -FD	1434719649	06/08/2003	1231739507	94
5	9078044456	ATTI -BC	1435104581	06/26/2003	1226902254	30506
6	9077742460	ATTI -CB	1435109506	07/02/2003	1239407855	93
7	9077078485	ATTI -AF	1435032433	07/26/2003	1231567944	373
8	9077742460	ATTI -DD	1435109506	07/31/2003	1226859630	93
9	9077745206	ATTI -FB	1435147808	08/30/2003	1227052025	396
10	9077105547	ATTI -DD	1434714874	09/16/2003	1211774448	117
11	9077745206	ATTI -EE	1435147808	09/26/2003	1239195535	396
12	9077107856	ATTI -DB	1434723773	10/06/2003	1210826627	94
13	9077104492	ATTI -BB	1434730232	10/14/2003	1231589956	737
14	9077108226	ATTI -DA	1434725027	10/15/2003	1231303951	254
15	9077108226	ATTI -AB	1434725027	10/27/2003	1210433693	93
16	9078054456	ATTI -ED	1435104581	11/03/2003	1239153363	39338
17	9077109927	ATTI -DC	1434731250	11/05/2003	1231681405	709
18	9077106815	ATTI -DB	1434719649	11/16/2003	1210622786	141
19	9077105547	ATTI -CF	1434714874	11/25/2003	1231330690	141
20	9077742464	ATTI -FE	1435131450	12/16/2003	1239381137	93

PROC SORT Along with Other Options

You can rename, keep, drop, and use a where condition in a proc sort procedure. The following example sorts the data based on bill amount and keeps only the accounts where the bill is more than 100,000.

```
proc sort data=MYDATA.bill out=mydata.bill_top100k;
by descending Bill_Amount ;
where Bill_Amount>100000;
run;

proc print data=mydata.bill_top100k;
run;
```

Table 4-18 lists the output.

Table 4-18. *Output of proc sort with where Clause on the Bill Data Set (Bill_Amount>100000)*

Obs	Cust_id	Cust_type	ACCOUNT_ catid_NO	cust_account_ start_date	Bill Id	Bill_Amount
1	2019163645	ATTI -ED	1443831475	12/12/2009	1208136629	2498886
2	6019124803	ATTI -FA	1443792516	10/17/2008	1229767424	1264144
3	9077649354	ATTI -FF	1442268246	02/16/2007	1208060015	842793
4	2019163654	ATTI -AF	1443831587	02/21/2010	1208133216	637458
5	9077692456	ATTI -AA	1439417474	11/22/2005	1227041576	611285
6	10254469134	ATTI -FC	2146374084	02/14/2012	1234675885	582857
7	10254646912	ATTI -AB	2146412749	08/18/2012	1234570950	506827
8	2019163646	ATTI -EE	1443831477	12/28/2009	1208069539	493230
9	10254658023	ATTI -CE	2146412750	01/15/2012	1234688311	490434
10	2019165914	ATTI -EC	1443836616	01/26/2010	1208155818	449101
11	9077692456	ATTI -CB	1439417474	10/31/2006	1239190652	435840
12	2019145910	ATTI -FC	1443832534	01/07/2010	1208101338	354218
13	10254315887	ATTI -DF	2146895608	11/08/2012	1234667240	331287
14	9077583456	ATTI -CC	1441012525	06/21/2007	1229999711	329104
15	2019145904	ATTI -EC	1443832496	07/18/2009	1208054195	324730
16	9078289456	ATTI -AC	2142893560	10/30/2011	1239181341	275988
17	6019123940	ATTI -FF	1443795885	01/16/2009	1208032704	252900
18	9077651540	ATTI -FE	1442286993	12/19/2007	1208104760	236049
19	2019134657	ATTI -BE	1443814020	11/15/2009	1238345134	233339
20	2019134656	ATTI -FF	1443814018	12/20/2009	1238346617	233339
21	2019163655	ATTI -FE	1443832487	08/23/2009	1208057193	221453
22	9078289456	ATTI -FB	2142893560	06/26/2011	1227030890	219704
23	4011125545	ATTI -DD	1443797530	01/11/2009	1208034685	211467
24	2019204479	ATTI -EF	1443855060	09/22/2010	1208105777	210767
25	9077583456	ATTI -BC	1441012525	04/05/2007	1207790642	203463
26	6019189714	ATTI -FC	1443870327	04/12/2011	1207981367	202345
27	9077317566	ATTI -EB	1443768772	12/04/2008	1225081608	196723
28	9077320429	ATTI -BF	1443765739	08/18/2009	1229715387	193601
29	2019145134	ATTI -DC	1443833256	09/06/2010	1208058102	187826
30	9077314727	ATTI -CD	1443257150	07/18/2008	1208104221	182678

(continued)

Table 4-18. (*continued*)

Obs	Cust_id	Cust_type	ACCOUNT_catid_NO	cust_account_start_date	Bill Id	Bill_Amount
31	2019126967	ATTI -DD	1443819722	04/20/2009	1208043167	180400
32	2019133923	ATTI -FF	1443812511	08/29/2009	1208140810	180398
33	8019133567	ATTI -EE	1443832612	10/21/2009	1208086678	176451
34	8019133565	ATTI -EF	1443832615	11/24/2009	1208123022	176438
35	9077319490	ATTI -DC	1443716173	09/29/2008	1207983440	173663
36	9078431556	ATTI -AB	2144787600	02/24/2012	1235745034	151896
37	9078431556	ATTI -CE	2144787600	07/08/2012	1215806763	148594
38	10254646912	ATTI -FF	2146412749	03/10/2012	1214536083	140433
39	9078848456	ATTI -EA	2144373403	06/13/2012	1214398775	139701
40	2019147405	ATTI -EA	1443836711	01/24/2010	1208170734	131113
41	2019150164	ATTI -DF	1443850596	02/27/2010	1226457535	130443
42	2019163640	ATTI -CE	1443831580	06/12/2010	1208123432	129329
43	2019163641	ATTI -EB	1443831583	08/02/2009	1208136399	129329
44	2019163642	ATTI -CC	1443831582	01/29/2010	1208011444	129329
45	2019163644	ATTI -DF	1443831476	09/05/2009	1208160793	129329
46	2019184355	ATTI -CD	1443834259	09/22/2009	1208137492	129307
47	9078740456	ATTI -CC	1443822918	06/01/2010	1239713907	122922
48	9078740456	ATTI -AC	1443822918	01/22/2010	1225547457	122379
49	2019136337	ATTI -BC	1443824559	11/06/2009	1208083564	118071
50	8830333572	ATTI -AE	1443833396	09/05/2009	1208095895	111322
51	10254469134	ATTI -CE	2146374084	11/30/2012	1214478721	110284
52	2019144773	ATTI -AB	1443828846	08/07/2010	1208102132	109644
53	10254658023	ATTI -AD	2146412750	05/01/2012	1214634671	108807
54	9077970856	ATTI -EC	1440365282	01/31/2007	1225081133	105935
55	2019136664	ATTI -FE	1443823171	06/15/2010	1208104946	105932
56	2019163647	ATTI -AC	1443832461	01/30/2010	1208083716	103660
57	9077970856	ATTI -FD	1440365282	09/22/2007	1239333478	102318
58	2019185719	ATTI -AD	1443837684	03/13/2010	1208168642	100200

PROC SORT for Removing Duplicates

Apart from simply sorting the data, proc sort can also be used for removing duplicate records. There are several occasions where you want to remove the duplicates. Imagine salary records of a set of employees. In any case, you don't want to credit salary more than once for any single employee. Using proc sort, you can very well remove the duplicate records. You use the nodup option in the proc sort code to remove the duplicate records.

Just as an example, you want to sort some data based on customer ID and remove the duplicate records in the same step. The following code sorts the whole data based on customer ID and also removes the duplicates from the input data before storing it in a new data set:

```
proc sort data=MYDATA.bill out=mydata.bill_wod nodup;
by cust_id  ;
run;
```

Here is the log message from SAS:

```
NOTE: There were 31183 observations read from the data set MYDATA.BILL.
NOTE: 19 duplicate observations were deleted.
NOTE: The data set MYDATA.BILL_WOD has 31164 observations and 6 variables.
NOTE: PROCEDURE SORT used (Total process time):
      real time           0.06 seconds
      cpu time            0.05 seconds
```

The log message clearly states that there are 19 duplicate observations and they are deleted.

You can see the preceding code removes the duplicate records. When a complete observation is repeated, then it will be removed using the nodup option. What if you want to remove the records based on a few variables? For example, say you want to remove all the repetitions based on bill ID. Every bill should have one ID, and it should not be repeated. In this case, the nodup option is not sufficient. You have to use the nodupkey so that the duplicates will be deleted based on the key, in other words, the bill ID.

This code removes the duplicates based on bill ID:

```
proc sort data=MYDATA.bill out=mydata.bill_cust_wod nodupkey;
by Bill_Id ;
run;
```

Here is the log message from SAS:

```
NOTE: There were 31183 observations read from the data set MYDATA.BILL.
NOTE: 93 observations with duplicate key values were deleted.
NOTE: The data set MYDATA.BILL_CUST_WOD has 31090 observations and 6 variables.
NOTE: PROCEDURE SORT used (Total process time):
      real time           0.03 seconds
      cpu time            0.04 seconds
```

There are 93 entries in the bill data set with repeated bill IDs. They might have different bill values. Also, they may be of different account types, but the bill ID is still the same. Hence, they will be deleted.

Taking Duplicates into a New File

In the previous section, you saw how to remove duplicates. Many times you will want to put all these duplicate entries into a different file and use them in further analysis. The out option gives you the output file but not the duplicates. You need to use the dupout option to put the duplicates into a separate file.

The following code sends the duplicate entries into a new file. This code is using the same example as in the earlier section.

```
proc sort data=MYDATA.bill out=mydata.bill_wod nodup dupout=mydata.nodup_cust_id ;
by cust_id   ;
run;
proc print data=mydata.nodup_cust_id;
run;
```

Table 4-19 lists the output of this code.

Table 4-19. *Duplicate cust_id Records of the Bill Data Set Put into a Separate File*

Obs	Cust_id	Cust_type	ACCOUNT_catid_NO	cust_account_start_date	Bill_Id	Bill_Amount
1	2019129785	ATTI -DF	1443826763	01/14/2010	1208093336	89277
2	2019134442	ATTI -FA	1443817418	06/11/2009	1208121521	80394
3	2019134963	ATTI -EF	1443821926	04/27/2010	1207962346	76932
4	2019136596	ATTI -BB	1443822377	01/25/2010	1208154630	89393
5	2019137840	ATTI -CD	1443824086	01/04/2010	1208057924	75513
6	2019144899	ATTI -AF	1443829413	07/24/2010	1208115566	83972
7	2019148052	ATTI -AC	1443839399	07/29/2010	1207978398	76482
8	2019165851	ATTI -AC	1443836159	10/07/2009	1208043440	78155
9	2019167569	ATTI -ED	1443844753	06/08/2010	1226481397	89072
10	2019184898	ATTI -AD	1443834883	07/21/2009	1208066185	78522
11	6019168766	ATTI -AC	1443848706	01/11/2011	1225067938	77335
12	8019133564	ATTI -BB	1443832614	09/08/2009	1207956396	90781
13	8019133566	ATTI -AB	1443832611	09/06/2009	1208050388	90781
14	8019133568	ATTI -AF	1443832613	09/04/2009	1208056363	90781
15	8019133569	ATTI -FD	1443833390	05/23/2010	1208032121	90781
16	8094233570	ATTI -EC	1443833399	10/01/2009	1207996575	90781
17	8850333571	ATTI -CF	1443833863	09/25/2009	1208050518	90781
18	8858333573	ATTI -CD	1443833860	03/09/2010	1208056802	90781
19	9077315059	ATTI -CE	1443353495	04/01/2008	1226921413	85501

Now you repeat the same process for Bill_id and put the duplicate records into a separate file.

```
proc sort data=MYDATA.bill out=mydata.bill_cust_wod nodupkey dpout=mydata.nodupkeys_bill_id
;
by Bill_Id ;
run;
proc print data=mydata.nodupkeys_bill_id;
run;
```

Table 4-20 lists the output of this code.

Table 4-20. *Duplicate bill_id Records of the Bill Data Set Taken into a Separate File*

Obs	Cust_id	Cust_type	ACCOUNT_ catid_NO	cust_account_ start_date	Bill_Id	Bill_Amount
1	8019133564	ATTI -BB	1443832614	09/08/2009	1207956396	90781
2	2019134963	ATTI -EF	1443821926	04/27/2010	1207962346	76932
3	2019148052	ATTI -AC	1443839399	07/29/2010	1207978398	76482
4	9078135827	ATTI -EF	2138991994	12/30/2009	1207979140	205
5	2019125219	ATTI -DE	1443793994	03/22/2009	1207986127	85035
6	8094233570	ATTI -EC	1443833399	10/01/2009	1207996575	90781
7	2019141114	ATTI -CD	1443835683	07/09/2009	1208014658	10314
8	2019186225	ATTI -DD	1443844885	01/29/2010	1208020055	3397
9	2019186412	ATTI -FC	1443854447	10/28/2010	1208020303	21575
10	2019184834	ATTI -FF	1443834947	08/18/2010	1208027910	2865
11	8019133569	ATTI -FD	1443833390	05/23/2010	1208032121	90781
12	2019165851	ATTI -AC	1443836159	10/07/2009	1208043440	78155
13	8019133566	ATTI -AB	1443832611	09/06/2009	1208050388	90781
14	8850333571	ATTI -CF	1443833863	09/25/2009	1208050518	90781
15	2019185979	ATTI -DA	1443843731	09/10/2010	1208056276	2865
16	8019133568	ATTI -AF	1443832613	09/04/2009	1208056363	90781
17	8858333573	ATTI -CD	1443833860	03/09/2010	1208056802	90781
18	2019137840	ATTI -CD	1443824086	01/04/2010	1208057924	75513
19	2019184898	ATTI -AD	1443834883	07/21/2009	1208066185	78522
20	2019139922	ATTI -EA	1443830204	02/26/2010	1208071592	2865

Graphs Using SAS

In this section, you will plot the data using SAS. SAS can produce most of the graphs that are used in day-to-day business. You use the GPLOT and GCHART procedures for drawing the graphs.

PROC gplot and Gchart

Using the gplot and Gchart procedures, you can draw one variable and two variable graphs. For example, you want to draw a graph that shows the relation between budget and reach. As the market campaign budget increases, what happens to reach? Does it increase unconditionally? You can infer information about these points looking at the scatter plot between budget and reach.

The generic code for drawing a scatter plot follows:

```
proc gplot data= <data set>;
plot y*x;
run;
```

Table 4-21 explains this code.

Table 4-21. *Code Listing for proc gplot Generic Code*

Code	Explanation
proc gplot	Calling gplot procedure
data=<data set>;	Data set name
plot y*x;	Scatter plot between Y and X

Here is the code for drawing a scatter plot between budget and reach:

```
symbol i=none;
proc gplot data= market_asset;
plot reach*budget;
run;
```

Figure 4-1 shows the results from this code.

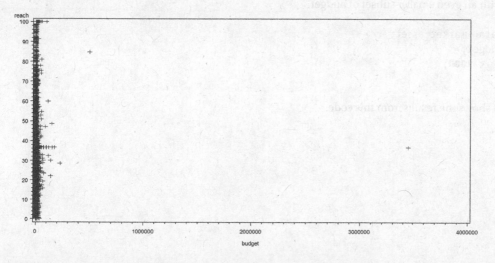

Figure 4-1. *A scatter plot for reach versus budget (market_asset data set)*

There is no clear conclusion that can be drawn from Figure 4-1. There are few extremely high values in the budget. Let's draw a scatter plot for a subset.

```
proc gplot data= market_asset;
plot reach*budget;
where budget  < 100000;
run;
```

Figure 4-2 shows the results from this code.

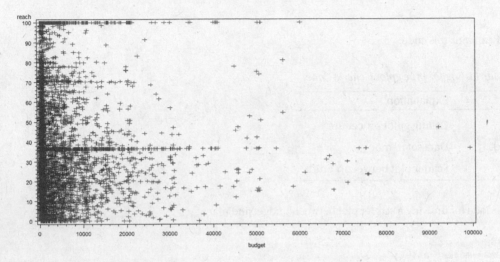

Figure 4-2. *A scatter plot for reach versus budget (market_asset data set, budget< 100000)*

Here also you can't see any clear evidence. It doesn't tell anything; if budget increases, what happens to reach?

Let's try with an even smaller subset of budget.

```
proc gplot data= market_asset;
plot reach*budget;
where budget  < 1000;
run;
```

Figure 4-3 shows the results from this code.

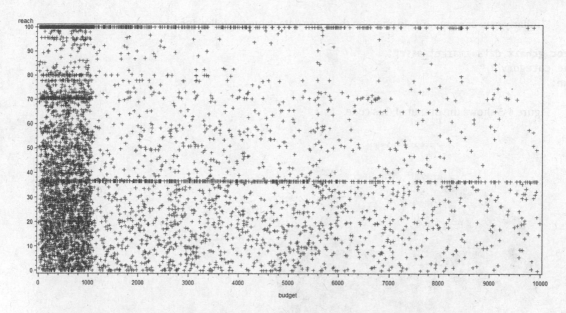

Figure 4-3. *A scatter plot for reach versus budget (market_asset data set, budget< 1000)*

Again, there is no clear trend on reach versus budget. So, you may conclude that budget and reach are not dependent on each other as far as this input data is concerned.

Bar Chart and Pie Chart Using SAS

You can also plot bar charts using SAS. For this you need to use proc Gchart. To explain this, you'll again use the same market campaign example. This time you want to see the number of market campaigns in each category. Here is the code:

```
proc gchart data= market_asset;
vbar category;
Run;
```

Figure 4-4 shows the output of this code.

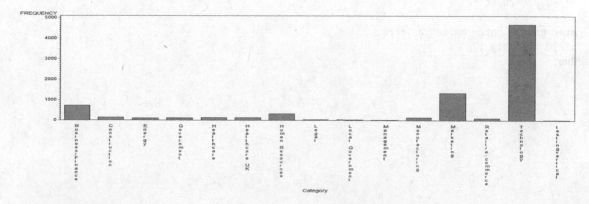

Figure 4-4. *Category-wise number market campaigns: a bar chart using proc gchart*

123

Similarly, you can create a pie chart on the same data using the following code:

```
proc gchart data= market_asset;
pie category ;
Run;
```

Figure 4-5 shows the output of this code.

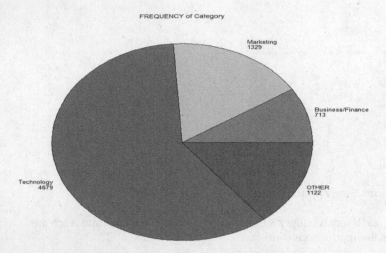

Figure 4-5. *Category-wise number market campaigns: a pie chart using proc gchart*

You can draw the same bar and pie charts in three dimensions (3D) using the following code. See Figures 4-6 and 4-7.

```
/* 3D bar chart */

proc gchart data= market_asset;
vbar3d category ;
Run;

/* 3D pie chart */

proc gchart data= market_asset;
pie3d category ;
Run;
```

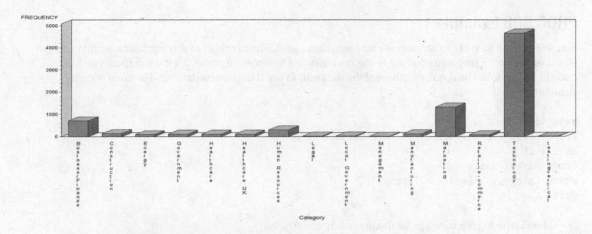

Figure 4-6. Category-wise number market campaigns: a 3D bar chart using proc gchart

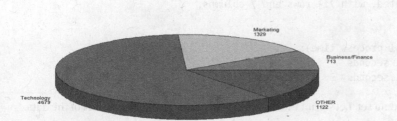

Figure 4-7. Category-wise number market campaigns: a 3D pie chart using proc gchart

PROC SQL

The PROC SQL procedure is a boon for those who are coming from a database and SQL background. SAS has a SQL engine, and you can turn it on by using PROC SQL. Once you use a PROC SQL statement in the code, then you can write the standard SQL commands to handle data. You need to finally use a QUIT statement to stop the SQL engine. A generic template for the code follows:

proc sql;

<SQL Statements> ;

Quit;

Note that you use QUIT instead of a usual RUN statement in proc sql. This is just to stop the execution of SQL statements. If you don't use QUIT, the SQL engine runs until you explicitly terminate it.

PROC SQL Example 1

You will use the same familiar market campaign data again. The manager of this market campaign analysis wants to see only campaigns that are in the business and finance categories. First you write some SQL code in SAS to take the required subset of the data and to put it into a new data set. The code is pretty straightforward.

```
proc sql;
create buss_fin /* This is the new dataset */
as select *
from  market_asset
where Category= 'Business/Finance';
Quit;
```

Here is the log file message for the preceding code:

```
87   proc sql;
88   create table buss_fin
89   as select *
90   from  market_asset
91   where Category= 'Business/Finance';
NOTE: Table WORK.BUSS_FIN created, with 713 rows and 7 columns.

92   Quit;
NOTE: PROCEDURE SQL used (Total process time):
      real time           0.01 seconds
      cpu time            0.01 seconds
```

There are 713 rows in the new data set. Let's print the first ten observations just to have a feel of the data.

```
proc print data=buss_fin(obs=10);
run;
```

Table 4-22 lists the output of the code.

Table 4-22. A Snapshot of the buss_fin Data Set

Obs	Camp_ID	Name	start_date	end_date	Category	reach	budget
1	5	DTI wp CPG	11/08/2008	11/16/2012	Business/Finance	100	2735
2	13	MTI WP CPG	11/28/2008	12/23/2013	Business/Finance	100	11085
3	14	Trade WPITHMay-August'11 CPG	11/29/2008	08/14/2010	Business/Finance	36.6	30180
4	50	Direct WindowsNov2012 CPG	01/19/2009	12/31/2012	Business/Finance	100	1517
5	56	MARI Windows2012 CPG	01/25/2009	04/09/2009	Business/Finance	36.6	6878
6	73	Excel Whitepapers CPG	02/06/2009	02/15/2016	Business/Finance	100	40588
7	75	DOI Whitepaper CPG	02/08/2009	06/27/2009	Business/Finance	36.6	54532
8	77	NexGen whitepaper CPG	02/12/2009	07/24/2010	Business/Finance	36.6	30774
9	89	Direct White90711 CPG	02/22/2009	06/27/2009	Business/Finance	36.6	21034
10	118	Lone Website CPG	03/23/2009	06/20/2010	Business/Finance	36.6	1540

PROC SQL Example 2

If the manager wants to see the total budget spend in each category, the following SQL code can be used:

```
proc sql;
select  Category, sum(budget) as total_budget
from  market_asset
group by  Category;
Quit;
```

Table 4-23 lists the output of this code.

Table 4-23. *Category-wise Budget Spends in the Market Campaign Data*

Category	total_budget
Business/Finance	5388826
Construction	174274.5
Energy	66633
Government	124867
Healthcare	522192
Healthcare UK	692567
Human Resources	1288243
Legal	18624.5
Local Government	32442
Management	5102
Manufacturing	194594
Marketing	4920834
Retail/e-commerce	235812.9
Technology	20757224
testingvertical	1234

■ **Note**   The SQL code used for left, right, and inner joins remains the same in `proc sql`. It's not repeated here. You may want to refer to any standard SQL text or Google it. Google search strings like *left join in SAS* should give you the required answer.

Data Merging

Analysts might have to work with multiple data sets and join or merge them to prepare one final data set for analysis. Imagine a scenario in the market campaign data. The information on leads is in a different table. You may want to merge these two data sets to prepare a final data set for analysis.

Let's try to understand these concepts using examples, which has been the usual approach throughout this book. The following code uses two sets of students. Each set has five students. The first set of students has mathematics as a subject in the exam, and the second set has science. The first three students appear in both the data sets. You will consider these two simple data sets to understand the concept of various types of data merging:

```
data students1;
input name $ maths;
cards;
Andy 78
Bill 90
Mark 80
Jeff 75
John 60
;
data students2;
input name $ science;
cards;
Andy 56
Bill 75
Mark 78
Fred 86
Alex 77
;
```

Appending the Data

Appending is simply affixing the second data set after the first data set. It's not really a big deal. In the resultant data set, the rows in the second data set will simply appear after the first data set. Here is the code for appending the data sets. You simply use a SET statement.

```
Data Students_1_2;
set students1 students2;
run;
```

Here is the log file when you execute this code:

```
NOTE: There were 5 observations read from the data set WORK.STUDENTS1.
NOTE: There were 5 observations read from the data set WORK.STUDENTS2.
NOTE: The data set WORK.STUDENTS_1_2 has 10 observations and 3 variables.
NOTE: DATA statement used (Total process time):
      real time           0.03 seconds
      cpu time            0.01 seconds
```

Let's take a snapshot of the resultant data set (Table 4-24).

```
proc print data=Students_1_2;
run;
```

Table 4-24. *A Snapshot of the Resultant Students_1_2 Data Set, Using the SET Keyword*

Obs	name	maths	science
1	Andy	78	.
2	Bill	90	.
3	Mark	80	.
4	Jeff	75	.
5	John	60	.
6	Andy	.	56
7	Bill	.	75
8	Mark	.	78
9	Fred	.	86
10	Alex	.	77

If there are some common columns in the data sets, then they will be populated (as a single column) in the new data set; otherwise, the columns will be left blank. In this data set, the name column is populated automatically, but the mathematics and science columns are left blank in the data rows, which are not applicable.

Table 4-24 shows that Andy, Bill, and Mark are repeated twice each. You see that the SET statement simply appends the data. You may want to do more than that. The SET statement doesn't work if you want to see both science and math marks against Andy, Bill, and Mark, without repeating their names (Table 4-25). For this, you need to use the MERGE option instead of SET.

Table 4-25. *Data Sets with SET and MERGE Options in SAS Code*

Obs	name	maths	science	Obs	name	maths	science
1	Andy	78	.	1	Andy	78	56
2	Bill	90	.	2	Bill	90	75
3	Mark	80	.	3	Mark	80	78
4	Jeff	75	.	4	Jeff	75	.
5	John	60	.	5	John	60	.
6	Andy	.	56	6	Fred	.	86
7	Bill	.	75	7	Alex	.	77
8	Mark	.	78				
9	Fred	.	86				
10	Alex	.	77				

From SET to MERGE

To get the result shown on the right side of Table 4-25, you need to replace set with a merge statement in your SAS code. Sorting is necessary before merging data sets; otherwise, SAS will throw an error. The following is the code for merging two students' data sets:

```
proc sort data=students1;
by name;
run;
proc sort data=students2;
by name;
run;

data studentmerge;
Merge students1 students2;
by name;
run;

proc print data=studentmerge;
run;
```

The log file after execution looks like this:

```
182   proc sort data=students1;
183   by name;
184   run;

NOTE: There were 5 observations read from the data set WORK.STUDENTS1.
NOTE: The data set WORK.STUDENTS1 has 5 observations and 2 variables.
NOTE: PROCEDURE SORT used (Total process time):
      real time            0.03 seconds
      cpu time             0.03 seconds

185   proc sort data=students2;
186   by name;
187   run;

NOTE: There were 5 observations read from the data set WORK.STUDENTS2.
NOTE: The data set WORK.STUDENTS2 has 5 observations and 2 variables.
NOTE: PROCEDURE SORT used (Total process time):
      real time            0.03 seconds
      cpu time             0.01 seconds

188
189   data studentmerge;
190   Merge students1 students2;
191   by name;
192   run;
```

```
NOTE: There were 5 observations read from the data set WORK.STUDENTS1.
NOTE: There were 5 observations read from the data set WORK.STUDENTS2.
NOTE: The data set WORK.STUDENTMERGE has 7 observations and 3 variables.
NOTE: DATA statement used (Total process time):
      real time            0.03 seconds
      cpu time             0.03 seconds
```

Table 4-26 lists the output.

Table 4-26. *A Snapshot of the Resultant studentmerge Data Set, Using the MERGE Keyword*

Obs	name	maths	science
1	Alex	.	77
2	Andy	78	56
3	Bill	90	75
4	Fred	.	86
5	Jeff	75	.
6	John	60	.
7	Mark	80	78

Blending with Condition

Sometimes the situation demands that you have all the observations from one data set and just the matching observations from the other. In other words, you want to have all the observations from data set 2 and only matching values from data set 1.

Table 4-27 shows the example for data set 1 with the math entries completely filled in and some values blank in the science marks column.

Table 4-27. *Example Data Set 1 with Two Blanks in the Science Column*

Obs	name	maths	science
1	Andy	78	56
2	Bill	90	75
3	Mark	80	78
4	Jeff	75	
5	John	60	

Table 4-28 shows the example for data set 2 with some math entries blank.

Table 4-28. *Example Data Set 2 with Two Blanks in the maths Column*

Obs	name	maths	science
1	Andy	78	56
2	Bill	90	75
3	Mark	80	78
4	Fred	.	86
5	Alex	.	77

At first you get all observations from data set 1 and only the matching observations from data set 2. Here is the code:

```
data final;
Merge data1(in=a) data2(in=b);
by var;
if a;
run;
```

The if a statement will keep all the entries from data set 1 and only the matching records from data set 2. Similarly, you can do the reverse by using the following code:

```
data final1;
Merge data1(in=a) data2(in=b);
by var;
if b;
run;
```

The if b statement will keep all the entries of data set 2 and keep only the matching records from data set 1.

If you want to keep the records, which appear in both the data sets, you can use an if a and b statement, as given here:

```
data final2;
Merge data1(in=a) data2(in=b);
by var;
if a and b;
run;
```

As an exercise, you may want to print the data sets final, final1, and final2 to validate the results.

Matched Merging

To demonstrate the concepts of matched merging in this section, you use the students1 and students2 data sets from the data merging section. You may recall that both had only two columns. The following is the code to keep all records from students1, along with matching records from students2:

```
/*matched merging example-1 */

proc sort data=students1;
by name;
run;
proc sort data=students2;
by name;
run;

data studentmerge1;
Merge students1(in=a) students2(in=b);
by name;
if a;
run;
```

Here is the log file for the preceding code:

```
NOTE: There were 5 observations read from the data set WORK.STUDENTS1.
NOTE: There were 5 observations read from the data set WORK.STUDENTS2.
NOTE: The data set WORK.STUDENTMERGE1 has 5 observations and 3 variables.
NOTE: DATA statement used (Total process time):
      real time            0.03 seconds
      cpu time             0.01 seconds
```

Table 4-29 is the result of this code.

Table 4-29. *Matched Merging Example 1*

Obs	name	maths	science
1	Andy	78	56
2	Bill	90	75
3	Jeff	75	.
4	John	60	.
5	Mark	80	78

The following code keeps all records from students2, along with matching records in students1:

```
/*Matched merging example-2 */

data studentmerge3;
Merge students1(in=a) students2(in=b);
by name;
if a and b;
run;
```

Here is the log file for the preceding code:

```
NOTE: There were 5 observations read from the data set WORK.STUDENTS1.
NOTE: There were 5 observations read from the data set WORK.STUDENTS2.
NOTE: The data set WORK.STUDENTMERGE2 has 5 observations and 3 variables.
NOTE: DATA statement used (Total process time):
      real time          0.03 seconds
      cpu time           0.01 seconds
```

Table 4-30 shows the output of this code.

Table 4-30. *Matched Merging Example 2*

Obs	name	maths	science
1	Alex	.	77
2	Andy	78	56
3	Bill	90	75
4	Fred	.	86
5	Mark	80	78

The following code keeps all matching records in `students1` and `students2`:

```
/* Matched merging example-3 */

data studentmerge3;
Merge students1(in=a) students2(in=b);
by name;
if a and b;
run;
```

The log file shows the following:

```
NOTE: There were 5 observations read from the data set WORK.STUDENTS1.
NOTE: There were 5 observations read from the data set WORK.STUDENTS2.
NOTE: The data set WORK.STUDENTMERGE3 has 3 observations and 3 variables.
NOTE: DATA statement used (Total process time):
      real time          0.03 seconds
      cpu time           0.03 seconds
```

Table 4-31 is the output of this code.

Table 4-31. *Matched Merging Example 3*

Obs	name	maths	science
1	Andy	78	56
2	Bill	90	75
3	Mark	80	78

Matched Merging: A Brief Case Study

You have two data sets from a telecom case study: bill data and complaints data. The billing data contains customer- and billing-related variables, whereas complaints data contains the type of complaints given by the customers for the period of the last six months. The complaints data also includes details of the disconnected customers, which are not part of the billing data. You want to do the following to prepare the data for further analysis:

1. Sort the billing data by customer ID and remove the duplicates.

2. Create a consolidated data set that contains only the existing billing customers who gave complaints. You will also attach the type of comment in the resultant data set; name this data set as `active_complaints` data.

3. Attach the billing details of customers to those who made a complaint. You may want to use it for analyzing the complaints and usage information together.

4. Attach the complaints details of customers to those who are actively paying the bill. You may want to use it to get the usage pattern along with the type of complains.

Here is the code and the results for executing these tasks:

1. Sort the billing data by customer ID and remove the duplicates.

```
proc sort data= bill nodupkey;
by cust_id;
run;
```

The log file shows the following:

```
NOTE: There were 31183 observations read from the data set WORK.BILL.
NOTE: 13259 observations with duplicate key values were deleted.
NOTE: The data set WORK.BILL has 17924 observations and 6 variables.
NOTE: PROCEDURE SORT used (Total process time):
      real time            0.06 seconds
      cpu time             0.03 seconds
```

As an exercise, you may want to print a snapshot of the resultant data set.

2. Create a consolidated data set that contains only the existing billing customers who made a complaint. Attach the type of comment in the resultant data set. Name the resultant data set `active_complaints`.

```
proc sort data=    bill nodupkey;
by cust_id;
run;

proc sort data=    complaints nodupkey;
by cust_id;
run;
```

```
data   active_complaints;
merge  bill(in=a)  complaints(in=b);
by     cust_id;
if a and b;
run;

proc print data= active_complaints(obs=10) noobs;
run;
```

The following is the log file for the previous code:

```
NOTE: There were 17924 observations read from the data set WORK.BILL.
NOTE: There were 29687 observations read from the data set WORK.COMPLAINTS.
NOTE: The data set WORK.ACTIVE_COMPLAINTS has 12638 observations and 11 variables.
NOTE: DATA statement used (Total process time):
      real time            0.03 seconds
      cpu time             0.03 seconds
```

Table 4-32 lists the output snapshot of this code set.

Table 4-32. *Consolidated Data Set That Contains Only the Existing Billing Customers Who Gave Complaints*

Cust_id	Cust_type	ACCOUNT_catid_NO	cust_account_start_date	Bill_Id	Bill_Amount	Network_complaints	Data_pack_complaints	bill_complaints	SMS_pack_Complaints	Other_Complaints
2019123897	ATTI-CC	1443789382	11/16/2009	1229877873	64332	No	No	Yes	No	No
2019124254	ATTI-EA	1443793896	11/08/2009	1238191219	55430	No	No	Yes	No	No
2019124521	ATTI-CE	1443793879	01/21/2009	1229772481	4868	No	No	No	Yes	No
2019124885	ATTI-EF	1443812409	01/29/2009	1208124832	11350	Yes	No	No	No	No
2019125217	ATTI-DD	1443793992	01/04/2009	1215443283	12416	Yes	No	No	No	No
2019125219	ATTI-DE	1443793994	03/22/2009	1207986127	85035	Yes	No	No	No	No
2019125220	ATTI-DF	1443793995	12/24/2008	1229768364	10040	Yes	No	No	No	No
2019125239	ATTI-FE	1443794102	02/25/2009	1215444698	11306	No	No	No	No	Yes
2019125336	ATTI-EA	1443808857	07/13/2009	1229866699	24500	No	Yes	No	No	No
2019125337	ATTI-AA	1443808858	08/27/2009	1229821661	24500	Yes	No	No	No	No

3. Attach the billing details of customers to the customers who made complaints.

```
data complaints_bill;
merge  bill(in=a)  complaints(in=b);
by     cust_id;
if b;
run;

proc print data= complaints_bill(obs=10) noobs;
run;
```

The log file message for this code is as follows:

```
NOTE: There were 17924 observations read from the data set WORK.BILL.
NOTE: There were 29687 observations read from the data set WORK.COMPLAINTS.
NOTE: The data set WORK.COMPLAINTS_BILL has 29687 observations and 11 variables.
NOTE: DATA statement used (Total process time):
      real time          0.10 seconds
      cpu time           0.11 seconds
```

Table 4-33 shows the output snapshot resulting from this code.

Table 4-33. *Billing Details of Customers to the Customers Who Gave Complaints*

Cust_id	Cust_ type	Account_ catid_NO	cust_ account_ start_date	Bill_ Id	Bill_ Amount	Network_ complaints	Data_pack_ complaints	bill_ complaints	SMS_pack_ Complaints	Other_ Complaints
2019115772	.	.	.	.		No	No	Yes	No	No
2019116058	.	.	.	.		Yes	No	No	No	No
2019119805	.	.	.	.		No	No	Yes	No	No
2019119981	.	.	.	.		No	Yes	No	No	No
2019120532	.	.	.	.		No	Yes	No	No	No
2019120656	.	.	.	.		No	No	No	No	Yes
2019120937	.	.	.	.		No	No	No	No	Yes
2019120969	.	.	.	.		No	No	Yes	No	No
2019121082	.	.	.	.		Yes	No	No	No	No
2019121188	.	.	.	.		No	No	No	Yes	No

It looks like the first ten observations don't have any bill details. So, let's sort the resultant data set based on the bill amount.

```
proc sort data= complaints_bill;
by descending Bill_Amount;
run;

proc print data= complaints_bill(obs=10) noobs;
run;
```

Table 4-34 shows the snapshot of output when this code is executed.

Table 4-34. Output of proc sort on complaints_bill with Descending Order of Bill Amount

Cust_id	Cust_type	ACCOUNT_catid_NO	cust_account_start_date	Bill_Id	Bill_Amount	Network_complaints	Data_pack_complaints	bill_complaints	SMS_pack_Complaints	Other_Complaints
2019163645	ATTI-ED	1443831475	12/12/2009	1208136629	2498886	No	No	Yes	No	No
6019124803	ATTI-FA	1443792516	10/17/2008	1229767424	1264144	No	Yes	No	No	No
9077649354	ATTI-FF	1442268246	02/16/2007	1208060015	842793	Yes	No	No	No	No
2019163654	ATTI-AF	1443831587	02/21/2010	1208133216	637458	No	No	No	Yes	No
9077692456	ATTI-AA	1439417474	11/22/2005	1227041576	611285	No	Yes	No	No	No
10254469134	ATTI-FC	2146374084	02/14/2012	1234675885	582857	No	No	Yes	No	No
10254646912	ATTI-AB	2146412749	08/18/2012	1234570950	506827	No	No	Yes	No	No
2019163646	ATTI-EE	1443831477	12/28/2009	1208069539	493230	No	No	No	No	Yes
10254658023	ATTI-CE	2146412750	01/15/2012	1234688311	490434	Yes	No	No	No	No
2019165914	ATTI-EC	1443836616	01/26/2010	1208155818	449101	No	No	No	Yes	No

The sorted output in Table 4-34 makes sense now.

4. Attach the complaints details of customers to the customers who are actively paying bill.

```
data  bill_with_complaints;
merge  bill(in=a)  complaints(in=b);
by     cust_id;
if a;
run;

proc print data= bill_with_complaints(obs=10) noobs;
run;
```

The log file when this code is executed is as follows:

```
NOTE: There were 17924 observations read from the data set WORK.BILL.
NOTE: There were 29687 observations read from the data set WORK.COMPLAINTS.
NOTE: The data set WORK.BILL_WITH_COMPLAINTS has 17924 observations and 11 variables.
NOTE: DATA statement used (Total process time):
      real time            0.10 seconds
      cpu time             0.07 seconds
```

Table 4-35 lists the output snapshot.

Table 4-35. Complaints Details of Customers to the Customers Who Are Actively Paying Bill

Cust_id	Cust_type	ACCOUNT_catid_NO	cust_account_start_date	Bill_Id	Bill_Amount	Network_complaints	Data_pack_complaints	bill_complaints	SMS_pack_Complaints	Other_Complaints
2019123897	ATTI -CC	1443789382	11/16/2009	1229877873	64332	No	No	Yes	No	No
2019124254	ATTI -EA	1443793896	11/08/2009	1238191219	55430	No	No	Yes	No	No
2019124521	ATTI -CE	1443793879	01/21/2009	1229772481	4868	No	No	No	Yes	No
2019124885	ATTI -EF	1443812409	01/29/2009	1208124832	11350	Yes	No	No	No	No
2019125217	ATTI -DD	1443793992	01/04/2009	1215443283	12416	Yes	No	No	No	No
2019125219	ATTI -DE	1443793994	03/22/2009	1207986127	85035	Yes	No	No	No	No
2019125220	ATTI -DF	1443793995	12/24/2008	1229768364	10040	Yes	No	No	No	No
2019125239	ATTI -FE	1443794102	02/25/2009	1215444698	11306	No	No	No	No	Yes
2019125267	ATTI -DD	1443803173	12/06/2009	1208100364	14561					
2019125335	ATTI -EB	1443808856	01/26/2009	1229750007	24500					

Conclusion

In this chapter, you learned some commonly used SAS functions and procedures along with options, which make them more useful for analysis. The chapter also discussed creating charts and data merging using SAS. In addition, you learned about the basics of analytics and basic programming in SAS, which is undoubtedly the most widely used analytical tool under the sun.

Next, in Chapter 5, you will enter the world of business analytics and use the concepts you have learned so far. You will learn more programming and analysis concepts. You are going to learn about basic descriptive statistics, correlation, and predictive modeling in coming chapters. The SAS knowledge gained in the past few chapters is vital in executing the analytics algorithms in the coming chapters. Good luck! And get ready for more fun and excitement.

Using SAS for Business Analytics

CHAPTER 5

■ ■ ■

Introduction to Statistical Analysis

This and subsequent chapters will delve into the details of business analytics techniques. It has already been established in the previous chapter that statistics forms a major portion of this art. This chapter will begin with the basic definition of statistics. It will also refer to a few web sites to access data sets, which you can use for the examples. By the end of this chapter, you will be able to comprehend the following concepts that are essential for proceeding with business analytics techniques:

- The difference between population and sample

- Different types of sampling

- The difference between variable and parameter

- The differences between descriptive, inferential, and predictive statistics

- The steps involved in solving a business analytics problem

- A complete business analytics example

What Is Statistics?

Statistics can be defined simply as the science of gathering, organizing, summarizing, and analyzing information. Statistics is a vital part of our daily lives. For example, the sports player career summaries that are displayed regularly on television are statistical summarizations of the player's career data. A good deal of such information that we encounter is dominated by numbers. Here are some such examples:

- Census statistics (www.census.gov/)

- World Bank statistics (http://data.worldbank.org/)

- Cricket Game statistics (www.espncricinfo.com/ci/content/stats/index.html)

- Stock market statistics (http://finance.yahoo.com/)

Table 5-1 shows the statistics on the percentage of the world population with access to electricity in the years 2009 and 2010. This is just a snapshot of the complete table, available on the World Bank data page (http://data.worldbank.org).

Table 5-1. *Statistics on Percentage of Population with Access to Electricity*

Country Name	Country Code	Percent in 2009	Percent in 2010
United Arab Emirates	ARE	100.0	100.0
Kuwait	KWT	100.0	100.0
Singapore	SGP	100.0	100.0
Lebanon	LBN	99.9	99.9
Libya	LBY	99.8	99.8
Brunei Darussalam	BRN	99.7	99.7
China	CHN	99.4	99.7
Israel	ISR	99.7	99.7
Egypt, Arab Rep.	EGY	99.6	99.6
Tunisia	TUN	99.5	99.5
Venezuela, RB	VEN	99.0	99.5
Bahrain	BHR	99.4	99.4
Chile	CHL	98.5	99.4
Jordan	JOR	99.9	99.4
Mauritius	MUS	99.4	99.4
Malaysia	MYS	99.4	99.4
Algeria	DZA	99.3	99.3
Costa Rica	CRI	99.3	99.2
Saudi Arabia	SAU	99.0	99.0
Trinidad and Tobago	TTO	99.0	99.0

A close look at Table 5-1 will show that this is only raw data with no statistical operation performed on it. But some useful inferences or insights about the availability of electricity for the listed countries can still be drawn from it. In these 20 countries, the smallest value in the 2009 column is 98.5, which means that at least this percentage of population has access to electricity. An access rate of 98.5 percent or above says something about the state of development in the listed countries. In the year 2010, this percentage increased to 99.3 percent, a definite increase in lifestyle. For countries like United Arab Emirates, Kuwait, and Singapore, it is a perfect 100 percent.

Even raw data narrates a story. It might not be possible to open raw data that takes up gigabytes and petabytes with commonly available tools. The data set size, in some cases, might be too large for any system to handle it. In such cases, visualization and advanced analytics techniques can be applied to gain any useful business insights. We may not be able to gain insights with a naked eye. We need some statistical techniques and tools to find the hidden patterns in the data.

Basic Statistical Concepts in Business Analytics

This section will discuss some basic terms such as *population*, *sample*, *variable*, and *parameters*, which will be useful as you learn more about business analytics techniques.

Population

Population is the complete set of objects or data records that are available for an analytics project or data analysis. For example, in a countrywide marketing campaign, a narrowed-down list of the country's citizens will form the *population* for the analytics problem. Generally it might not be possible to analyze the entire population because of the sheer size of the data, availability of time, funding, or limited processing power of available computing machines. These reasons may compel you to consider only a subset of the population. This subset is usually referred to as a *sample* in statistical terminology. If properly chosen, analyzing with a sample can be as good as analyzing the full population.

For increased clarity, here are a couple of examples in detail:

- *Example 1*: If a retailer like Wal-Mart undertakes to analyze its worldwide customer buying patterns, the entire customer base of Wal-Mart across the globe will form the population. The Wal-Mart business analytics team may decide to work either on the entire population or only on a representative sample, based on the resources available for the project.

- *Example 2*: A telecom company of the size Vodafone or AT&T, with a customer base across the country, may want to decide strategies to decrease the number of customers switching over to seemingly cheaper mobile phone plans offered by the competition. In such a case, all the mobile network users of Vodafone across the country will be considered as the population. As always, the analytics team will decide to use either the entire population or only a representative sample. As discussed, in most of the cases, a well-crafted sample is good enough to get the required business insights.

Sample

A *sample* can be formally defined as the subset of a population that is selected for analysis. The procedure of creating or collecting this subset is called *sampling*. Sometimes, it might be necessary to manually collect some records from the overall population. There are several types of sampling techniques. The following are the ones that are most commonly used in business analytics projects.

Simple Random Sampling

Simple random sampling is the most commonly used sampling method. Randomly choosing some records from a population (denoted by n) is called simple random sampling. There are several methods for deciding on the right sample size. Sometimes the business problem that we are handling gives us an idea of the sample size. Once the sample size (n) has been decided based on one of the methods, records are randomly selected from the population. Convenient functions are available in SAS for this purpose.

A classic example of random sampling is of a blindfolded man picking up ten apples from a basket full of apples. All the apples have an equal probability of being picked from the basket.

Stratified Sampling

Consider an example population, which has preexisting segments of same or different sizes. Segments are the population records that are already classified into a distinct number of subgroups. In such a case, it is best to do a random sampling from each segment; as such, a sample will truly represent the nature of such population.

149

The size of each segment can be based upon the proportion of that segment in the entire population. Such segments are usually referred to as *strata*. The process of simple random sampling from each strata is called *stratified sampling*. Segments can be manually created, and stratified sampling can be performed even when there are no obvious segments in the population.

For example, if 1,000 random candidates are to be picked from across the country for a sporting event, it might be a good idea to pick them proportionately from each state.

Systematic Sampling

Systematic sampling is based on a fixed rule, like picking every fifth or seventh observation from a given population. It is different from random sampling, wherein any random values are picked. This type of sampling is generally done if testing is a continuous process. Recording the room temperature every 60 minutes or measuring the blood pressure of a patient every 10 minutes are examples of systematic samples.

- *Example*: Consider a mass manufacturing machine that produces simple bolts to be used in a chemical plant erection project. Every 30th bolt manufactured by the machine can be collected as sample. This may look like a random sample from the whole lot, but you are not actually waiting for the whole lot to form; instead, you are collecting your sample much before creating the heap.

Variable

Simply put, a *variable* in a statistical data table is nothing but a column or a field in the table, a feature that may change its value from one record to another. It may well be a numeric, which can be measured for each record, or a non-numeric such as city, gender, or a status field containing Yes or No entries. Other examples are age, monthly income, daily sales, and cost data. The following are the major types of variables that a population or a sample may contain.

Table 5-1 has four columns: Country Name, Country Code, 2009 and 2010. Each one of these four columns represents a variable. So, in this data set, you have a total of four variables. The variable Country Name is taking the values of United Arab Emirates, Kuwait, and Singapore as you proceed from the row 1 to 3.

This data set is small because it has only four variables, but in the banking applications such as credit scoring, there may be hundreds of such variables, even as high as 500. Modeling with this kind of data set may be a challenging task and even unmanageable at times. So, when using proper statistical techniques, it may be required to limit the number of variables to, say, 50, which will make most sense for the analysis under consideration.

Numeric Variables

Numeric or quantitative variables are measurable, comparable, and orderable. Height, weight, expenses, and distance are a few examples. There are two types of numerical variables.

Continuous

A continuous variable can take any value between two limits.

For example, a height variable can be anything between 1 and 7 feet in most cases. It can take continuous values such as 5.1, 5.12, 5.6, 5.6134, and 6.5 feet.

Discrete

These numeric variables take values in steps only. They can take only an integer or some predefined values between the given limits.

For example, the number of children in a family can only be 1, 2, 3, 4, and so on. It can never be 1.5 or 2.34.

A continuous variable is like an analogue clock's hand. It can take any position between two given time points.

In this example, you can stop the clock at 4:25:25.5.

A discrete variable is like a digital clock's display. It can display only certain predefined numbers.

Here, you can either stop only at 4:25:24 or 4:25:26.

Categorical or Non-numeric Variables

Non-numeric, qualitative, and categorical variables are the type of variables that represent quality or a characteristic field.

Examples are shirt sizes expressed as S, M, L, XL, and XXL, or distance, which is expressed as near and far. It can as well be a Boolean value like a pass or a fail or a yes or no field.

Variable Types in Predictive Modeling Context

The aim of any data scientist or an analytics professional is to get useful business insights from a given set of data, which may be historical in nature. Forecasting based on the historical data into the foreseeable future (or predictive modeling) is the purpose of many data modeling exercises done in business analytics. Predictive modeling techniques such as regression have a target variable, which is the final outcome of the whole data modeling exercise. In other words, predictive modeling forecasts can predict a target variable using some other variables, which are known at the time of modeling in the form of historical data. This target variable is termed as a *dependent variable*; the other variables that are used for prediction are called *independent variables*.

The examples of dependent variables are sales in a month, probability of fraud, final grades of students, and effect of fertilizer on a crop.

Here are some more examples of independent and dependent variables: customer income while predicting expenses (dependent), hours of study while predicting grades (dependent), and fertilizer concentration while forecasting the crop yield (dependent). The variable other than what is denoted as "dependent" is an independent variable.

Parameter

A *parameter* is a measure that is calculated on the entire population. Any summary measure that gives information of population is called a parameter. Remember it this way: it's simply "P for P," meaning parameter is for population.

For example, take the data on electricity utility bills of an entire state like California. It will be huge by any standards because it represents the variables such as name, address, type of connection, month, units consumed, and the bill amount for all households in the state. Now for planning purposes, that is, to forecast the electricity demand for the next five years in the state, if you calculate the averages on all the state's households for the variables like units consumed and bill amount, it will be termed as parameters. So, two example parameters, that is, the entire state's average units consumed per household and the average bill amount may look like 650 units and $100, respectively. These parameters are calculated on the entire population, which might be really large at times. So, it's not hard to predict that it may require huge amount of computational effort.

Statistic

A *statistic* is the measure that is calculated over a sample. Going by this definition, a summary measure that gives information of the sample is called a statistic. Similar to "P for P," a statistic is "S for S," meaning statistic is for sample.

As an example, consider the same data set that you used to explain the concept of parameter in the previous section. For parameters, the average electricity consumed and the average bill amount were calculated on the entire population of the state of California. If the computations are not possible on the entire population, you may prefer to take a representative sample of, say, only 10,000 randomly chosen households from across the state. On this sample of 10,000 households, you calculate the average electricity consumed and the average bill amount per household. These two averages on the sample data will be termed as a statistic. If the samples are chosen properly, they closely represent the entire population. And in business analytics, in many cases you prefer to work with samples only rather than handling the huge computations involved with entire population.

Example Exercise

Consider the prdsale data set. It is available in the SAS help library. Answer these questions to get clarity on the concepts covered so far in this chapter:

1. Print the contents of Prdsale data and write your observations.

2. Print the first 20 observations of Prdsale data and write your observations.

3. What is the size of population?

4. Filter the data and take a sample (where country=Canada).

5. Take a random sample of size 30.

6. Identify the continuous, discrete, and categorical variables.

7. What are cause variables (independent)? What are effect variables (dependent)?

8. Calculate a parameter (mean actual sales of the population).

9. Calculate a statistic (mean actual sales of the sample).

10. How close is the statistic to a parameter? Is it a good estimate?

It's now the time for some hands-on examples. In the following section, you will take each of the previous ten exercises and solve them using a sample data set available from the SAS help. You will need to open the SAS environment to execute the code given. You are expected to have some basic understanding of how SAS works.

1. Print the contents of prdsale data and write your observations.

The following SAS code prints the metadata on the prdsale data set:

```
proc contents data=sashelp.prdsale varnum;
run;
```

Table 5-2 is the SAS output.

Table 5-2. *Output of PROC CONTENTS on prdsale dataset*

The CONTENTS Procedure

Data Set Name	SASHELP.PRDSALE	Observations	1440
Member Type	DATA	Variables	10
Engine	V9	Indexes	0
Created	Thursday, January 31, 2008 09:40:56 PM	Observation Length	96
Last Modified	Thursday, January 31, 2008 09:40:56 PM	Deleted Observations	0
Protection		Compressed	NO
Data Set Type		Sorted	NO
Label	Furniture sales data		
Data Representation	WINDOWS_32		
Encoding	us-ascii ASCII (ANSI)		

Engine/Host Dependent Information

Data Set Page Size	8192
Number of Data Set Pages	18
First Data Page	1
Max Obs per Page	84
Obs in First Data Page	62
Number of Data Set Repairs	0
Filename	C:\Program Files\SAS\SASFoundation\9.2\core\sashelp\prdsale.sas7bdat
Release Created	9.0201M0
Host Created	XP_PRO

(continued)

Table 5-2. (*continued*)

Alphabetic List of Variables and Attributes

#	Variable	Type	Len	Format	Label
1	ACTUAL	Num	8	DOLLAR12.2	Actual Sales
2	PREDICT	Num	8	DOLLAR12.2	Predicted Sales
3	COUNTRY	Char	10	$CHAR10.	Country
4	REGION	Char	10	$CHAR10.	Region
5	DIVISION	Char	10	$CHAR10.	Division
6	PRODTYPE	Char	10	$CHAR10.	Product type
7	PRODUCT	Char	10	$CHAR10.	Product
8	QUARTER	Num	8	8.	Quarter
9	YEAR	Num	8	4.	Year
10	MONTH	Num	8	MONNAME3.	Month

Observations from Table 5-2:

- Prdsale data has ten variables.

- At first glance, it looks like this is furniture sales data.

- A closer look at the labels tells you that this is monthly product sales data.

2. **Print the first 20 observations of Prdsale data and write your observations.**

```
proc print data = sashelp.prdsale(obs=20);
run;
```

Table 5-3 lists the output of this code.

Table 5-3. *Output of PROC PRINT on prdsale Dataset – 1ˢᵗ 20 observations*

Obs	ACTUAL	PREDICT	COUNTRY	REGION	DIVISION	PRODTYPE	PRODUCT	QUARTER	YEAR	MONTH
1	$925.00	$850.00	CANADA	EAST	EDUCATION	FURNITURE	SOFA	1	1993	Jan
2	$999.00	$297.00	CANADA	EAST	EDUCATION	FURNITURE	SOFA	1	1993	Feb
3	$608.00	$846.00	CANADA	EAST	EDUCATION	FURNITURE	SOFA	1	1993	Mar
4	$642.00	$533.00	CANADA	EAST	EDUCATION	FURNITURE	SOFA	2	1993	Apr
5	$656.00	$646.00	CANADA	EAST	EDUCATION	FURNITURE	SOFA	2	1993	May
6	$948.00	$486.00	CANADA	EAST	EDUCATION	FURNITURE	SOFA	2	1993	Jun
7	$612.00	$717.00	CANADA	EAST	EDUCATION	FURNITURE	SOFA	3	1993	Jul
8	$114.00	$564.00	CANADA	EAST	EDUCATION	FURNITURE	SOFA	3	1993	Aug
9	$685.00	$230.00	CANADA	EAST	EDUCATION	FURNITURE	SOFA	3	1993	Sep
10	$657.00	$494.00	CANADA	EAST	EDUCATION	FURNITURE	SOFA	4	1993	Oct

(*continued*)

Table 5-3. (*continued*)

Obs	ACTUAL	PREDICT	COUNTRY	REGION	DIVISION	PRODTYPE	PRODUCT	QUARTER	YEAR	MONTH
11	$608.00	$903.00	CANADA	EAST	EDUCATION	FURNITURE	SOFA	4	1993	Nov
12	$353.00	$266.00	CANADA	EAST	EDUCATION	FURNITURE	SOFA	4	1993	Dec
13	$107.00	$190.00	CANADA	EAST	EDUCATION	FURNITURE	SOFA	1	1994	Jan
14	$354.00	$139.00	CANADA	EAST	EDUCATION	FURNITURE	SOFA	1	1994	Feb
15	$101.00	$217.00	CANADA	EAST	EDUCATION	FURNITURE	SOFA	1	1994	Mar
16	$553.00	$560.00	CANADA	EAST	EDUCATION	FURNITURE	SOFA	2	1994	Apr
17	$877.00	$148.00	CANADA	EAST	EDUCATION	FURNITURE	SOFA	2	1994	May
18	$431.00	$762.00	CANADA	EAST	EDUCATION	FURNITURE	SOFA	2	1994	Jun
19	$511.00	$457.00	CANADA	EAST	EDUCATION	FURNITURE	SOFA	3	1994	Jul
20	$157.00	$532.00	CANADA	EAST	EDUCATION	FURNITURE	SOFA	3	1994	Aug

Observations from Table 5-3:

- The data set contains monthly actual and predicted sales figures along with product types and regions.

- The product type in the first 20 records is furniture, and the product is sofa.

- It looks like the data is from 1993 and onward.

3. **What is the size of population?**

```
proc contents data=sashelp.prdsale varnum;
run;
```

The size of the population in this example is the total number of records. Proc contents show there are 1,440 records in total. Refer to Table 5-4 for the output.

Table 5-4. *Results of PROC CONTENTS on predsale Dataset With varnum Option*

Data Set Name	SASHELP.PRDSALE	Observations	1440
Member Type	DATA	Variables	10
Engine	V9	Indexes	0
Created	Thursday, January 31, 2008 09:40:56 PM	Observation Length	96
Last Modified	Thursday, January 31, 2008 09:40:56 PM	Deleted Observations	0
Protection		Compressed	NO
Data Set Type		Sorted	NO
Label	Furniture sales data		
Data Representation	WINDOWS_32		
Encoding	us-ascii ASCII (ANSI)		

4. **Filter the data and take a sample (where country=Canada).**

```
data prod_sample;
set  sashelp.prdsale;
where country='CANADA';
run;
```

Here is the log file for the previous code:

```
NOTE: There were 480 observations read from the data set SASHELP.PRDSALE.
      WHERE country='CANADA';
NOTE: The data set WORK.PROD_SAMPLE has 480 observations and 10 variables.
NOTE: DATA statement used (Total process time):
      real time          0.67 seconds
      cpu time           0.10 seconds
```

5. **Take a random sample of size 30.**

```
proc surveyselect data = sashelp.prdsale
method = SRS
rep = 1
sampsize = 30 seed = 12345 out = prod_sample_30;
id _all_;
run;
```

■ **Note**   The method SRS represents the type of sampling. SRS stands for Simple random Sampling. Seed will make sure that we will have the same random sample drawn again. To refer to same sample, we need an index. Seed will make sure that same sample of 30 observations is drawn again.

Table 5-5 lists the output.

Table 5-5. *Result of PROC SURVEYSELECT on prdsale*

The SURVEYSELECT Procedure	
Selection Method	Simple Random Sampling
Input Data Set	PRDSALE
Random Number Seed	12345
Sample Size	30
Selection Probability	0.020833
Sampling Weight	48
Output Data Set	PROD_SAMPLE_30

The following SAS code prints the sample prod_sample_30.

```
proc print data=prod_sample_30;
run;
```

Table 5-6 lists the output of this code.

Table 5-6. *Output of PROC PRINT on prod_sample_30 Dataset*

Obs	ACTUAL	PREDICT	COUNTRY	REGION	DIVISION	PRODTYPE	PRODUCT	QUARTER	YEAR	MONTH
1	$670.00	$679.00	CANADA	EAST	EDUCATION	OFFICE	CHAIR	1	1993	Mar
2	$768.00	$948.00	CANADA	EAST	EDUCATION	OFFICE	DESK	1	1994	Mar
3	$511.00	$402.00	CANADA	EAST	CONSUMER	OFFICE	DESK	1	1993	Feb
4	$379.00	$819.00	CANADA	EAST	CONSUMER	OFFICE	DESK	2	1994	Apr
5	$190.00	$969.00	CANADA	WEST	EDUCATION	FURNITURE	SOFA	3	1994	Jul
6	$112.00	$263.00	CANADA	WEST	EDUCATION	FURNITURE	SOFA	3	1994	Sep
7	$638.00	$145.00	CANADA	WEST	EDUCATION	FURNITURE	BED	4	1994	Dec
8	$760.00	$17.00	CANADA	WEST	EDUCATION	OFFICE	CHAIR	3	1994	Aug
9	$586.00	$363.00	CANADA	WEST	EDUCATION	OFFICE	DESK	1	1993	Jan
10	$956.00	$149.00	CANADA	WEST	CONSUMER	FURNITURE	BED	3	1993	Jul
11	$911.00	$318.00	CANADA	WEST	CONSUMER	FURNITURE	BED	2	1994	May
12	$749.00	$852.00	GERMANY	EAST	EDUCATION	FURNITURE	SOFA	4	1994	Nov
13	$394.00	$262.00	GERMANY	EAST	EDUCATION	FURNITURE	BED	3	1993	Sep
14	$104.00	$68.00	GERMANY	EAST	EDUCATION	FURNITURE	BED	3	1994	Aug
15	$798.00	$215.00	GERMANY	EAST	EDUCATION	OFFICE	CHAIR	3	1993	Sep
16	$33.00	$901.00	GERMANY	EAST	CONSUMER	OFFICE	TABLE	3	1993	Jul
17	$458.00	$309.00	GERMANY	EAST	CONSUMER	OFFICE	CHAIR	4	1994	Oct
18	$100.00	$285.00	GERMANY	WEST	EDUCATION	OFFICE	DESK	1	1993	Jan
19	$505.00	$59.00	GERMANY	WEST	CONSUMER	OFFICE	TABLE	1	1993	Jan
20	$58.00	$998.00	U.S.A.	EAST	EDUCATION	FURNITURE	BED	3	1994	Aug
21	$229.00	$524.00	U.S.A.	EAST	EDUCATION	OFFICE	CHAIR	1	1993	Jan
22	$390.00	$617.00	U.S.A.	EAST	EDUCATION	OFFICE	CHAIR	4	1993	Oct
23	$790.00	$229.00	U.S.A.	EAST	EDUCATION	OFFICE	CHAIR	3	1994	Sep
24	$119.00	$427.00	U.S.A.	EAST	CONSUMER	FURNITURE	SOFA	2	1993	May
25	$198.00	$652.00	U.S.A.	EAST	CONSUMER	OFFICE	TABLE	3	1994	Aug
26	$602.00	$497.00	U.S.A.	EAST	CONSUMER	OFFICE	CHAIR	4	1994	Nov
27	$512.00	$933.00	U.S.A.	WEST	EDUCATION	FURNITURE	BED	1	1993	Feb
28	$208.00	$382.00	U.S.A.	WEST	CONSUMER	OFFICE	CHAIR	1	1993	Mar
29	$739.00	$656.00	U.S.A.	WEST	CONSUMER	OFFICE	DESK	4	1993	Nov
30	$87.00	$145.00	U.S.A.	WEST	CONSUMER	OFFICE	DESK	4	1993	Dec

Observations:

This sample data output clearly gives a better picture of the overall population rather than printing the first few observations. We can see various product types, various countries, and so on.

6. **Identify the continuous, discrete, and categorical variables.**

Here are the continuous variables:

- Actual (Actual Sales)
- Predicted (Predicted Sales)

These two variables are continuous as they can take any real values. For example, the sample values can be $200, $201, or $201.5.

Here are the numerical discrete variables:

- QUARTER
- YEAR

These variables can take a set of values only. Quarter can take 1 or 2; it can't be equal to 1.5. Hence it is a discrete variable. Year is also a discrete variable.

Here are the categorical variables:

- COUNTRY
- REGION
- DIVISION
- PRODTYPE
- PRODUCT

These are not numeric type of variables.

7. **Which are independent variables? What are dependent variables?**

    a. Actual and Predicted sales is the effect or the dependent variables.

    b. The independent variables make up the rest of the list.

    c. The sales here depend on country, region, product month, year, quarter, and so on. These all are independent variables.

8. **Calculate a parameter (mean actual sales of the population).**

```
proc means data=sashelp.prdsale ;
var actual;
run;
```

Means is a SAS procedure name. In this example we have given the variable name as actual. The PROC MEANS procedure will act on this variable to find the mean.

Table 5-7 lists the output of this code.

Table 5-7. *The Result of PROC MEANS on prdsale*

Analysis Variable : ACTUAL Actual Sales

N	Mean	Std Dev	Minimum	Maximum
1440	507.1784722	287.0313065	3.0000000	1000.00

The mean sale of the population is 507.17. Details about averages will be presented later in this chapter.

9. **Calculate a statistic (mean actual sales of the sample).**

```
proc means data=prod_sample_30 ;
var actual;
run;
```

The output of this code is listed in Table 5-8.

Table 5-8. *Mean and Standard Deviation on prod_sample_30; Variable = actual*

The MEANS Procedure

Analysis Variable : ACTUAL Actual Sales

N	Mean	Std Dev	Minimum	Maximum
30	451.8000000	286.8939970	33.0000000	956.0000000

The mean sale of the sample is 451.8.

10. **How close is the statistic to a parameter? Is it a good estimate?**

- The statistic is not very close to parameter. The sample average sale does not really represent the overall population average sales.

- There is almost an 11% difference between parameter and statistic.

- An increase in sample size might help get a good estimate. Increasing the sample size to 100 might be a better option.

The following SAS code extracts a sample of size 100 from prdsale dataset and then calculates the mean.

```
/* Simple Random Sample; Size is 100 */
proc surveyselect data = sashelp.prdsale
method = SRS
rep = 1
sampsize = 100 seed = 12345 out = prod_sample_100;
id _all_;
run;

proc means data=prod_sample_100 ;
var actual;
run;
```

Table 5-9 lists the output of SAS code for calculating the mean and standard deviation of variable actual.

Table 5-9. *Mean and Standard Deviation on prod_sample_100; Variable = actual*

Analysis Variable : Actual Sales

N	Mean	Std Dev	Minimum	Maximum
100	527.7000000	297.0428837	25.0000000	988.0000000

The average sales estimate (the statistic) looks closer to the overall sales of the population. The new difference is less than 5 %

Statistical Analysis Methods

There are three methods of Statistical analysis: descriptive, inferential, and predictive. In descriptive statistics methods, the data is simply summarized using statistical central tendencies and variations. In inferential statistics, a sample is drawn from the population to infer on the full set of data or population. Predictive statistics, as expected, can predict the dependent variable using methodologies such as linear and logistic regression. Some of these terms might appear strange at this point, but they will be explained in detail in the coming chapters.

Descriptive Statistics

Descriptive statistics is the right solution for presenting an overall picture about a set of data. These methodologies represent summaries, which give a fair picture about a given population. The business insights earned from the analysis of data will be in the form of tables and charts. Descriptive statistics output might help draw useful inferences, but this output in itself is not an inference.

Here is a sales example. Descriptive statistics, applied on one year's worth of sales data, might include average sales in the last 12 months, maximum and minimum sales of the past 12 months, and so on. The business expert can infer on the performance on his or her business, based on this analysis.

The following are the important measures or outcomes of descriptive statistics:

- *Measures of central tendency*: Mean, median, mode, and midrange

- *Measures of variation*: Range, variance, standard deviation, z-scores

The next chapter will examine these terms in more detail.

Inferential Statistics

You now know it is not always easy to analyze a whole population every time because of various constraints. It is therefore better to analyze a representative sample and draw inferences on the entire population. If done carefully, it can be the right decision most of the time. But since only a portion of the data is analyzed, inferences drawn on the whole data set, factors such as sample size and errors, become very important.

- *Example 1*: A bottling machine is supposed to fill 300 ml of soft drink in every bottle. The population size is a few thousand per day. You took a representative sample of 400 bottles and found that the sample average is 300.5 ml. Is the machine performing well?

- *Example 2*: Two notebook computer buyers, among a sample of 300, reported a problem with the screen. Is that a reason for concern? Based on this, can you draw the conclusion about the screen quality on a population size of about 10,000 at least?

Predictive Statistics

Predictive statistics is the science of predicting future results, based on historical events. Predictive modeling or model building is like driving a car while looking into the rear-view mirror, expecting that the road ahead will be same as the road that has already been traveled on. Predictive statistics tries to predict a dependent variable or outcome by using a combination of independent variables or predictors. The historical data needs to be developed into a mathematical equation between Y (dependent variable) and Xi (independent variables).

Various techniques can be used to build the predictive model. Regression, logistic regression, and time-series analysis are good representative samples. A classic example of using predictive statistics is a bank building a credit risk model to decide whether to offer a loan to an individual.

Solving a Problem Using Statistical Analysis

This section discusses the main steps involved in statistical data analysis, also known as business analytics problem solving.

Setting Up Business Objective and Planning

Any data analysis problem begins with the business objective. Why am I doing this analysis? What is in and out of scope? Do I have enough expertise available to perform the analysis? What are the challenges? How can I implement the results? Is all relevant data available? If it is available, is it reliable? Several questions such as these abound.

The scope and analysis design or approach is the major outcome of this process.

The Data Preparation

The data that is required for analysis can be gathered after deciding the objective and scope. The data might not be perfect for analysis, whether in format, quality, or completeness. There are bound to be some missing values, outliers, and other noise factors in the data set. Starting basic data exploration, validating it for accuracy and consistency and identifying all the issues are important tasks in building a model. All the issues must be resolved before going to analysis. The following are the major steps involved:

1. Collect the data.
2. Explore the data.
3. Validate the data.
4. Clean the data.

Descriptive Analysis and Visualization

Understanding and gaining insights into the data before moving on to building a model are important tasks. Understanding each variable and their relationships is also necessary. The following are the major steps:

1. Visualize the data and gain simple insights.
2. Perform simple descriptive statistics.
3. Perform univariate analysis for each variable.
4. Create derived variables and metrics if necessary.
5. Find the correlation between the variables.

Predictive Modeling

This step deals with building a model to address the objective. You need to identify the most appropriate analysis technique, identify the dependent variables, and remove any redundancies. The final step is to build the best fit. Model iterations are sometimes necessary. Interpreting the model helps to know the most impacting predictor variables. The following points summarize this step:

1. Identify the modeling technique.

2. Build the predictive model.

3. Iterate the model and find the best fit.

4. Interpret the model.

Model Validation

The following are the questions to be answered in this step. How good is the model you have just built? Is it good only on the sample data that is used to build this model? Can you get a similar but different sample and test the accuracy of the model? How robust is my model with regard to the accuracy of prediction? Understand the following terms in connection with this step:

- *In-time validation*: This involves taking a sample that looks the same as the development sample from a similar time period. Sometimes, the in-time sample is 20 to 30 percent of the overall records considered for analysis; the rest is used in model building. The validation using an in-time sample is in-time validation.

- *Out-of-time validation*: This method takes a sample from a different time period for model validation. Sometimes, the model is perfect only on the development and in-time validation samples. Cross-validation of sample from a different time period helps test the real robustness of the model.

Model Implementation

The model is ready when it has passed all the tests of validation. It can now be used for the business objective it was built for. The model can be implemented and applied into the production systems for trial runs in the actual business environment. Finally, it needs to be documented and handed over to the end users.

Figure 5-1 shows all the steps used in the model building. This book will touch upon most of them.

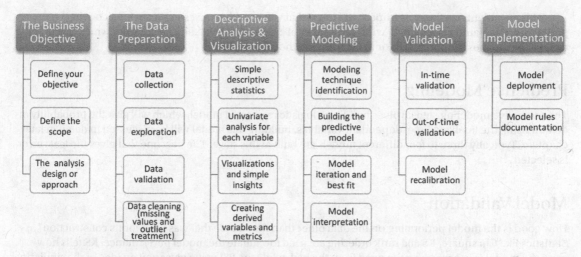

Figure 5-1. *Steps Used in Model Building*

An Example from the Real World: Credit Risk Life Cycle

Consider an example of credit risk model building. Every bank wants to analyze how risky it is to offer a loan or approve a credit card to an applicant. The approval or rejection is normally based on a predictive model, which may contain 15 to 20 variables that are filtered from 400 independent variables. The following are the basic steps in building a credit risk model.

Business Objective and Planning

The objective is to predict the risk factor on every potential applicant. Determining the portfolio size, creating growth plans, and simulating the competitive and business environment can be other objectives. Detailed project management plans can also be made at this stage.

Data Preparation

The required data is collected from customer historical records with the bank, in-house applications, and customer facing retail applications or any other federated data marts. Data from these multiple sources is likely to have heterogeneous data formats. Hence data cleansing and transformation is a must. Correction for outliers, missing value treatment, and other data cleansing steps are done at this stage. Creating dummy or derived variables is also done if required.

Descriptive Analysis and Visualization

The performance window is decided, which tells us how much historical data should be considered for analysis. The usual performance window for this kind of analysis is 12 months. Defining the good and bad accounts that will be used in model building later is done at this stage. Analysts must decide whether the population needs to be divided into various segments. If yes, then the segmentation variables are also identified.

This step identifies the most important factors that will have the most significant effect on the probability of default. For example, with the number of current loans versus age, consider which one of these two will be the most important factor in determining the credit risk.

Predictive Modeling

The predictive modeling step involves building a logistic regression model, which will give the probability of default using the finalized list of dependent variables. Building this model will be discussed in detail in later chapters. Typically three to four different models are built at this stage. The best one that gives the least error is selected.

Model Validation

How good is the model performing on the data other than the sample that was used for its construction? Statistics like Chi square, KS and rank ordering are used to validate the model performance. KS tells how good the model is in separating the good from the bad, while the PSI value gives you an idea of the similarity of the current population (the real production data) versus that of the development sample.

- *KS test*: KS stats for Kolmogorov-Smirnov. This is used to find whether the model is efficiently separating good from bad. The higher the KS, the better the model

- *PSI*: PSI stands for Population Stability Index. Banks have to make sure the current population on which the credit scoring model will be used is same as the development time population. If there is a drastic shift in the population, then the scorecard (model) might not be valid anymore.

Model Implementation

The final step is to start using the credit risk model, which will convert the risk into a score. The model, at this stage, is given to the IT developers for implementation. Whenever there is a new credit card application, the system, equipped with this model, will automatically calculate the credit score. Using this credit score the bank's front end staff will approve or decline the application

Conclusion

This chapter defined some basic statistical terms. They will be useful in all the chapters that follow. It also covered the basic steps that are followed in solving business analytics problems. This was done with the help of a real-life bank scenario. The next few chapters will discuss descriptive statistics, followed by data preparation and predictive modeling. All the steps mentioned in Figure 5-1 will be discussed along with its terminology, which at this stage might look somewhat alien.

CHAPTER 6

∎ ∎ ∎

Basic Descriptive Statistics and Reporting in SAS

The first step in statistical data analysis is to define the business objectives, which determines the need of the project. This step will require some initial planning. Once the data is gathered in the required format, the next step is to explore the data.

A raw data set is all that is available at this stage. The next step is to get a basic understanding of the data. If the data set is too large, only the first few records can be printed. The business analyst must then visualize the data, highlight the outliers, identify the caveats, find interesting patterns in the data, or build a predictive model that will help forecast the result, which is still unknown at this stage.

This chapter discusses basic descriptive statistics steps, which will help in data exploration and also help in data-cleansing operations, which is the topic of the next chapter.

As discussed earlier, basic descriptive statistics gives an overall picture of the data set on hand. The actors of this picture are discussed later in this chapter. The topics of advanced statistical modeling techniques, namely, inferences and predictions on the data, will be dealt with in the next few chapters.

Rudimentary Forms of Data Analysis

To get a feel of the data, complex statistical analysis is not always required. Sometimes trivial techniques such as printing the first few rows of the data set or visually inspecting the computer screen can do the job. In fact, before attempting any analysis on the data, we strongly recommend you inspect the first few rows, just to get a feel for the data.

Simply Print the Data

A simple visual inspection of the data in hand can tell a lot and help you to understand it. If the data set is small, it can be printed without trouble, but if it is a large data set, it is recommended that only a few rows (such as 1 to 100) are printed. A simple sorting might help sometimes. If you have a lot of columns or variables, you should print only important variables that deserve careful inspection.

The first few pages of the previous chapter contained the snapshot of a big data set, which showed the statistics on the percentage of population with access to electricity in 2009 and 2010. Recall that we discussed a few simple observations on this data set that helped give you a feel of the complete data set.

Print and Various Options of Print in SAS

SAS gives you easy-to-use print options for any kind of data set. You can print all the variables or select a few important ones to print. We will discuss the SAS print procedure and its options using an example.

The following code outputs a snapshot of some data generated by an online store over a period of one month:

```
Proc print data = online_sales (obs=20);
run;
```

Table 6-1 lists the output of this code, and the explanation of the variables involved in the online_sales data set follows:

- *brand*: The brand name
- *listPrice*: The price of the item as listed on web site
- *shippingPeriod*: The shipping period in days
- *date*: The date of the order
- *category*: The item category

The data in Table 6-1 belongs to only a few mobile phone orders placed by the customers.

Table 6-1. *Results of proc print on online_sales Data Set*

Obs	tr_id	Brand	listPrice	shippingPeriod	date	category
1	1	Samsung	39900	4	16OCT12:01:03:00	Mobiles
2	2	Samsung	16990	6	16OCT12:01:03:00	Mobiles
3	3	Samsung	7090	5	16OCT12:01:03:00	Mobiles
4	4	AirTyme	1549	3	16OCT12:01:04:00	Mobiles
5	5	AirTyme	1549	3	16OCT12:01:04:00	Mobiles
6	6	Arise	859	7	16OCT12:01:04:00	Mobiles
7	7	Karbonn	8399	6	16OCT12:01:04:00	Mobiles
8	8	Motorola	19999	4	16OCT12:01:04:00	Mobiles
9	9	Samsung	34900	4	16OCT12:01:04:00	Mobiles
10	10	Airtyme	2338	4	16OCT12:01:04:00	Mobiles
11	11	HTC	8049	6	16OCT12:01:04:00	Mobiles
12	12	HTC	8099	3	16OCT12:01:04:00	Mobiles
13	13	HTC	10399	6	16OCT12:01:04:00	Mobiles
14	14	Micromax	1584	4	16OCT12:01:04:00	Mobiles
15	15	Micromax	1584	4	16OCT12:01:04:00	Mobiles
16	16	Micromax	3449	7	16OCT12:01:04:00	Mobiles
17	17	Motorola	13199	3	16OCT12:01:04:00	Mobiles
18	18	Motorola	9729	6	16OCT12:01:04:00	Mobiles
19	19	Nokia	1559	6	16OCT12:01:04:00	Mobiles
20	20	Samsung	11400	3	16OCT12:01:04:00	Mobiles

We'll now discuss some useful options in the SAS print procedure. The following is the generic code and its explanation:

```
proc print data=<<data set>> label noobs heading=vertical;
var <<variable-list>>;
by var1; run;
```

- **Label**: This option prints variable labels as column headings instead of variable names.

- **Noobs**: This option removes the OBS column from output.

- **Heading=vertical**: This option prints the column headings vertically. *This is useful when the names are long but the values of the variable are short.*

- **By**: The by statement produces output grouped by values of the mentioned variables.

Let's use an example. The following is the code for sorting the data based on list price:

```
Proc sort data=online_sales ;
by listPrice  ;
run;
```

You can use the descending option to sort from high price to low price. SAS sorts in ascending order by default.

```
Proc sort data=online_sales ;
By descending listPrice  ;
run;
```

The following is the SAS log for the preceding code:

```
There were 5129 observations read from the data set WORK.ONLINE_SALES.
The data set WORK.ONLINE_SALES has 5129 observations and 6 variables.
PROCEDURE SORT used (Total process time):
real time            0.04 seconds
cpu time             0.04 seconds
```

Now let's try to print this sorted data set.

```
proc print data =  online_sales (obs=20);
run;
```

Refer to Table 6-2 for the output of this code.

Table 6-2. *Results of proc print on online_sales Data Set After Sorting on listPrice*

Obs	tr_id	Brand	listPrice	shippingPeriod	date	category
1	156	BlackBerr	139990	3	16OCT12:12:56:00	Mobiles
2	2937	Apple	59500	4	30OCT12:07:02:00	Mobiles
3	3456	Apple	59500	4	31OCT12:07:02:00	Mobiles
4	3777	Apple	59500	5	01NOV12:07:02:00	Mobiles
5	4021	Apple	59500	6	02NOV12:07:02:00	Mobiles
6	4283	Apple	59500	6	03NOV12:07:02:00	Mobiles
7	4461	Apple	59500	7	04NOV12:07:02:00	Mobiles
8	4830	Apple	59500	5	05NOV12:07:02:00	Mobiles
9	157	Apple	57500	3	16OCT12:12:57:00	Mobiles
10	2936	Apple	52500	7	30OCT12:07:02:00	Mobiles
11	3455	Apple	52500	7	31OCT12:07:02:00	Mobiles
12	3776	Apple	52500	3	01NOV12:07:02:00	Mobiles
13	4020	Apple	52500	7	02NOV12:07:02:00	Mobiles
14	4282	Apple	52500	4	03NOV12:07:02:00	Mobiles
15	4337	Apple	52500	6	04NOV12:03:00:00	Mobiles
16	4460	Apple	52500	3	04NOV12:07:02:00	Mobiles
17	4552	Apple	52500	6	05NOV12:02:41:00	Mobiles
18	4561	Apple	52500	6	05NOV12:03:00:00	Mobiles
19	4637	Apple	52500	6	05NOV12:03:20:00	Mobiles
20	4829	Apple	52500	6	05NOV12:07:02:00	Mobiles

The output in Table 6-2 shows that Apple has most of the expensive items on the web site. One BlackBerry item tops the list.

Summary Statistics

When the data set is large, opt for summaries or aggregated measures to describe the data. The following are some summary measures that can be applied on each variable of the data set:

- *Central tendencies*: What is the average value?

- *Dispersion*: What is the spread in a particular variable? We will discuss more about this in the following sections.

- *Other summary statistics such as correlation coefficient*: This will be covered in coming chapters.

Next, we describe the summary statistics measures individually and use examples to show how to use them.

Central Tendencies

Answering questions such as the following will help you understand how to use aggregated measures or summary statistics to describe a data set:

- What were the sales figures last year?

- What is the cost of air travel from Dubai to the United Kingdom?

- How many hours do you sleep in a day?

- How much time does it take to reach the office?

The answers are usually average values rather than definite numbers; for example, answers might be along the following lines: the average sale is worth $5 million a year; it takes 30 minutes to reach the office on an average, and so on. This average is simply a measure of *central tendency*, a measure that gives an idea about the middle value of a variable containing a number of data points. There are three measures of central tendency.

- Mean

- Median

- Mode

■ **Note**   All central tendencies are averages.

We will discuss them individually.

Mean

When a variable is numeric, it is a simple arithmetic mean. This is the most commonly measured central tendency. A simple aggregate sum of values divided by a count of those values gives the arithmetic mean.

For example, the average age of four people aged 24, 24, 27, and 25 is 25. The average age was derived by finding the sum and dividing it by the count.

```
Average age= (24+24+27+25)/ (4) = 25
```

$$\bar{x} = \frac{\sum_{i=1}^{n} x_i}{n}$$

Arithmetic mean is a good measure of the central tendency, but it is not perfect. The arithmetic mean might not indicate the middle value of the variable if there are any outliers in the data. It will tend to deflect toward the outlier.

The Outlier Effect

An *outlier* is the value of a set or a column that is entirely different from other values in the same set. It can be an extraordinarily high or low value, compared to the other entries in that variable column.

For example, Table 6-3 contains the monthly incomes of 14 individuals.

Table 6-3. *Monthly Income of 14 Individuals*

Id	Monthly Income
1	7,400
2	9,800
3	6,500
4	9,100
5	9,000
6	718,900
7	9,400
8	9,300
9	7,800
10	8,900
11	8,600
12	9,600
13	10,000
14	6,800

The monthly incomes of 13 individuals out of 14 are less than or equal to $10,000 in the data set. However, there is one individual whose income is $718,900. It is not apparent whether it is an error or a deliberate inclusion. This value is completely different and much higher than the other values. This particular value in the monthly income data is the outlier. Subsequent sections in this chapter will discuss how to identify the outliers.

When the mean of the preceding sample is calculated, the result is $59,364; whereas 13 out of 14 individuals earn less than $59,364. Only one individual's salary is more than the arithmetic mean. The central tendency is expected to be around the midpoint of the data, so the average value can be determined, but the mean in this instance has changed drastically because of the outlier. Outliers generally pull mean values toward them, whether they are high-side outliers (as in this example) or low-side ones. For example, on the lower side, if the 15th entry in Table 6-3 is 1,000, the average will be lowered to $55,473. Figure 6-1 shows a plot of this example.

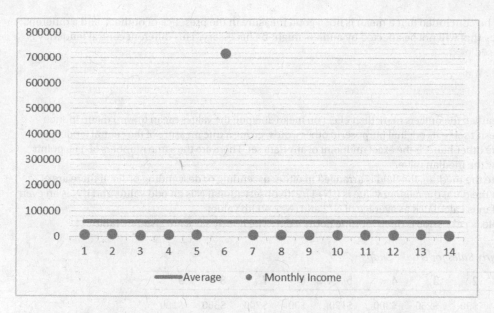

Figure 6-1. *Monthly income of individuals with outlier and calculated average*

Here, the mean does not give the true central tendency of the variable because of just one outlier. What is the impact when the outlier is removed (Figure 6-2)?

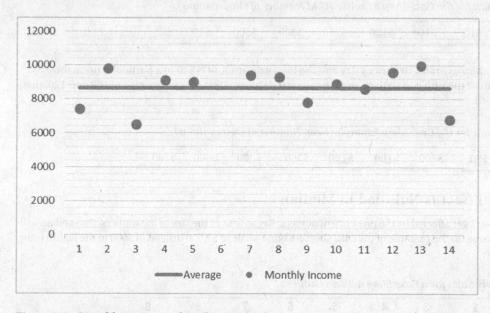

Figure 6-2. *Monthly income with outlier removed*

Once the outlier is removed, the new mean, $8,631, is in the middle of all the values, the true central tendency or average value. This is called the *truncated mean*. To arrive at this value, the outlier must be eliminated or replaced with the second highest value for the given variable and then calculated. In general, it might not be a good idea to fiddle with the original data.

It can now be concluded that a mean is not a good measure in the presence of outliers, and a different central tendency that will not be affected by outliers might be necessary. This different central tendency is median, discussed next.

Median

As established earlier, the outliers mark their effect on mean and pull the values mean toward them. In such cases, mean doesn't serve its original purpose, in other words, giving a true measure of the central tendency. A median, on the other hand, is the exact midpoint of any data set. There are the same numbers of data points above and below the median value.

To calculate the median, the field is arranged in either ascending or descending order. If there are N observations in the data, the median value is ((N+1)/2)th observation if N is an odd value, and if N is an even value, the median is calculated as average of N/2 and (N/2 + 1)th value.

For example, a cake shop is open for nine hours a day. The hourly sales are given in Table 6-4.

Table 6-4. *Hourly Sales for a Cake Shop*

Hour	1	2	3	4	5	6	7	8	9
Sales	$100	$90	$250	$300	$120	$30	$700	$500	$100

To find the median, the sales are simply arranged in ascending order, and the fifth value is noted. The median in this case is $120 (Table 6-5). It is independent of the order, either ascending or descending.

Table 6-5. *Median for the Cake Shop Example (Odd Number of Observations)*

$30	$90	$100	$100	**$120**	$250	$300	$500	$700

If the same shop operates 10 hours a day and the least sale amount is $30 in a particular hour, the median is a mean of the fifth and sixth records in the sorted values. Here, it is an average of 100 and 120, that is, $110 (Table 6-6).

Table 6-6. *Median for the Cake Shop Example (Even Number of Observations)*

$30	$90	$90	$100	**$100**	**$120**	$250	$300	$500	$700

The Outlier Effect Is Nullified in Median

The median doesn't get affected by the presence of outliers. Regardless of the size of the outliers, the median will still remain the same. In the example of the cake shop, imagine a sale of $1,500 instead of $500 in the eighth hour (Table 6-7).

Table 6-7. *Hourly Sales for a Cake Shop with an Outlier*

Hour	1	2	3	4	5	6	7	8	9
Sales	$100	$90	$250	$300	$120	$30	$700	$1500	$100

The median in Table 6-8 is still the same.

Table 6-8. *Median for the Cake Shop Example with an Outlier*

$30	$90	$100	$100	**$120**	$250	$300	$700	$1500

Median is not affected by the low-side outliers either. The way median works is simple: it takes the position of the record instead of actually considering the value. Its position remains the same from either the top or the bottom, regardless of the size of the value.

■ **Note**   Is it safe to say that there are no outliers in the data when the mean is close to the median?

Yes, if the mean and median are close, it can be concluded that there are no outliers in the variable. However, If there are balancing outliers on the either side of median (both extreme low and high values), then also the mean and median can be close.

Mode

Mode is the value that occurs most frequently in the data set. Sometimes, calculating mean or median might not make much sense, particularly if a particular value occurs multiple times in a variable. In such a case the recurring value is likely to be quoted as the average.

For example, consider ten families residing in an apartment complex. The numbers of family members are 3, 4, 4, 4, 4, 4, 4, 2, and 4, respectively. What is the average family size in that apartment? The answer is most likely to be 4 since most of the families consist of 4 members and is the mode in this case. Mode is not exactly in the central tendency, but it makes better sense rather than the mean, which is 3.7. A family size of 3.7 is improbable. Mode is a good measure in cases where a mean is not meaningful. Examples are average shoe size of a city, average number of loans, average family size of a country, and so on. As you have already learned, mode is the most frequent value of a variable in a given data set. You just need to look at the frequency table to find the mode as the most frequent occurring value.

Calculating Central Tendencies in SAS

The following code will help find the average list price in the online sales example:

```
Proc means data=online_sales;
var listPrice;
run;
```

Refer to Table 6-9 for the output of this code.

Table 6-9. *Output of proc Means on online_sales data base (listPrice)*

Analysis Variable : listPrice				
N	Mean	StdDev	Minimum	Maximum
5129	11598.14	10583.31	849.0000000	139990.00

SAS gives the mean value by default and additional information on other measures like minimum, maximum, and so on. The following code helps in printing only the mean value:

```
Proc means data=online_sales mean;
var listPrice;
run;
```

Refer to Table 6-10 for the output of this code.

Table 6-10. *Mean of listPrice*

Analysis Variable: listPrice
Mean
11598.14

This code calculates the mean list price by brand:

```
Proc means data=online_sales mean;
var listPrice;
class brand;
run;
```

Refer to Table 6-11 for the output of this code.

Table 6-11. *Output of proc means on online_sales (Mean Values for Every Brand)*

Analysis Variable : listPrice		
brand	N Obs	Mean
Acer	1	8499.00
AirTyme	90	1549.00
Airtyme	89	2551.10
Apple	146	42854.80
Arise	161	1215.92
BlackBerr	201	14374.20
Blackberr	18	21221.78
Canon	3	5964.67
Fujifilm	11	6512.09
HTC	469	16816.29
HTC	38	11511.00
Huawei	8	8780.25
IBall	108	5491.60

(*continued*)

Table 6-11. (*continued*)

Analysis Variable : listPrice		
brand	N Obs	Mean
Intel	1	17500.00
Intex	26	887.30
Karbonn	299	7172.62
LAVA	4	2100.00
LG	361	13164.46
Lava	25	1289.12
Lemon	1	999.00
MICROMAX	19	4950.00
MOTOROLA	19	22368.00
Micromax	579	3550.90
Motorola	259	13035.01
Nikon	23	9319.78
Nokia	683	9919.18
Olympus	4	4399.00
Samsung	956	13313.62
Sony	355	17242.56
Sony Eric	62	16899.82
Soyer	7	1050.00
Spice	80	3967.20
Videocon	13	2840.92
Xolo	6	17937.50
Zen	4	1999.00

The output in Table 6-10 shows the average price of each brand. We will concentrate on the brands that have more than 30 orders (>30). The overall average list price is $11,598. The list price for Apple is way above it, and the average list price for Micromax is much below the overall average.

This code helps find the median:

```
Proc means data=online_sales median;
var listPrice;
run;
```

Refer to Table 6-12 for the output of this code.

Table 6-12. *Median for listPrice*

Analysis Variable: listPrice
Median
8399.00

The median is much lower than the mean, which indicates that there are some high-side outliers. The data indicates that the outliers in this case are the list prices of Apple and BlackBerry.

The following is the code for the mode:

```
Proc means data=online_sales mode;
var listPrice;
run;
```

Refer to Table 6-13 for the output of this code.

Table 6-13. *Mode for listPrice*

Analysis Variable: listPrice
Mode
990.00

A mode value for a continuous variable may not make a lot of sense. Let's find the mode of shippingPeriod using the following code:

```
Proc means data=online_sales mode;
var shippingPeriod;
run;
```

Refer to Table 6-14 for the output of this code.

Table 6-14. *Mode for shippingPeriod*

Analysis Variable: shippingPeriod
Mode
5

The shipping period for most of the items is 5. This is simply the most occurring value. A shipping period of 5 days sounds more probable than 4.98 days, which is the mean value. Take a look at the code that follows:

```
Proc means data=online_sales mean;
var shippingPeriod;
run;
```

Refer to Table 6-15 for the output of this code.

Table 6-15. Mean for shippingPeriod

Analysis Variable: shippingPeriod
Mean
4.98

The central tendencies of different variables were discussed in the preceding examples. Central tendencies give us a unique value that represents that variable. But the question remains: is it sufficient, or are other measures needed to better understand that variable?

ANDERSON WANTS TO CROSS A RIVER

Mr. Anderson, who can't swim, wants to cross a small waterway. He asked a neighbor to describe the depth of that river, and the neighbor said its depth is 4 feet on average. Mr. Anderson is happy and starts to cross it. His happiness does not last long. The reason is that although the average is 4 feet, the depth at some places might have been 7 feet, which is more than Mr. Anderson's height. If he had inquired about the deviation from average depth or the inconsistency of depth at various points, or at least the range of depth apart from the average depth of the river, it would have saved Mr. Anderson from drowning.

Therefore, merely knowing the average or the center value may not be sufficient in all cases. The deviation from center (or the dispersion) or the spread of a variable is also important. Given next are a few measures of dispersion.

What Is Dispersion?

Dispersion is the variation in data—the anomaly or inconsistency in the values of a variable. The measures of dispersion indicate nothing about the middle value of the data. Rather, they give you an idea about the spread.

Next, we discuss a few measures that quantify the dispersion.

Range

Range is a basic measure that explains the dispersion, or the spread, in the data. The calculation of range is simple: it's the difference between the maximum and minimum values of a variable.

Range = Maximum – Minimum

Range is a good measure of dispersion when dealing with a small data set and a quick estimate of the range is required. It is better to mention the maximum and minimum values while quoting the range.

For example, you are looking for a smartphone with certain features. There are nine brands that produce such a phone. Table 6-16 gives their respective prices.

Table 6-16. Smartphone Brand and Prices

Id	Brand-1	Brand-2	Brand-3	Brand-4	Brand-5	Brand-6	Brand-7	Brand-8	Brand-9
Price	$ 4,700	$ 4,800	$ 8,900	$ 6,000	$ 2,900	$ 7,400	$ 3,300	$ 7,800	$ 3,900

The maximum price is $8,900, and the minimum is $2,900; the range of this price variable is $8,900 – $2,900 = $ 6,000

Although range is a good measure for small samples, a stronger, more reliable measure for quantifying the actual dispersion in the data is necessary. A more granular measure that considers the spread in each record, instead of the overall range, will serve this purpose. Variance is such a measure.

Variance

The overall spread in the data needs to be quantified. For example, consider the closing stock prices of a startup company, in dollars (Table 6-17, Figure 6-3 and Figure 6-4).

Table 6-17. *Stock Prices for a Startup Company*

Day	Stock Price
1	66
2	55
3	60
4	60
5	77
6	70
7	79
8	82
9	77
10	72
11	75
12	67
Mean	**70**

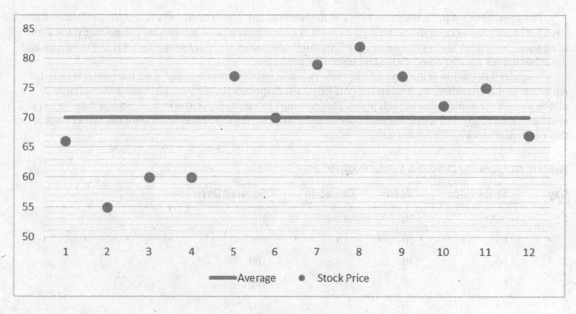

Figure 6-3. *Stock prices for a startup company (Table 6-17)*

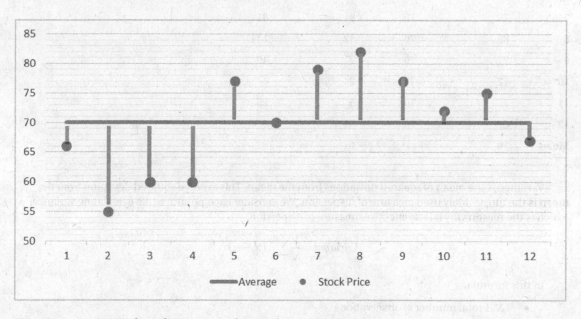

Figure 6-4. *Deviations from the mean; stock prices for a startup company*

The average stock price is almost $70. The deviation of day 1 from average is 4, and the deviation of day 2 is 15. This deviation from the mean at each point is a good indicator of dispersion. Interestingly, the sum of all such deviations always comes to zero. This inference is obvious because the mean is in the middle and the other values are dispersed above and below the mean line.

A squared deviation is the next best option. The squared deviation average of all the points is called *variance*, and it quantifies the dispersion perfectly. Less dispersion in the variable means the variable is taking almost the same value at each point. This amounts to a deviation close to zero. Squared deviation will also be close to zero, and it can be concluded that variance is very low. The variance for the last 12 days in this particular stock is 67 (Table 6-18).

Table 6-18. *Squared Deviation for the Stock Prices*

Day	Stock Price	Mean	Deviation	Squared Deviation
1	66	70	-4	16
2	55	70	-15	225
3	60	70	-10	100
4	60	70	-10	100
5	77	70	7	49
6	70	70	0	0
7	79	70	9	81
8	82	70	12	144
9	77	70	7	49
10	72	70	2	4
11	75	70	5	25
12	67	70	-3	9
Mean	70		0	67

Variance is the mean of squared deviations from the mean. This average squared deviation from the mean is the most widely used measure of dispersion. We consider each point x_i while calculating variance, which is the reason why it is an effective measure of dispersion.

$$Variance = \frac{1}{N} \sum_{i-1}^{N} (x_i - \mu)^2$$

In this formula,

- N is total number of observations.
- x_i is the ith value of x.
- μ is the mean of variable x.

Similarly, the variance for a different stock (Table 6-19) with daily values is 413.

Table 6-19. *Stock Prices for a Startup Company, Another Example*

Day	Stock Price	Average	Deviation	Squared Deviation
1	32	41	-9	81
2	28	41	-13	169
3	75	41	34	1156
4	22	41	-19	361
5	41	41	0	0
6	75	41	34	1156
7	63	41	22	484
8	25	41	-16	256
9	59	41	18	324
10	19	41	-22	484
11	21	41	-20	400
12	32	41	-9	81
Mean	41		0	413

■ **Note** Although variance is a good measure for quantifying the dispersion, it poses a challenge with regard to applying units of measures to the calculated numeric values.

In the example in Table 6-18, variance is 67. Is it 67 square dollars? Would a businessperson understand it? Several other variables such as number of customers (square customers), age (year square), and so on, do not make sense. They have to be brought back to the original form of units after calculating variance. Since variance is the average squared deviation, a square root of that value will be in the same units as the original variable. That is called *standard deviation*.

Standard Deviation

Standard deviation is the square root of variance, as expressed by this formula:

$$\sigma = \sqrt{\frac{1}{N}\sum_{i=1}^{N}(x-\mu)^2}$$

This measure also helps in comparisons because the unit is the same as in the original unit of the variable. The standard deviation in the stock price example is sqrt(67) = 8.2. This means there is an $8.2 deviation from the mean stock price of $70.

Calculating Dispersion Using SAS

SAS gives convenient procedures to calculate the measures of dispersion. We will show you some examples to help you understand. What follows is the code for calculating variation on the list price:

```
Proc means data=online_sales var;
var listPrice;
run;
```

Refer to Table 6-20 for the variance given by this code.

Table 6-20. *Variance for online_sales*

Analysis Variable : listPrice
Variance
112006391

The following code calculates the standard deviation on the variable list price:

```
Proc means data=online_sales std ;
var listPrice;
run;
```

Refer to Table 6-21 for the variance given by this code.

Table 6-21. *Variance for online_sales*

Analysis Variable: listPrice
StdDev
10583.31

Determining whether the variance is high or low is not possible simply by looking at variance. The magnitude of the variable will make sense only when it's compared with the original variable units. For example, a variance of 1,000 may be insignificant if the average variable value is in millions. The following code calculates the standard deviation on the list price for every brand:

```
/* SD in list price for each brand */
Proc means data=online_sales std ;
var listPrice;
class brand;
run;
```

Refer to Table 6-22 for the output of this code.

Table 6-22. *Output of proc means on online_sales Data Set (Brandwise Standard Deviation)*

Analysis Variable : listPrice

Brand	N Obs	StdDev
Acer	1	.
AirTyme	90	0
Airtyme	89	133.04
Apple	146	7290.86
Arise	161	493.21
BlackBerr	201	10270.17
Blackberr	18	6220.25
Canon	3	351.48
Fujifilm	11	3579.65
HTC	469	9039.04
Htc	38	0.00
Huawei	8	2127.74
IBall	108	4485.59
Intel	1	.
Intex	26	0.74
Karbonn	299	1866.15
LAVA	4	52.00
LG	361	7024.66
Lava	25	198.36
Lemon	1	.
MICROMAX	19	0.00
MOTOROLA	19	0.00
Micromax	579	1919.67
Motorola	259	5992.82
Nikon	23	4891.98
Nokia	683	8026.02
Olympus	4	0.00
Samsung	956	11870.12

(*continued*)

Table 6-22. (*continued*)

| Analysis Variable : listPrice | | |
Brand	N Obs	StdDev
Sony	355	7297.27
Sony Eric	62	1909.80
Soyer	7	0.00
Spice	80	871.76
Videocon	13	786.43
Xolo	6	479.26
Zen	4	0.00

It can be inferred by Table 6-22 that Samsung has sold mobile phones of different price ranges, and hence the standard deviation is high for this brand. This is partly because of the number of items sold by that brand. Micromax has a relatively low (zero in fact) standard deviation when compared to other brands.

EXAMPLE OF A HEALTHCARE CLAIM

Consider a healthcare insurance company that has collected claims data over a period of time. Here are the variables:

- *Patient_id*: Patient ID
- *Patient_Age*: Patient age
- *Days_admitted_hosp*: Number of days spent in the hospital
- *Num_medical_bills_submitted*: Number of medical bills submitted
- *Claim_amount*: The claim amount

The following is the SAS code and analysis on central tendencies and dispersion of the claim amount. Refer to Table 6-23 for the output.

```
Proc means data= health_claim  ;
var Claim_amount;
run;
```

Table 6-23. *Output of proc means on health_claim data set*

| Analysis Variable : Claim_amount | | | | |
N	Mean	StdDev	Minimum	Maximum
50000	1537.48	3287.49	0	256058.00

The mean claim amount is $1,537. The maximum claim amount is $256,058. The standard deviation in claim amount is $3,287.5.

The following is the code for the mean and median:

```
Proc means data= health_claim  meanmedian;
var  Claim_amount;
run;
```

Refer to Table 6-24 for the output of this code.

Table 6-24. Mean and Median for claim_amount data set

Analysis Variable : Claim_amount	
Mean	Median
1537.48	438.00

There are definitely some outliers. The mean value is far away from median. Chapter 7 will discuss how to deal with outliers.

This chapter so far discussed some of the measures that give an overall picture of variables. Mean indicates the average value, whereas the variance and standard deviation indicate dispersion. There are a few additional measures, such as quantiles, that help you get a comprehensive understanding of a variable. Quantiles help you better understand the distribution of a variable. In SAS, you use univariate analysis to get quantiles and box plots, which are covered in the following sections.

Quantiles

Quantiles are simply identical fragments. Examples of quantiles are percentiles, quartiles, and deciles, which will be discussed in this section. A percentile is a result of breaking the variable values into 100 pieces.

For example, consider the marks of 20 students in ascending order (Table 6-25).

Table 6-25. Marks for 20 Students

S1	S2	S3	S4	S5	S6	S7	S8	S9	S10	S11	S12	S13	S14	S15	S16	S17	S18	S19	S20
37	58	40	55	67	75	35	40	56	37	60	52	57	66	43	74	75	63	72	54

Here, the median is 56.5, and the student who is close to average is S9. This means that S9 is almost at the 50 percent mark of the population (Table 6-26).

Table 6-26. The Median for Population

S7	S1	S10	S3	S8	S15	S12	S20	S4	S9	S13	S2	S11	S18	S14	S5	S19	S16	S6	S17
35	37	37	40	40	43	52	54	55	56	57	58	60	63	66	67	72	74	75	75

- Which student is at 100 percent? Remember that it is the 100 percent position value that is relevant here, not the 100 marks. The whole variable is divided into 100 parts, and each part is called a *percentile*. Now consider Tables 6-27 and 6-28 and the following list to better understand percentiles, and two other terms, *deciles* and *quartiles*:

  - What value is at the 100th percentile? Is it the maximum value, that is, 75?

  - What value is at the 50th percentile? Is it the median value, 56? This means 50% of the cases are less than 56 and rest 50% are more than 56.

  - What value is at the 10th percentile? Since there are 20 pieces here, each piece contributes 5% percent in the overall variable, so the 10th percentile is the second one, that is, i.e., 37. This means 10% of the cases are less than 37 and rest 90% are more than 37.

  - The 70th percentile is the 14th value, that is, i.e., 63. This means 70% of the cases are less than 63 and rest 30% are more than 63.

  - A decile is 1/10th part of the variable, and in this example the first decile equals the 10th percentile (37), and the second decile equals the 20th percentile (40).

  - A quartile is a 1/4th part of a variable. In this example, the first quartile equals the 25th percentile (40). The second quartile equals the 50th percentile, which is the median, and it equals the 5th decile (56). The third quartile equals the 75th percentile (66).

Table 6-27. *Different Percentile Values for the Population*

S7	S1	S10	S3	S8	S15	S12	S20	S4	S9	S13	S2	S11	S18	S14	S5	S19	S16	S6	S17
35	37	37	40	40	43	52	54	55	56	57	58	60	63	66	67	72	74	75	75

Table 6-28. *Different Quartile Values for the Population*

S7	S1	S10	S3	S8	S15	S12	S20	S4	S9	S13	S2	S11	S18	S14	S5	S19	S16	S6	S17
35	37	37	40	40	43	52	54	55	56	57	58	60	63	66	67	72	74	75	75

The first and the second quartiles are also known as the lower and upper quartiles, respectively. The difference between lower and upper quartiles is called *inter quartile range* (Table 6-29). It gives you an idea about the middle 50 percent values range.

Table 6-29. *Inter Quartile Range*

S7	S1	S10	S3	S8	S15	S12	S20	S4	S9	S13	S2	S11	S18	S14	S5	S19	S16	S6	S17
35	37	37	40	40	43	52	54	55	56	57	58	60	63	66	67	72	74	75	75

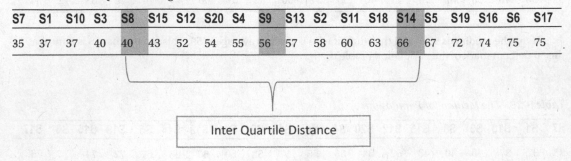

Inter Quartile Distance

Inter quartile range = 66 – 40 = 26

This interquartile range is a useful measure in identifying outliers. In the example in Table 6-29, the overall range is 75 – 35 =40, whereas the interquartile range, that is, the middle 50 percent range, is 26, which looks right. Now consider the example given in Table 6-30.

Table 6-30. *Another Example of Inter Quartile Range*

S7	S1	S10	S3	S8	S15	S12	S20	S4	S9	S13	S2	S11	S18	S14	S5	S19	S16	S6	S17
35	37	37	45	46	48	52	50	52	55	57	58	58	59	59	66	67	69	96	99

Inter quartile range = 59 – 46 =13

This indicates that the middle 50 percent of the values are within the range of 13. The overall range is 99 – 35 = 64. The following additional inferences can be drawn for Table 6-30:

- The least 25 percent values are in the range of 11 (46 – 35).

- The middle 50 percent values are in the range of 13.

- The highest 25 percent values are in the range of 99 – 59 = 40, which indicates that there are some unusually high values in the top 25 percent.

What would the scenario be if there were just ten students (Table 6-31)? This is how the first quartiles can be found: The first quartile falls between the second-ranked and third-ranked values. In such a case the quartile is calculated by taking the average. The third quartile falls between the seventh-ranked and eighth-ranked values.

Table 6-31. *Quartiles with Only Ten Observations*

S7	S1	S10	S3	S8	S15	S12	S20	S4	S9
35	37	37	40	40	43	52	54	55	56

First quartile = (37 + 37) / 2 = 37
Third quartile = (52 + 54) /2 = 53

Calculating Quantiles Using SAS

The following code yields some of the important quantiles. Box plots, which are discussed in the next section, can also be plotted using univariate analysis, which you use to get quantiles.

Here is the code for obtaining quantiles for the list price in the online sales example:

```
Proc univariate data= online_sales ;
var  listPrice ;
run
 ;
```

Proc univariate offers several other measures about a variable. This section will consider only the quantiles tables in the output (Table 6-32).

Table 6-32. *Quartiles from the Output of Proc univariate on online_sales*

Quantiles (Definition 5)	
Quantile	**Estimate**
100% Max	139990
99%	44500
95%	35198
90%	28590
75% Q3	15302
50% Median	8399
25% Q1	4110
10%	1559
5%	999
1%	849
0% Min	849

The minimum list price is $849, and the maximum is $139,990. The 25th percentile, or the first quartile, is $4,110. The third quartile value is $15,302. The interquartile range is 15,302 – 4,110 = 11,192. The quantiles also show that 99 percent of the items list price is less than or equal to $44,500. The maximum value of $139,990 looks much bigger than 99% value i.e $44,500. There is a clear indication of outliers. In later sections we will introduce a measure to identify the outliers.

HEALTH CLAIM EXAMPLE

Given in Table 6-33 is the output taken from a healthcare data set. The code to perform a univariate analysis on health claim data follows:

```
Proc univariate data= health_claim ;
var  Claim_amount ;
run;
```

Table 6-33. *Quartiles for health_claim Example*

Quantiles (Definition 5)	
Quantile	Estimate
100% Max	256058
99%	11428
95%	6149
90%	4368
75% Q3	2044
50% Median	438
25% Q1	0
10%	0
5%	0
1%	0
0% Min	0

The output in Table 6-33 shows that 25 percent of the customers claimed nothing, and 50 percent of the customers claimed $438. Ninety-five percent of the customers claimed $6,149 or less, so the rest of the customers, who make up 5 percent, seem to have an unusual claim amount. Performing an excess scrutiny of these 5 percent customers might be in order. Elements such as their number of medical bills, type of disease, and so on, might need to be examined.

Box Plots

Box plots indicate outlier values in a variable. A box plot is a graphical representation of the quartiles and minimum and maximum values. Box plots make it easier to interpret the distribution of a variable. A box plot is illustrated in Table 6-33 using the example data given in Table 6-34.

Table 6-34. *An Example Population*

S7	S1	S10	S3	S8	S15	S12	S20	S4	S9	S13	S2	S11	S18	S14	S5	S19	S16	S6	S17
35	37	37	40	40	43	52	54	55	56	57	58	60	63	66	67	72	74	75	75

A box plot shows these values:

- Minimum value
- First quartile value
- Median value
- Second quartile value
- Maximum value

Figure 6-5 illustrates a box plot based on the data shown in Table 6-34.

S7	S1	S10	S3	S8	S15	S12	S20	S4	S9	S13	S2	S11	S18	S14	S5	S19	S16	S6	S17
35	37	37	40	40	43	52	54	55	56	57	58	60	63	66	67	72	74	75	75

Figure 6-5. *The box plot for Table 6-34*

If there are rare values in the variable very different from the others, they will appear at either the upper or lower end of the box. The outliers are glaringly visible in a box plot. The plot given in Figure 6-6 shows that all the values are well distributed. There are no outliers in this variable.

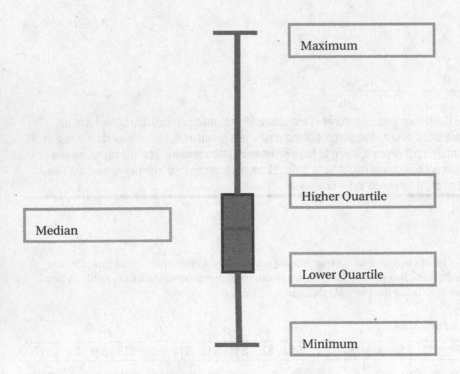

Figure 6-6. *A Boxplot without outliers*

For example, in Table 6-35 a few of the values are much beyond the median and very far from the third quartile. The box plot looks like Figure 6-7.

Table 6-35. *An Example Population with Outliers*

S7	S1	S10	S3	S8	S15	S12	S20	S4	S9	S13	S2	S11	S18	S14	S5	S19	S16	S6	S17
35	37	37	45	46	48	52	50	52	55	57	58	58	59	59	66	67	69	98	99

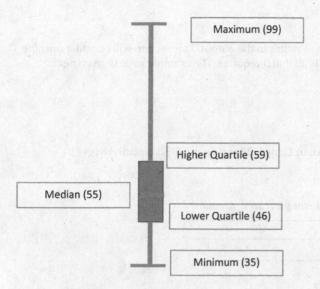

Figure 6-7. The box plot for Table 6-35

An easy way to identify the outliers is to look at the interquartile range. If there are values beyond three times the interquartile range, they can be considered as outliers (Figure 6-8). In this example, the interquartile range is 59 – 46 = 13, but there are values beyond 59 + 3 * (13) = 59 + 39 = 98. Therefore, the values 98 and 99 are outliers in the data.

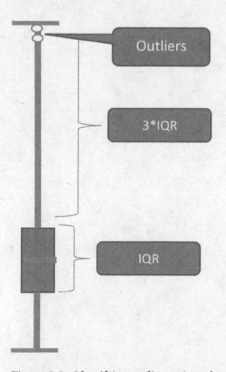

Figure 6-8. Identifying outliers using a box plot

Creating Boxplots Using SAS

As discussed earlier, univariate analysis gives several values in the output. This section will consider only the box plots. Adding a plot option to univariate code is all that it requires. The example code is given next:

```
Proc univariate data= health_claim  plot;
var  Claim_amount ;
run;
```

The resulting quantiles and box plot are shown in Table 6-36 and Figure 6-9, respectively.

***Table 6-36.** Quantiles Table for the Health Claim Example (claim_amount)*

Quantiles (Definition 5)	
Quantile	Estimate
100% Max	256058
99%	11428
95%	6149
90%	4368
75% Q3	2044
50% Median	438
25% Q1	0
10%	0
5%	0
1%	0
0% Min	0

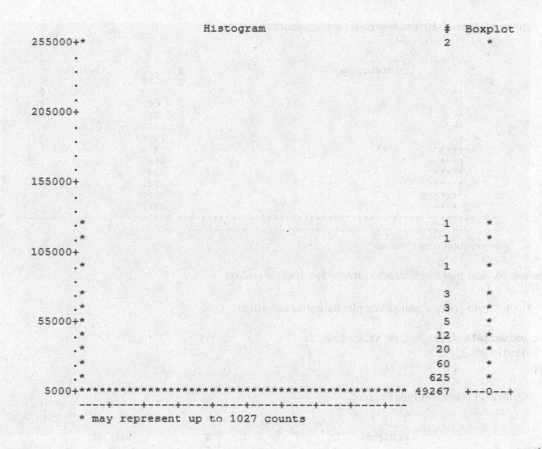

Figure 6-9. *Box plot for the health_claim data set (claim_amount)*

The graph in Figure 6-9 shows that the whole box (first quartile and median and third quartiles) is right at the bottom. This is because there are some high-side outliers. The first step is to remove all values that are high-side outliers and redraw the box plot. Consider only the first 95 percent of the population, that is, all the customers with a claim amount less than $6,149. The 95 percent value can be found from the quantiles table (Table 6-36).

This code shows a condition on claim amount:

```
Proc univariate data= health_claim  plot;
var  Claim_amount ;
where  Claim_amount<6149;
run;
```

The box (Figure 6-10) can now be seen after ignoring outliers.

```
         Histogram                                #          Boxplot
   6250+*                                         154           0
      . **                                        553           0
      . **                                        702           0
      . **                                        857           0
      . **                                       1042           0
      . ***                                      1298           |
   3250+***                                      1573           |
      . ****                                     1826           |
      . *****                                    2247           |
      . ******                                   2701        +------+
      . *******                                  3520        |  +  |
      . **********                               5222        |     |
    250+**************************************** 25804       *-----*
      ----+----+----+----+----+----+----+----+----+----+----+---
```
* may represent up to 538 counts

Figure 6-10. *Box plot for claim amount after ignoring the outliers*

The following code creates a box plot for online sales data:

```
Proc univariate data= online_sales plot ;
var  listPrice ;
run;
```

Figure 6-11 is the resulting box plot.

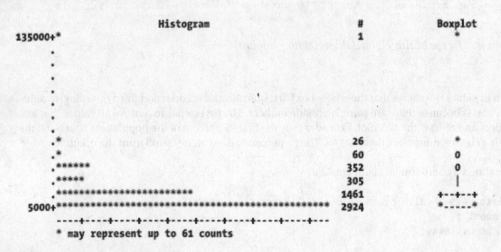

```
         Histogram                                #          Boxplot
 135000+*                                         1             *
      .
      .
      .
      .
      .
      .
      .
      .*                                          26            *
      .*                                          60            0
      .*****                                     352            0
      .*****                                     305            |
      .*********************                    1461         +--+--+
   5000+************************************** 2924         *-----*
      ----+----+----+----+----+----+----+----+----+----+---
```
* may represent up to 61 counts

Figure 6-11. *Box plot for listPrice*

There are some outliers again (Table 6-37).

Table 6-37. *Quantiles Table for the Health Claim Example (listPrice)*

Quantiles (Definition 5)	
Quantile	Estimate
100% Max	139990
99%	44500
95%	35198
90%	28590
75% Q3	15302
50% Median	8399
25% Q1	4110
10%	1559
5%	999
1%	849
0% Min	849

Consider the values up to 99 percent to see the box. The elimination and treatment of the outliers are discussed in detail in Chapter 7. This example simply removes outliers and redraws the box plot (Figure 6-12) to show the box clearly. Consider the following code:

```
Proc univariate data= online_sales plot ;
var  listPrice ;
where listPrice<44500;
run;
```

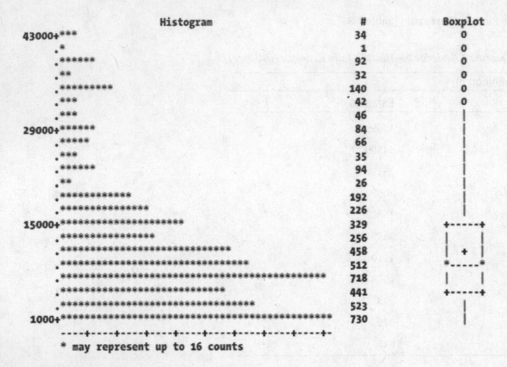

Figure 6-12. *Box plot for listPrice after removing outliers*

The box now looks better distributed around the population after removing outliers.

Bivariate Analysis

You have examined one variable at a time so far. Bivariate analysis analyzes two variables together. It answers questions such as what the association between two variables is, what happens to the other variable when one increases, and so on. Here are some examples:

- If the number of study hours increases, do the exam grades increase too?

- If the number of church buildings increases, does the crime rate decrease?

- If the numbers of cars increase, does the average age decrease?

Correlation is used to quantify the association between two variables. Correlation is one of the most used methods to carry out bivariate analysis. Subsequent chapters will discuss correlation.

Conclusion

In this chapter, you learned how to do some basic reporting on data and analyze variables using some very basic statistical techniques. You also see how univariate analysis can be useful in analyzing variables. Until recently, even this much of statistical analysis was considered good enough. Things have changed a lot in the past few years. Now much more advanced and sophisticated data analysis tools and techniques are employed as an aid to executive decision making. In the coming chapters, we will take you through some classical and advanced data analysis techniques. You will appreciate how easy these otherwise complex techniques become when you deploy SAS as your primary data analysis tool.

CHAPTER 7

■ ■ ■

Data Exploration, Validation, and Data Sanitization

Preparing the data for the actual analysis is an important portion of any analytics project. The raw data comes from a variety of sources such as classical relational databases, flat files, spreadsheets, and unstructured data from sources such as social media text. A project may contain both structured and unstructured data, and to add to the complexity, there can be numerous data sources. As you would expect, the data will have a lot of challenges—both in quality and in quantity. An analyst needs to first read the data from its sources, which itself can be a challenging task, and then parse it to be useful for any further analysis. SAS needs data to be in its own datasets before you can use any of its routines for analysis. In short, the raw data is not always ready for the analysis; it needs to be validated and cleaned before the analysis.

Considering the importance of the topic, we have treated this topic in sufficient detail. This chapter first makes you aware of the general issues that you may face while preparing data for analysis. We then cover the topics of data validation and cleaning. Finally, we take a very detailed case study in the banking domain to demonstrate the concepts with actual data.

Data Exploration Steps in a Statistical Data Analysis Life Cycle

The previous chapters dealt with applications of basic descriptive statistics. It has already been established that data analysis can yield great results if used effectively. Figure 7-1 recaps the steps followed in statistical data analysis.

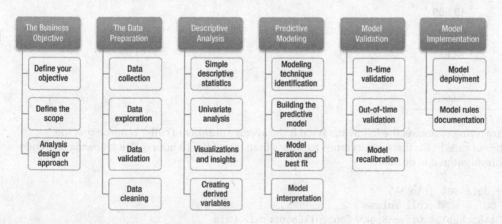

Figure 7-1. Steps in the data analysis project

Once the data business objectives have been established, a complete understanding of the data is necessary before proceeding with the analysis process. The data preparation step in the second column in the order, shown in Figure 7-1, raises the following questions:

- Why is understanding and exploring data such an important step in statistical analysis?

- Why does this step need to be mentioned along with other important steps such as descriptive analysis and predictive modeling?

The following example will address these two questions.

Example: Contact Center Call Volumes

Consider the call volume data for a typical customer contact center of a large organization. The data snapshot in Table 7-1 has just three columns. The first column is the day, the second column is the hour of the day, and the third column is the call volume (number of calls) in a given hour. The data is recorded for five consecutive days.

Table 7-1. *Call Volume Data*

Day	Hour	Volume
1	1	3,504
1	2	3,378
1	3	6,872
1	4	5,993
1	5	3,093
1	6	3,512
1	7	4,142
1	8	6,441
1	9	61,906
1	10	43,175
1	11	49,989
1	12	9,862
1	13	18,231
1	14	46,282
1	15	36,665

Say you are trying to answer the following: What is the average number of calls? Is it true that the average number of calls in the first 8 hours tends to be less than rest of the 16 hours? The following SAS code needs to be commissioned to do this:

```
/*Import the data set into SAS */
PROC IMPORT OUT = WORK.call_volume
DATAFILE= "D:\Backup\SkyDrive\Books\Content\Chapter-12 Data
Exploration and Cleaning\Data sets\Call_volume.csv"
```

```
DBMS=CSV REPLACE;
    GETNAMES=YES;
    DATAROW=2;
RUN;

/*Get the contents of the SAS data set */
Proc contents data=WORK.call_volume varnum;
run;
```

The varnum option in the preceding code lists the variables in the order in which they appear in the original data set. SAS usually prints the variables based on alphabetical order.

Refer to Table 7-2 for the output of this code.

Table 7-2. *The Variable List from SAS Output For PROC CONTENTS*

Variables in Creation Order					
#	Variable	Type	Len	Format	Informat
1	Day	Num	8	BEST12.	BEST32.
2	Hour	Num	8	BEST12.	BEST32.
3	Volume	Num	8	BEST12.	BEST32.

The PROC MEANS needs to be executed in order to find the average values of the call volume per hour. The following SAS code finds out the mean on the variable call_volume.

```
Title'Overall mean of the call volume';
Proc means data=WORK.call_volume ;
var volume ;
run;
```

Proc means data gives the output in Table 7-3 by default.

Table 7-3. *Output Of PROC MEANS On call_volume Data set*

Analysis Variable : Volume				
N	Mean	StdDev	Minimum	Maximum
120	168812.73	862472.38	0	7260664.00

Since you require only the mean, you can use the mean option in the code, as follows:

```
Title'Overall mean of the call volume';
proc means data=WORK.call_volume mean;
var volume ;
run;
```

Table 7-4 shows the output of the preceding code.

Table 7-4. *The Mean for Variable volume*

Mean
168,812.73

Following is the code to find the mean call volume for the first eight hours.

```
Title 'Mean of the call volume in 1 to 8 Hours';
proc means data=WORK.call_volume mean;
var volume;
where Hour <9;
run;
```

Table 7-5 shows the output of the preceding code.

Table 7-5. *Mean of the Call Volume in Hours 1 to 8*

Mean
417,169.30

Following is the code to find the mean call volume for the later 16 hours, that is, mean call volume between 9 to 24 hours.

```
Title 'Mean of the call volume in 9 to 24 Hours';
procmeans data=WORK.call_volume mean;
var volume;
where  Hour ge 9;
run;
title;
```

Table 7-6 shows the output of the preceding code.

Table 7-6. *Mean of the Call Volume in Hours 9 to 24*

Mean
44,634.44

The overall mean of all call numbers across the file was calculated in the SAS code preceding Table 7-4. The same mean for the first 8 hours and the same data for the subsequent 16 hours were then calculated (Tables 7-5 and 7-6). So, a simple proc means the SAS script helps find that the overall average number of calls per hour is 168,812. The assumption that the average number of calls per hour in the first 8 hours is less than the rest of 16 hours is not true. The call rate in the first 8 hours is almost 10 times higher than the next 16 hours of data.

The contact center can now deploy resources based on the average number of calls per hour. However, the manager reviews these results and rejects them.

The rationale behind the rejection is given next.

Need for Data Exploration and Validation

To explain the need for data exploration and validation, we will continue by extending the same contact center call volumes example.

- The manager knows by experience that the average number of calls per hour is generally between 50,000 to 100,000.

- He also questions the likelihood of there being more calls during the first 8 hours of the day (12 a.m. to 8 a.m.) compared to the next 16 hours. His experience also tells him that the call rate should be considerably higher in the later 16 hours.

His confidence necessitates a reexamination of the results from the raw call volume data for all five days. Table 7-7 lists the data for the number of calls every hour, for 24 hours each day, for 5 days. Each data row contains the observation number, the hour, and the number of calls received in that hour.

Table 7-7. *Hourly Call Volume Data for 5 Days*

Day=1			Day=2			Day=3		
Obs	Hour	Volume	Obs	Hour	Volume	Obs	Hour	Volume
1	1	3504	25	1	6288	49	1	2135
2	2	3378	26	2	6091	50	2	1392
3	3	6872	27	3	2546	51	3	4502
4	4	5993	28	4	6989	52	4	2969
5	5	3093	29	5	1028	53	5	5580128
6	6	3512	30	6	2287	54	6	2576968
7	7	4142	31	7	3256	55	7	1116094
8	8	6441	32	8	2332	56	8	7260664
9	9	61906	33	9	59691	57	9	17587
10	10	43175	34	10	65386	58	10	51723
11	11	49989	35	11	43383	59	11	30032
12	12	9862	36	12	49614	60	12	28268
13	13	18231	37	13	47330	61	13	44800
14	14	46282	38	14	53819	62	14	72084
15	15	36665	39	15	44540	63	15	53854
16	16	29240	40	16	26274	64	16	18097
17	17	46201	41	17	0	65	17	0
18	18	61044	42	18	20817	66	18	79203
19	19	59205	43	19	69010	67	19	23298
20	20	49223	44	20	78719	68	20	29236
21	21	44375	45	21	24440	69	21	71985
22	22	17617	46	22	48211	70	22	48338
23	23	35117	47	23	67518	71	23	12168
24	24	63976	48	24	13732	72	24	39269

Day=4			Day=5		
Obs	Hour	Volume	Obs	Hour	Volume
73	1	5780	97	1	4217
74	2	6932	98	2	7534
75	3	2897	99	3	1093
76	4	4553	100	4	2485
77	5	6533	101	5	1470
78	6	4807	102	6	6045
79	7	7920	103	7	6519
80	8	1929	104	8	3454
81	9	57631	105	9	38552
82	10	15706	106	10	76981
83	11	33103	107	11	65421
84	12	62481	108	12	78585
85	13	48118	109	13	54074
86	14	75616	110	14	68203
87	15	16769	111	15	11175
88	16	58803	112	16	62702
89	17	0	113	17	20033
90	18	76676	114	18	65875
91	19	51608	115	19	29631
92	20	76464	116	20	61052
93	21	34958	117	21	25397
94	22	51244	118	22	25505
95	23	79254	119	23	70465
96	24	52602	120	24	21537

Note that there are a few unusually high values in the Volume column on day 3. They may be true entries, but the high call rates might be because of some incidents that do not occur every day. The other entries look to be in order. Also, note that the value for the 17th hour is zero on days 2, 3, and 4. These zeros might be because of technical mistakes or any other unknown reason. These rarely occurring values are the outliers. You need to remove them and calculate the same three averages again. Restrict the sample to the call volumes between 1,000 and 100,000 for doing so.

```
/* Call volume subset */
Data call_Volume_subset;
Set WORK.call_volume;
if volume >1000 and volume<100000;
run;
```

When the preceding SAS code is executed, the log file shows the following:

```
NOTE: There were 120 observations read from the data set WORK.CALL_VOLUME.
NOTE: The data set WORK.CALL_VOLUME_SUBSET has 113 observations and 3 variables.
NOTE: DATA statement used (Total process time):
real time              0.01 seconds
cpu time               0.01 seconds
```

The new average values are as follows:

```
Title 'Overall mean of the call volume';
proc means data=WORK. call_Volume_subset mean;
var volume ;
run;
```

Table 7-8 shows the output of the preceding code.

Table 7-8. Overall Mean of call_volume

Mean
32,952.86

The following code is for calculating the mean of the first 8 hours, after removing the outliers.

```
Title 'Mean of the call volume in 1 to 8 Hours';
proc means data=WORK. call_Volume_subset mean;
var volume;
where Hour <9;
run;
```

Table 7-9 shows the output of the preceding code.

Table 7-9. Mean of the Call Volume in 1 to 8 Hours

Mean
4,247.72

The following code is for calculating the mean of the later 16 hours, after removing the outliers.

```
Title 'Mean of the call volume in 9 to 24 Hours';
proc means data=WORK. call_Volume_subset mean;
var volume;
where  Hour ge 9;
run;
title;
```

Table 7-10 shows output of the preceding code.

Table 7-10. *Mean of the Call Volume in Hours 9 to 24*

Mean
46,373.44

The modified data set indicates that the average number of calls per hour is 32,952. The number of calls in the first 8 hours is almost 10 times less than the remaining 16 hours. These numbers make business sense and are in accordance with the manager's experience. The analytics inferences need not always match the experience. When analyzed, the opposite may be true. Whatever the case, the analytics results should always be rational and make business sense. The blind averages in the call center example on the raw data resulted in the opposite of what the business was experiencing. The careful examination of data revealed the presence of outliers.

Although 7 out of 120 records were troublesome, the results had large differences. It is always a good idea to inspect data routinely before starting the analysis.

Issues with the Real-World Data and How to Solve Them

Real-world data is often crude and cannot be analyzed in its given form. The following are some of the frequent challenges encountered in data made available for analytics projects:

- Real-world data often contains missing values, which may be the data points where data is not captured or is not available.

- The data sets often contain outliers. For example, the monthly income of a few selected individuals in a sample is recorded in millions.

- A few fields in some data sets are sometimes constant, such as seeing age = 11 and number of personal loans = 6 in all the records.

- The data is simply erroneous, such as number of dependents = –3.

- The data may contain default values such as 999999, 1111111, #VALUE, NULL, or N/A.

- The data is incomplete or is not available at the transaction level but only at the aggregate level, such as most economic indicators. If you are performing analysis at the transaction level, then you may incur a lot of error.

- The data is insufficient. The vital variables required for analysis may not be present in the data set.

Missing Values

Missing values and the outliers are the two most common issues in data sets. The nonavailability of data is also a frequent issue. Sometimes a considerable size of the population or sample might have missing values. The following are a couple of examples:

- There might be a lot of missing values in the credit bureau data related to developing and underdeveloped countries.

- A marketing survey with optional fields will definitely have missing values by design.

The Outliers

The outliers are not exactly errors, but they can be the cause for misleading results. An outlier is typically a value that is very different from the other values of that variable. An example is when, in a given data column, 95 percent of the data is between certain limits but the rest of the 5 percent is completely different, which may significantly change the overall results. This issue occurred with the call volumes example. A couple of other examples of outliers are as follows:

- Insurance claim data, where just 2 percent of claims are more than $5 million and the rest are less than $800,000

- Superstore data on customer spend, where in a month just 4 customers spend $50,000 and the rest of the customers spend less than $6,000

These are just two examples of the issues that may be present in the raw data that is made available for analysis. There may be several other issues at the data set level, such as format-related challenges with variable types such as dates, numbers, and characters. The variable lengths may not make business sense. In addition, several other data issues might arise while transferring the data from one database platform to other. This chapter concentrates only on the data analysis–related issues.

Manual Inspection of the Dataset Is Not a Practical Solution

The small amount of data in the call volume example made it possible to manually look for abnormalities in the data. This is not possible with larger data sets with many variables. A data set with more than 100 variables and 1 million records does not allow for it to be examined manually for the outliers and other data-related issues.

Therefore, you need a sound, scientific method to look at the data before the analysis. The data exploration process should highlight all the hidden problems that are hard to identify manually.

Removing Records Is Not Always the Right Way

Can the troubled records simply be removed? Is that the right way? If not, what is the best possible way to deal with data errors? After identifying issues such as outliers and missing values, it is tempting to drop these *erroneous records* and proceed with the rest of the analysis. But doing so also gets rid of some precious information.

For example, consider the employee profile data given in Table 7-11. The data is divided into seven columns:

> *EmpID*: Employee ID
>
> *Name*: Name of the employee
>
> *Age*: Age of the employee
>
> *Qualification*: Educational qualification of the employee
>
> *Experience*: Years of experience of employee
>
> *Gender*: Gender of the employee
>
> *Team*: The team in which the employee is working

Table 7-11. Employee Profile Data

Emp ID	Name	Age	Qualification	Experience	Gender	Team
EMP001	-	45	Masters	20	M	Analytics
EMP002	Will	34	Masters	12	M	-
EMP003	White	35	Phd	10	-	IT
EMP004	Phoebe	-	Masters	9	F	IT
EMP005	Jack	32	Bachelors	-	M	Analytics
EMP006	Mike	45	-	10	M	Accounts

It's apparent from the columns that every record has at least one field missing, and hence dropping records with missing values is not a solution. You need a robust and consistent solution.

Understanding and Preparing the Data

Chapter 6 used the box-plot technique to identify the outliers in a variable. A few simple descriptive statistics techniques and checkpoints can be used in data analysis as well in order to completely understand the data. The data needs to be understood and cleaned first, as discussed in the earlier call volume example. Only then will it be ready for analysis. The following sections examine data exploration, validation, and cleaning in detail.

Data Exploration

The first data inspection step is to get a complete understanding of the data and the minute details of its structure. The following questions need to be answered:

- How is the data structured?
- What are the variables?
- What are their types?
- How many variables are there?
- How many records are there?
- Are there any missing values?

Data Validation

Data validation will answer the following questions:

- Are all the values correct?
- Are there any outliers?
- Are the variable types correct?
- Does the data match the data dictionary, which describes all the variables involved in the data set?

All these questions deal with the accuracy checks on the data. This step can be executed during data exploration or can be carried out as a separate step. Simple descriptive statistics techniques are brought into play at this stage.

Data Cleaning

You can identify almost all the issues with the data set using data exploration and data validation. Next, the data needs to be cleaned and made ready for the analysis. The outliers and missing values along with any other errors need to be fixed by appropriate substitutions. But is the process of data cleaning the same if the variable is discrete or continuous? What if just 1 percent of the values are missing in the data? What if 30 percent of the values are missing? Is the treatment the same for the 10 percent, 30 percent, and 90 percent missing values cases?

The steps of exploring, validating, and cleaning data take a considerable amount of time in the project life cycle, sometimes as high as 70 percent of the total person hours available for the project. There are no shortcuts either. You need to explore all the variables to understand and resolve the issues. The subsequent topics in this chapter will discuss ways of understanding, validating, and cleaning the data using a credit risk case study.

Data Exploration, Validation, and Sanitization Case Study: Credit Risk Data

What follows is a credit risk case study, which we use throughout this chapter to demonstrate various steps and concepts pertaining to data preparation prior to the start of the actual analysis process.

THE CREDIT SCORING SYSTEM

Reference: The data set used in this case study is based on and used with permission from Kaggle's web site at https://www.kaggle.com/c/GiveMeSomeCredit.

It is common knowledge that banks offer several products, such as personal loans, credit cards, mortgages, and car loans. Every bank seeks to evaluate the risks associated with a customer before issuing a loan or a credit card. Each customer is assessed based on a few crucial parameters, such as the number of previous loans, average income, age, number of dependents, and so on. The banks use advanced analytics methodologies and build predictive models to find the probability of a default before issuing the card or a loan.

Historical customer payments and usage data is used for building a predictive model, which will quantify the risk associated with each customer. The bank decides a cutoff point, and any customer with a higher risk than the cutoff is rejected. This methodology of quantifying the risk is called *credit scoring*. Each customer gets a credit score as a result of this model, which is built on historical data. The higher the credit score, the better a customer's probability of availing a credit is. In general, a credit score is between 0 and 1000, but this is not mandatory. For example, if a customer gets a credit score of 850, her application for a loan or credit card may be approved. On the other hand, if a customer scores 350, her probability of getting credit becomes almost zero.

One needs to be careful about data errors such as missing values and outliers while building credit risk models. An error might negate the credibility of the entire model if not handled well, and one major default on a loan may negate the profits earned from 100 good cases. The model has to be robust under all circumstances because no bank wants to give a high score to a bad customer. Hence, data exploration, validation, and cleaning become important in models involving financial transactions.

DATA DICTIONARY

Reference: The base data set is taken from Kaggle and modified.

Given in Tables 7-12 and 7-13 is the data set for building the predictive model. A data dictionary that usually accompanies the data explains all the variable details. Each variable needs to be examined in order to understand, validate, and clean if necessary. The following are the details given by the bank's data team.

The historical data for 250,000 borrowers, collected over a two-year performance window, is provided. Part of this data is used for building the models, and some data is set aside for testing and validation purposes.

The data set file name is `Customer_loan_data.csv`, and the data dictionary (the variable details) is given in Table 7-12.

Table 7-12. *Data Dictionary for the Example*

Variable Name	Description	Type
SeriousDlqin2yrs	Person experienced 90 days past due delinquency or worse	Y/N
RevolvingUtilizationOfUnsecuredLines	Total balance on credit cards and personal lines of credit except real estate and no installment debt such as car loans divided by the sum of credit limits	Percentage
Age	Age of borrower in years	Integer
NumberOfTime30-59DaysPastDueNotWorse	Number of times borrower has been 30 to 59 days past due but no worse in the last 2 years	Integer
DebtRatio	Monthly debt payments, alimony, living costs divided by monthly gross income	Percentage
MonthlyIncome	Monthly income	Real
NumberOfOpenCreditLinesAndLoans	Number of open loans (installment loans such as a car loan or mortgage) and lines of credit (such as credit cards)	Integer
NumberOfTimes90DaysLate	Number of times borrower has been 90 days or more past due	Integer
NumberRealEstateLoansOrLines	Number of mortgage and real estate loans including home equity lines of credit	Integer
NumberOfTime60-89DaysPastDueNotWorse	Number of times borrower has been 60 to 89 days past due but no worse in the last 2 years	Integer
NumberOfDependents	Number of dependents in family excluding themselves (spouse, children, and so on)	Integer

All the variables in Table 7-12 are related to credit risk. Table 7-13 provides a detailed explanation of the variables.

Table 7-13. *Detailed Explanation of the Variables*

Variable Name	Description	Type
SeriousDlqin2yrs	Person experienced 90 days past due delinquency or worse. These accounts are also known as *bad* accounts. The bad definition changes from product to product. For example, a serious delinquency for credit cards is 180 days delinquent. This might be a loan or a mortgage, and hence 90 days past due is serious delinquency. In general, this is the target variable that needs to be predicted.	Y/N
RevolvingUtilizationOf UnsecuredLines	Total balance on credit cards and personal lines of credit except real estate and number of installment debt such as car loans divided by the sum of credit limits. Consider a credit card with $100,000 as the credit limit. If $25,000 is used on average every month, the utilization percentage is 25 percent. If $10,000 is used on an average, the utilization is 10 percent. So, utilization takes values between 0 and 1 (0 to 100 percent).	Percentage
Age	Age of borrower in years.	Integer
NumberOfTime30-59Days PastDueNotWorse	Number of times borrower has been 30 to 59 days past due but no worse in the last 2 years. This data spans 2 years. How many times was a customer 30 days late but not later than 59 days? Once, twice, three, or six times?	Integer
DebtRatio	Monthly debt payments, alimony, living costs divided by monthly gross income. Debt to income ratio. With an income of $50,000, debt is $10,000, and debt ratio is 20 percent. Hence, the debt ratio can take any value between 0 and 100 percent. It can also be slightly more than 100 percent.	Percentage
MonthlyIncome	Monthly income.	Real
NumberOfOpenCredit LinesAndLoans	Number of open loans (an installment loan such as car loan or mortgage) and lines of credit (such as credit cards).	Integer
NumberOfTimes90DaysLate	Number of times borrower has been 90 days or more past due. This data is for 2 years. How many times a customer was is 90 days late? Once, twice, three times, or five times?	Integer
NumberRealEstate LoansOrLines	Number of mortgage and real estate loans including home equity lines of credit.	Integer
NumberOfTime60-89Days PastDueNotWorse	Number of times borrower has been 60 to 89 days past due but no worse in the last 2 years. This data is for 2 years. How many times was a customer 60 to 89 days late but not worse? Once, twice, three, or five times?	Integer
NumberOfDependents	Number of dependents in family excluding applicant (spouse, children, and so on).	Integer

The following sections will use simple descriptive statistics techniques to explore, validate, and sanitize the credit risk data.

Importing the Data

The following SAS code imports the .csv file into SAS:

```
/*Import the customer raw data into SAS */
PROCIMPORT OUT= WORK.cust_cred_raw
DATAFILE= "C:\Users\Google Drive\Training\Books\Content\Chapter-12 Data Exploration and
Cleaning\Datasets\Customer_loan_data.csv"
DBMS=CSV REPLACE;
    GETNAMES=YES;
    DATAROW=2;
RUN;
```

Here are the main notes from the SAS log file of the preceding code, when executed:

```
NOTE: The infile 'C:\Users\Google Drive\Training\Books\Content\Chapter-12 Data
      Exploration and Cleaning\Datasets\Customer_loan_data.csv' is:

      Filename=C:\Users\Google Drive\Training\Books\Content\Chapter-12 Data Exploration
and Cleaning\Datasets\Customer_loan_data.csv,
      RECFM=V,LRECL=32767,File Size (bytes)=14516824,
      Last Modified=05Mar2014:14:55:27,
      Create Time=05Mar2014:12:54:18

NOTE: 251503 records were read from the infile 'C:\Users\Google
      Drive\Training\Books\Content\Chapter-12 Data Exploration and
      Cleaning\Datasets\Customer_loan_data.csv'.
      The minimum record length was 33.
      The maximum record length was 64.
NOTE: The data set WORK.CUST_CRED_RAW has 251503 observations and 13 variables.
NOTE: DATA statement used (Total process time):
real time          1.82 seconds
cpu time           1.52 seconds


251503 rows created in WORK.CUST_CRED_RAW from C:\Users\Google
Drive\Training\Books\Content\Chapter-12 Data Exploration and
Cleaning\Datasets\Customer_loan_data.csv.
NOTE: WORK.CUST_CRED_RAW data set was successfully created.
NOTE: PROCEDURE IMPORT used (Total process time):
real time          2.07 seconds
cpu time           1.76 seconds
```

According to the preceding analysis, there are no major warnings or errors. The name of the SAS data set is cust_cred_raw. The following is the process of data exploration.

Step 1: Data Exploration and Validation Using the PROC CONTENTS

PROC CONTENTS data is used to get basic details about the data, such as the number of records variables and variable types. Because the intention is to understand the data completely without actually opening the data file, PROC CONTENTS can be the starting point to get metadata (data about data) and related details.

The SAS code for contents looks like this:

```
title'Proc Contents on raw data';

proc contents data= WORK.cust_cred_raw varnum ;
run;
```

The preceding SAS code generates the output in Table 7-14.

Table 7-14. Output of proc contents on cust_cred_raw Data Set

The CONTENTS Procedure

Data Set Name	WORK.CUST_CRED_RAW	Observations	251503
Member Type	DATA	Variables	13
Engine	V9	Indexes	0
Created	Tuesday, July 01, 2008 12:06:40 AM	Observation Length	104
Last Modified	Tuesday, July 01, 2008 12:06:40 AM	Deleted Observations	0
Protection		Compressed	NO
Data Set Type		Sorted	NO
Label			
Data Representation	WINDOWS_32		
Encoding	wlatin1 Western (Windows)		

Engine/Host Dependent Information

Data Set Page Size	12288
Number of Data Set Pages	2150
First Data Page	1
Max Obs per Page	117
Obs in First Data Page	93
Number of Data Set Repairs	**0**
Filename	**C:\Users\AppData\Local\Temp\SAS Temporary Files\ _TD6600\cust_cred_raw.sas7bdat**
Release Created	**9.0201M0**
Host Created	**W32_VSPRO**

(*continued*)

Table 7-14. (*continued*)

Variables in Creation Order

#	Variable	Type	Len	Format	Informat
1	Sr_num	Num	8	BEST12.	BEST32.
2	SeriousDlqin2yrs	Num	8	BEST12.	BEST32.
3	RevolvingUtilizationOfUnsecuredL	Num	8	BEST12.	BEST32.
4	Age	Num	8	BEST12.	BEST32.
5	NumberOfTime30_59DaysPastDueNotW	Num	8	BEST12.	BEST32.
6	DebtRatio	Num	8	BEST12.	BEST32.
7	MonthlyIncome	Char	6	$6.	$6.
8	NumberOfOpenCreditLinesAndLoans	Num	8	BEST12.	BEST32.
9	NumberOfTimes90DaysLate	Num	8	BEST12.	BEST32.
10	NumberRealEstateLoansOrLines	Num	8	BEST12.	BEST32.
11	NumberOfTime60_89DaysPastDueNotW	Num	8	BEST12.	BEST32.
12	NumberOfDependents	Char	3	$3.	$3.
13	obs_type	Char	8	$8.	$8.

Validations and Checkpoints in the Overall Contents

Although PROC CONTENTS is a simple procedure, it provides all the basic details and answers to some serious questions. The following is the checklist to keep in mind while reading the PROC CONTENTS output:

- Are all variables as expected? Are there any variables missing?

- Are the numbers of records as expected?

- Are there unexpected variables, say, x10, AAA, q90, r10, VAR1?

- Do the names of the variables match the data dictionary? Are there any variables whose names are trimmed?

- Are the data types as expected? Are there any date variables that are read as numbers? Are there any numeric variables that are read as characters?

- Is the length across variables correct? Are there any characters variables that are trimmed after certain characters?

- Have the labels been provided, and, if yes, are they sensible?

Examine the checklist (Table 7-15) for the customer loan data.

Table 7-15. *Checkpoints in the Customer Loan Data*

Number	Checkpoint	Observation	Issues?
1	Are all variables as expected?	All the 11 variables in the data dictionary are in the final data set.	No
2	Are the numbers of records as expected?	Yes. Around 250,000 records were expected, and the output shows 251,503 records. The extra 1,503 records can be inquired about, if required. But the number looks insignificant.	No
3	Are there unexpected variables?	Yes, there are two unexpected variables. There are only 11 variables mentioned in data dictionary. Sr_num and Obs_type are extra. It is apparent that Sr_num is a serial number, but what is obs_type?	Yes
4	Are the names of the variables the same?	No. Some variable names are trimmed in the dataset because of restrictions on the variable names in SAS. For example, RevolvingUtilizationOfUnsecuredLines is read as RevolvingUtilizationOfUnsecuredL;. Similarly, the following two also get trimmed. NumberOfTime30_59DaysPastDueNotW NumberOfTime60_89DaysPastDueNotW But these can be mapped to original variables. This looks fine.	No
5	Are the data types as expected?	No. The monthly incomes are expected to be numeric, but it is a character in the imported data. Similarly, the number of dependents should be a number. SeriousDlqin2yrs should be Y/N but is read as a number. There are serious issues with the monthly income and number of dependents.	Yes
6	Is the length across variables correct?	There seems to be no issue with the length.	No
7	Have labels been provided and are sensible?	There are no labels given to the variable in the current data set.	No

After going through the checklist of PROC CONTENTS, the following issues were found:

- There are two additional variables, Sr_num and obs_type.

- The monthly income and number of dependents should be numeric, but they are written as characters. This means that no arithmetic or numerical operations are possible on these two variables. This issue can be detrimental to the analysis and might yield disastrous results in the later phases of the project.

- SeriousDlqin2yrs should be Y/N but is read as a number. Y/N might have been converted to 0/1.

The rest of the data exploration steps will be explored. The discrepancies will be examined closely and resolved. If the answers to these issues are not found, check with the data team on whether they are because of manual or system errors.

Step 2: Data Exploration and Validation Using Data Snapshot

The PROC CONTENTS data gave an overview of observations and variables. The next step is to examine the values in each variable. As discussed earlier, printing the whole data set is not practical; only a snapshot of the data is to be printed. A snapshot is a small portion of the data. It may be random 100 observations, the first 30 observations, or the last 50. There is no rule on the snapshot size. Basically, it is nothing but the printing of data with some restrictions.

■ **Warning**   Be careful while printing the data; never try to print a large data set. The computer system and software may hang, and a reboot may be the only option to bring it back.

The SAS code to print the first 20 records follows:

```
title'Proc Print on raw data (first 20 observations only)';
proc print data=WORK.cust_cred_raw (obs=20) ;
run;
```

Table 7-16 lists the output for this code.

Table 7-16. The First 20 Observations in the cust_cred_raw Data Set

Obs	sr_num	Serious Dlqin2yrs	Revolving UtilizationOf UnsecuredL	age	Number OfTime 30_59Days PastDue NotW	DebtRatio	Monthly Income	Number OfOpen Credit Lines And Loans	Number OfTimes 90Days Late	Number RealEstate LoansOr Lines	Number OfTime 60_89Days PastDue NotW	NumberOf Dependents	obs_ type
1	1	1	0.766	45	2	0.80	9120	13	0	6	0	2	training
2	2	0	0.957	40	0	0.12	2600	4	0	0	0	1	training
3	3	0	0.658	38	1	0.09	3042	2	1	0	0	0	training
4	4	0	0.234	30	0	0.04	3300	5	0	0	0	0	training
5	5	0	0.907	49	1	0.02	63588	7	0	1	0	0	training
6	6	0	0.213	74	0	0.38	3500	3	0	1	0	1	training
7	7	0	0.306	57	0	5710.00	NA	8	0	3	0	0	training
8	8	0	0.754	39	0	0.21	3500	8	0	0	0	0	training
9	9	0	0.117	27	0	46.00	NA	2	0	0	0	NA	training
10	10	0	0.189	57	0	0.61	23684	9	0	4	0	2	training
11	11	0	0.644	30	0	0.31	2500	5	0	0	0	0	training
12	12	0	0.019	51	0	0.53	6501	7	0	2	0	0	training
13	13	0	0.010	46	0	0.30	12454	13	0	2	0	2	training
14	14	1	0.965	40	3	0.38	13700	9	3	1	0	2	training
15	15	0	0.020	76	0	477.00	0	6	0	1	0	0	training
16	16	0	0.548	64	0	0.21	11362	7	0	1	0	2	training
17	17	0	0.061	78	0	2058.00	NA	10	0	2	0	0	training
18	18	0	0.166	53	0	0.19	8800	7	0	0	0	0	training
19	19	0	0.222	43	0	0.53	3280	7	0	1	0	2	training
20	20	0	0.603	25	0	0.07	333	2	0	0	0	0	training

Validation and Checkpoints in the Data Snapshot

A snapshot is a cursory look at part of the actual data. This helps you understand the variables by looking at values they take. The following are the checkpoints that need to be observed in the PROC PRINT step:

- Look for the unique identifiers or primary key. Does some variable take unique values?

- Do all the text variables have relevant data? Or do they have some unexpected text in between, such as @!$# or &&&&.

- Do all the known numeric variables have appropriate values? Do they have non-numeric values also?

- Are there any coded values instead of the actual values? Examples are state codes like NY or ID in the state field.

- Do all the variables have data? Or is there any variable with missing values?

- Are the issues identified in the PROC CONTENTS output clear? This includes any information on variable lengths, variable types, number of variables, and so on.

Go through the checklist (Table 7-17) for the customer loan data.

Table 7-17. *Checkpoints in the Customer Loan Data*

Number	Checkpoint	Observation	Issues?
1	Look for the unique identifiers or primary key. Does some variable take unique values?	The Sr_num looks like a serial number and unique identifier.	No
2	Do all the text variables have relevant data? Or do they have some unexpected text in between, like @!$# or &&&&?	The variable monthly income and number of dependents has NA in some records. Obs_type takes only training as the column values. It might have something to do with training and testing samples in data description.	Yes
3	Do all the known numeric variables have appropriate values? Do they have any non-numeric values?	Yes, there are no major issues here. We already took a note of monthly income and number of dependents.	No
4	Are there any coded values instead of the actual values? For example, do they have state codes like NY or ID in the state field?	SeriousDlqin2yrs should be a Yes or No type according to the data dictionary. It looks like it has been coded to 0 and 1. But Y=1 or 0?	Yes
5	Do all the variables have data? Or are there any variable with missing values?	There are no clear missing values in the snapshot.	No
6	Can you get clarity on issues identified in the PROC CONTENTS output? Do you need any information on variable lengths, variable types, or number of variables?	Yes, there is additional information on monthly income, number of dependents, and Sr_num and obs_type.	No

The following are some open issue items from the contents:

1. There are two additional variables: Sr_num and obs_type.

2. The monthly income and number of dependents should be numeric but are written as characters.

3. SeriousDlqin2yrs should be Y/N but is read as a number. Y/N might have been converted to 0/1.

Resolving sr_num

From the snapshot, it was observed that Sr_num is a serial number, and the following code is a test to confirm this. A detailed analysis on Sr_num, using procunivariate, will reveal the facts.

```
Title 'Univariate on Sr_num';
proc univariate data=WORK.cust_cred_raw;
var sr_num;
run;
```

Table 7-18 provides a closer look at the quartiles and extreme values in the output.

Table 7-18. The Quartile and Extreme Observations in the cust_cred_raw Data Set

Quantiles (Definition 5)	
Quantile	Estimate
100% Max	251503
99%	248988
95%	238928
90%	226353
75% Q3	188628
50% Median	125752
25% Q1	62876
10%	25151
5%	12576
1%	2516
0% Min	1

(*continued*)

Table 7-18. (*continued*)

Extreme Observations			
Lowest		Highest	
Value	Obs	Value	Obs
1	1	251499	251499
2	2	251500	251500
3	3	251501	251501
4	4	251502	251502
5	5	251503	251503

The five highest values of sr_num are 251499, 251500, 251501, 251502, and 251503, and sr_num starts with 1, 2, 3, 4, and 5. Hence, it can be safely concluded that sr_num is the record number.

Resolving the Monthly Income and Number of Dependents Issue

Here are the burning open items:

1. For obs_type, proc print didn't give much information.

2. The monthly income and number of dependents should be numeric, but they are written as characters.

3. SeriousDlqin2yrs should be Y/N but is read as a number. Y/N might have been converted to 0/1.

Both monthly income and number of dependents are NA. As a result, the whole variable is stored as a character variable. Convert them to numerical missing values. Use if-then-else in SAS or the following code:

```
/* Monthly income & number of dependents Character issue*/
Data cust_cred_raw_v1;
Set cust_cred_raw;
MonthlyIncome_new= MonthlyIncome*1;
NumberOfDependents_new=NumberOfDependents*1;
run;
```

This creates two new variables, MonthlyIncome_new and NumberOfDependents_new, and creates a new data set called cust_cred_raw_v1 from old data set cust_cred_raw.

Some warning notes about NA multiplied by 1 are noticeable while executing the code. The log file will have messages similar to the following lines:

```
NOTE: Character values have been converted to numeric values at the places given by:
      (Line):(Column).
      71:21   72:24
NOTE: Invalid numeric data, MonthlyIncome='NA' , at line 71 column 21.
NOTE: Invalid numeric data, NumberOfDependents='NA' , at line 72 column 24.
```

Two new variables were created to resolve the character-related issues. The following is the PROC CONTENTS output of the new data set:

```
title'Proc Contents on data version1';
proc contents data=cust_cred_raw_v1 varnum ;
run;
```

Table 7-19 lists the output of this code.

Table 7-19. *Output of proc contents on the cust_cred_raw_v1 Data Set*

Variables in Creation Order					
#	Variable	Type	Len	Format	Informat
1	Sr_num	Num	8	BEST12.	BEST32.
2	SeriousDlqin2yrs	Num	8	BEST12.	BEST32.
3	RevolvingUtilizationOfUnsecuredL	Num	8	BEST12.	BEST32.
4	Age	Num	8	BEST12.	BEST32.
5	NumberOfTime30_59DaysPastDueNotW	Num	8	BEST12.	BEST32.
6	DebtRatio	Num	8	BEST12.	BEST32.
7	MonthlyIncome	Char	6	$6.	$6.
8	NumberOfOpenCreditLinesAndLoans	Num	8	BEST12.	BEST32.
9	NumberOfTimes90DaysLate	Num	8	BEST12.	BEST32.
10	NumberRealEstateLoansOrLines	Num	8	BEST12.	BEST32.
11	NumberOfTime60_89DaysPastDueNotW	Num	8	BEST12.	BEST32.
12	NumberOfDependents	Char	3	$3.	$3.
13	obs_type	Char	8	$8.	$8.
14	MonthlyIncome_new	Num	8		
15	NumberOfDependents_new	Num	8		

Two new numeric variables have been added, and they can now be used instead of character variables. The following are the open items:

1. For obs_type, PROC PRINT didn't give sufficient information.

2. SeriousDlqin2yrs should be Y/N but is read as a number. Y/N might have been converted to 0/1. What is 0 and what is 1 need to be decided.

PROC PRINT and the snapshot gave access to the real values, using certain issues that could be identified and resolved. The data issues may not be obvious in all the variables. For example, it is apparent that obs_type has something to do with test and training data, but PROC PRINT did not put forth a clear picture. Sometimes, increasing the print size will give a closer look at the values. Printing four or five parts of the data is usually sufficient to get a better picture.

The Continuous and Discrete Variables

The earlier chapters discussed continuous and discrete variables. A variable that can take any value between two limits is *continuous* (for example, height in feet). A variable that can take only limited values is called *discrete* (for example, number of children in a family). Time from an analog watch is a continuous variable, and time from a digital watch is a discrete variable. Both of them show time, but in an analog watch precision is not fixed, and a digital watch shows the time only to its lowest unit of measure, which can even be 100th of a second.

Table 7-20 shows the categorization of the variables into continuous and discrete in the given data.

Table 7-20. *Discrete and Continuous Variables in the Data*

#	Variable	Type	Variable Type	Further Inspection
1	Sr_num	Num	Discrete	No
2	SeriousDlqin2yrs	Num	Discrete	Yes
3	RevolvingUtilizationOfUnsecuredL	Num	Continuous	Yes
4	Age	Num	Discrete	Yes
5	NumberOfTime30_59DaysPastDueNotW	Num	Discrete	Yes
6	DebtRatio	Num	Continuous	Yes
7	MonthlyIncome	Char	NA	NA
8	NumberOfOpenCreditLinesAndLoans	Num	Discrete	Yes
9	NumberOfTimes90DaysLate	Num	Discrete	Yes
10	NumberRealEstateLoansOrLines	Num	Discrete	Yes
11	NumberOfTime60_89DaysPastDueNotW	Num	Discrete	Yes
12	NumberOfDependents	**Char**	**NA**	**NA**
13	obs_type	**Char**	**Discrete(Categorical)**	**Yes**
14	MonthlyIncome_new	**Num**	**Continuous**	**Yes**
15	NumberOfDependents_new	**Num**	**Discrete**	**Yes**

After running the basic contents and printing procedures, comes descriptive statistics. As discussed earlier, there are no shortcuts. Every variable needs to be examined individually. Everything needs to be explored and validated before analysis. The first step in basic descriptive analysis is to identify the continuous and discrete variables. Since the continuous variables can take infinite values, components like a frequency table will be really long. On the other hand, discrete variables take a countable number of values, so a frequency table makes some sense.

The next step is to perform a univariate analysis on the continuous variables and frequency tables for discrete variables. A frequency table for continuous variables is not a good idea because it can take an infinite number of values, but it can be useful for discrete variables.

Step 3: Data Exploration and Validation Using Univariate Analysis

A univariate analysis is almost like a complete report on a variable and covers details from every angle. The univariate analysis will be used here to explore and validate all continuous variables. This analysis gives the output on the following statistical parameters:

- N (a count of non-missing observations)

- Nmiss (a count of missing observations)

- Min, max, median, mean

- Quartile numbers and percentiles (P1, p5, p10, q1(p25), q3(p75), p90, p99)

- Stdde (standard deviation)

- Var (variance)

- Skewness

- Kurtosis

Skewness and kurtosis deal with variable distribution. These two measures will not be used in this analysis. Here is a quick explanation of them for the sake of completeness.

Skewness explains on which side the most variable distribution is skewed. A negative skewness means the left side tail is longer when a distribution plot of the variable is drawn. Consider a variable, say the age of people with natural deaths; for most of such cases age will be really high. Similarly, a right-skewed distribution may feature a tail on the right side. The number of accidents in one's life may be an example of right-skewed distribution. Look at Figures 7-2 and 7-3.

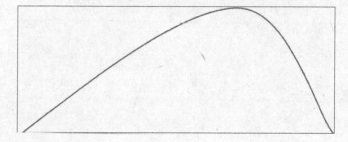

Figure 7-2. Left-skewed variable

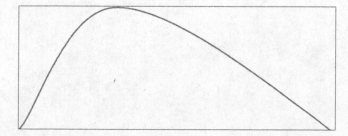

Figure 7-3. Right-skewed variable

Kurtosis details how well the variable is distributed in terms of the sharpness of the peak. The following figures show three different types of peaks. Figure 7-4 shows a sharp peak at one value. Figure 7-5 has almost no peak, and Figure 7-6 has a medium peak. The kurtosis measure value for the first graph will be in excess of 3; the kurtosis value will be much less than 3 for the second graph, and for the third variable, the kurtosis will be close to 3.

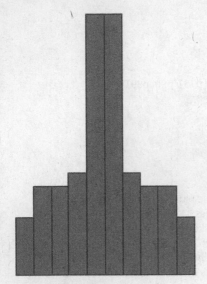

Figure 7-4. *High kurtosis*

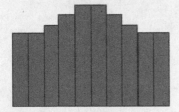

Figure 7-5. *Low kurtosis*

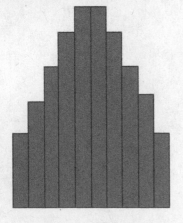

Figure 7-6. *Medium kurtosis*

The SAS code that follows is for conducting univariate analysis on variableRevolvingUtilizationOfUnsecuredL. More variables might be added if required, but you will start with one variable.

■ **Note** To recap, RevolvingUtilizationOfUnsecuredL is a total balance on credit cards and personal lines of credit except real estate and the number of installment debt such as car loans divided by the sum of credit limits.

Consider a credit card with $100,000 as a credit limit. If $25,000 is used on average every month, the utilization percentage is 25 percent. If $10,000 is used on average, the utilization is 10 percent. Hence, utilization takes values between 0 and 1 (that is, 0 to 100 percent).

```
Title 'Proc Univriate on utilization -RevolvingUtilizationOfUnsecuredL ';
Proc univariate data= cust_cred_raw_v1;
Var RevolvingUtilizationOfUnsecuredL;
run;
```

This code gives the output shown in Table 7-21.

Table 7-21. *Output of proc univariate on cust_cred_raw_v1 (Var RevolvingUtilizationOfUnsecuredL)*

The UNIVARIATE Procedure Variable: RevolvingUtilizationOfUnsecuredL
Moments

N	251503	Sum Weights	251503
Mean	5.75041518	Sum Observations	1446246.67
Std Deviation	229.63398	Variance	52731.7647
Skewness	89.5880819	Kurtosis	13043.6294
Uncorrected SS	1.32705E10	Corrected SS	1.32621E10
Coeff Variation	3993.3461	Std Error Mean	0.45789359

Basic Statistical Measures

Location		Variability	
Mean	5.750415	Std Deviation	229.63398
Median	0.153575	Variance	52732
Mode	0.000000	Range	50708
		Interquartile Range	0.53132

Tests for Location: Mu0=0

Test	Statistic		p Value	
Student's t	t	12.55841	Pr> |t|	<.0001
Sign	M	116657	Pr>= |M|	<.0001
Signed Rank	S	1.361E10	Pr>= |S|	<.0001

(continued)

223

Table 7-21. (*continued*)

Quantiles (Definition 5)

Quantile	Estimate
100% Max	5.07080E+04
99%	1.09129E+00
95%	1.00000E+00
90%	9.82102E-01
75% Q3	5.61298E-01
50% Median	1.53575E-01
25% Q1	2.99746E-02
10%	2.99276E-03
5%	0.00000E+00
1%	0.00000E+00
0% Min	0.00000E+00

Extreme Observations

Lowest		Highest	
Value	Obs	Value	Obs
0	251479	21821	189198
0	251455	22000	149161
0	251452	22198	16957
0	251438	29110	31415
0	251434	50708	85490

Validation and Checkpoint in Univariate Analysis

The univariate analysis gives almost all measures of central tendency and . Here is the checklist for univariate analysis:

- What are the central tendencies of the variable? What are the mean, median, and mode across each variable? Is the mean close to the median?

- Is the concentration of variable as expected? What are quartiles? Is the variable skewed toward higher values? Or is it skewed toward lower values? What is the quartiles distribution? Are there any indications of the presence of outliers?

- What is the percentage of missing values associated with the variable?

- Are there any outliers/extreme values for the variable?

- What is the standard deviation of this variable? Is it near 0, or is it too high? Is the variable within the limits or possible range?

Table 7-22 contains the observations on the output.

Table 7-22. *Observations on the Output of the Variable RevolvingUtilizationOfUnsecuredL*

Number	Checkpoint	Observation		Issues?
1	What are the central tendencies of the variable? What are the mean, median, and mode across each variable? Is the mean close to the median?	Mean: 5.750415 Median: 0.153575 Mode: 0.000000 The mean is 575 percent, and the median is 15 percent. Mean is far higher than the median. It is a clear indication of outliers and is a serious issue.		Yes
2	Is the concentration of variables as expected? What are quartiles? Are the variable skewed toward higher values or lower values?	Quantiles (Definition 5)		Yes
	What is the quantiles distribution? Are there any indications of the presence of outliers?	Quantile	Estimate	
		100% Max	5070800%	
		99%	109%	
		95%	100%	
		90%	98%	
		75% Q3	56%	
		50% Median	15%	
		25% Q1	3%	
		10%	0%	
		5%	0%	
		1%	0%	
		0% Min	0%	
		The difference between mean and median already indicated the values. Ten percent of the population almost never used their cards. Fifty percent of the population, that is, around 125,000 customers, uses only 15 percent of their credit limit.		
		Interestingly, a small percentage of the population uses more than 100 percent limit. Does that mean that around 5 percent of the population uses more than 100 percent of their credit limit? A few customers use 5000 times their credit limit, which looks impossible with regard to business logic.		

(*continued*)

Table 7-22. (*continued*)

Number	Checkpoint	Observation	Issues?
3	What is the percentage of missing value associated with the variable?	There are no missing values. Utilization is available for all customers.	No
4	Are there any outliers/extreme values for the variable?	Yes, identified. Around 5 percent values are outliers. It looks like utilization dollars are recorded directly instead of utilization percentages.	Yes
5	What is the standard deviation of this variable? Is it close to 0, or is it too high? Is the variable within the limits or possible range?	Standard deviation is not zero; it's slightly on the higher side, since there are abnormally huge outliers present.	No

The output clearly indicates the presence of outliers, and these 5 percent outliers are inducing drastic errors in all the measures. One such example is the mean utilization of overall customers of 500 percent, whereas the actual average is 15 percent. The outliers can also be verified using a box plot. The "plot" option needs to be mentioned in the PROC UNIVARIATE code to create the box-plot graph:

```
title'ProcUnivriate and boxplot on utilization';
Procc univariate data= cust_cred_raw_v1 plot;
Var RevolvingUtilizationOfUnsecuredL;
run;
```

Figure 7-7 shows the resulting box plot.

```
                    Histogram              #         Boxplot
     51000+*                               1            *



     .



     .




     41000+



     .





     .
```

Figure 7-7. Box plot on the variable RevolvingUtilizationOfUnsecuredL

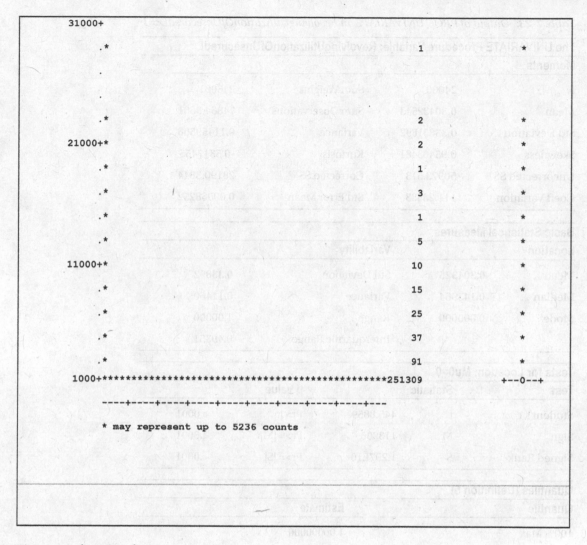

Figure 7-7. (*continued*)

The box cannot be seen in the preceding box-plot graph; recall the box-plot interpretation: a box has Q1, median, and Q3. It is evident from the graph that the variable is completely dominated by outliers. To verify further, consider values only below 100 percent. The SAS code and output will look like the following:

```
title'Proc Univariate on utilization less than or equal to 100%';

Proc univariate data= cust_cred_raw_v1 plot;
Var RevolvingUtilizationOfUnsecuredL;
Where RevolvingUtilizationOfUnsecuredL<=1;
run;
```

Table 7-23 and Figure 7-8 show PROC UNIVARIATE and the box plot on utilization less than or equal to 100 percent.

Table 7-23. *Output of PROC UNIVARIATE on RevolvingUtilizationOfUnsecuredL<=1*

The UNIVARIATE Procedure Variable: RevolvingUtilizationOfUnsecuredL

Moments

N	246001	Sum Weights	246001
Mean	0.30432543	Sum Observations	74864.3601
Std Deviation	0.33851892	Variance	0.11459506
Skewness	0.95463481	Kurtosis	-0.5317455
Uncorrected SS	50973.513	Corrected SS	28190.3844
Coeff Variation	111.23583	Std Error Mean	0.00068252

Basic Statistical Measures

Location		Variability	
Mean	0.304325	Std Deviation	0.33852
Median	0.143964	Variance	0.11460
Mode	0.000000	Range	1.00000
		Interquartile Range	0.49351

Tests for Location: Mu0=0

Test	Statistic		p Value	
Student's t	t	445.8859	Pr> |t|	<.0001
Sign	M	113906	Pr>= |M|	<.0001
Signed Rank	S	1.297E10	Pr>= |S|	<.0001

Quantiles (Definition 5)

Quantile	Estimate
100% Max	1.00000000
99%	0.99999990
95%	0.99999990
90%	0.93870941
75% Q3	0.52224629
50% Median	0.14396401
25% Q1	0.02873321
10%	0.00270126
5%	0.00000000
1%	0.00000000
0% Min	0.00000000

(*continued*)

Table 7-23. (*continued*)

Extreme Observations

Lowest		Highest	
Value	Obs	Value	Obs
0	251479	1	224166
0	251455	1	228890
0	251452	1	231077
0	251438	1	238536
0	251434	1	244591

```
                       Histogram                          #         Boxplot
      1.025+*                                            29          |

           .***************                           23431          |

      0.925+***                                        4807          |

           .***                                        4364          |

      0.825+***                                        4155          |

           .***                                        4052          |

      0.725+***                                        4137          |

           .***                                        4243          |

      0.625+***                                        4523          |

           .***                                        4772          |

      0.525+****                                       5398         +-----+

           .****                                       5611          |   |

      0.425+****                                       6283          |   |

           .*****                                      6931          |   |

      0.325+******                                     7473          | + |

           .******                                     8744          |   |

      0.225+******                                    10008          |   |

           .********                                  12257          |   |

      0.125+**********                                16663         *-----*

           .****************                          27497          |   |

      0.025+************************************************* 80623   +-----+

           ----+----+----+----+----+----+----+----+----+----+---

           * may represent up to 1680 counts
```

Figure 7-8. *The box plot of PROC UNIVARIATE on RevolvingUtilizationOfUnsecuredL<=1*

You can see the box now. Although there are some high-side values, there are no relentless outliers in this case. But simply removing the outliers is not a solution. This needs to be noted and resolved later. There are two open items already, and the revolving utilization variable here contains outliers, which need to be treated. That is the third issue on the list.

1. For obs_type, PROC PRINT didn't give sufficient information.

2. SeriousDlqin2yrs should be Y/N, but it is read as a number. It might be possible that Y/N is converted to 0/1. What is 0 and what is 1 need to be decided.

3. Revolving utilization has outliers, around 5 percent.

Similarly, univariate analysis is performed on monthly income using this code:

```
title' Univariate on monthly income ';
Proc univariate data= cust_cred_raw_v1 ;
Var MonthlyIncome_new;
run;
```

Table 7-24 shows the output of this code.

Table 7-24. Output of proc univariate on cust_cred_raw_v1 (Var MonthlyIncome_new)

The UNIVARIATE Procedure Variable: MonthlyIncome_new

Moments

N	201669	Sum Weights	201669
Mean	6626.1771	Sum Observations	1336294510
Std Deviation	8746.29438	Variance	76497665.3
Skewness	36.3109836	Kurtosis	2449.29916
Uncorrected SS	2.42817E13	Corrected SS	1.54271E13
Coeff Variation	131.996085	Std Error Mean	19.4762131

Basic Statistical Measures

Location		Variability	
Mean	6626.177	Std Deviation	8746
Median	5400.000	Variance	76497665
Mode	5000.000	Range	835040
		Interquartile Range	4812

Tests for Location: Mu0=0

Test	Statistic		p Value	
Student's t	t	340.219	Pr> |t|	<.0001
Sign	M	99507.5	Pr>= |M|	<.0001
Signed Rank	S	9.9018E9	Pr>= |S|	<.0001

(continued)

Table 7-24. (*continued*)

Quantiles (Definition 5)

Quantile	Estimate
100% Max	835040
99%	25083
95%	14583
90%	11600
75% Q3	8212
50% Median	5400
25% Q1	3400
10%	2044
5%	1300
1%	0
0% Min	0

Extreme Observations

Lowest		Highest	
Value	Obs	Value	Obs
0	251388	702500	93565
0	251374	729033	151133
0	251359	730483	123292
0	251305	772700	174163
0	251300	835040	122544

Missing Values

Missing Value	Count	Percent Of	
		All Obs	Missing Obs
.	49834	19.81	100.00

There are outliers in monthly income because the mean is slightly higher than the median, and these outliers are on the higher side of values taken by this variable. But a serious issue other than the mild outliers is that the variable has some missing values, almost 49,834. That is, 20 percent of the overall monthly income records are missing. This is detrimental to the analysis. SAS will ignore this 20 percent of the records whenever monthly income is used. This is the fourth issue on the following list:

1. For obs_type, Proc print didn't give sufficient information.

2. SeriousDlqin2yrs should be Y/N, but it is read as a number. It might be possible that Y/N is converted to 0/1. What is 0 and what is 1 need to be decided.

3. Revolving utilization has outliers, around 5 percent.

4. The monthly income is missing in nearly 20 percent of the cases.

Similarly, univariate analysis can be performed on all the continuous variables; they can be validated, and the issues can be recorded.

Step 4: Data Exploration and Validation Using Frequencies

Step 4 inspects all the discrete variables that take a countable number of values. A frequency table should be created only for discrete and categorical variables. A frequency table on a continuous variable can hang your system, since you expect continuous variables to take an almost infinite number of values. So, if you have too many records in the data, you need to be careful while choosing the variables for frequency distributions.

The following is the code for frequency tables for SeriousDlqin2yrs and obs_type. Once we show the frequency tables, we will explain how they can be used to your advantage.

```
Title 'Frequency table for Serious delinquency in 2 years ';
proc freq data= cust_cred_raw_v1;
table SeriousDlqin2yrs;
run;
```

Table 7-25 gives the output for this code.

Table 7-25. *Output of proc freq on the cust_cred_raw_v1 Data Set (Table SeriousDlqin2yrs)*

The FREQ Procedure				
SeriousDlqin2yrs	Frequency	Percent	Cumulative Frequency	Cumulative Percent
0	139974	93.32	139974	93.32
1	10026	6.68	150000	100.00

```
Frequency Missing = 101503
```

The output is missing in 101,503 values. The count of missing values is given in the output of PROC FREQ. It has already been mentioned by the data team that the serious delinquency values are not available for the testing population. If a proc frequency is run on obs_type, it should yield exactly 101,503 testing records.

```
title' Frquency table for  obs_type ';
proc freq data=    cust_cred_raw_v1;
table obs_type;
run;
```

Table 7-26 lists the output for this code.

Table 7-26. *Output of proc freq on the cust_cred_raw_v1 Data Set (Table obs_type)*

The FREQ Procedure				
obs_type	Frequency	Percent	Cumulative Frequency	Cumulative Percent
test	101503	40.36	101503	40.36
training	150000	59.64	251503	100.00

Hence, 150,000 observations will be used for model building, and the rest will be used for testing. Obs_type simply indicates the testing and training records.

A frequency table of the other discrete variable will also be created. The serious delinquency variable is 0 for 93.3 percent of the records and is 1 for 6.7 percent of the records. The second issue on the list can be resolved by applying some banking knowledge. Serious delinquencies are generally fewer than nondelinquencies. A good customer base of 93 percent and the bad customer base of 7 percent can be expected in a borrower population, but the other way around is almost impossible. So, it can safely be inferred that Y is coded as 1 and N is coded as 0.

That leaves two remaining issues on the list:

1. Revolving utilization has outliers of around 5 percent.

2. The monthly income is missing in 20 percent of the cases.

The following is a continuation of the frequency table for NumberOfTime30_59DaysPastDueNotW. The variable indicates how many times a customer is delinquent for one month but no later than 59 days. A customer can default possibly once, twice, and a maximum of 24 times in a 24-month period. A maximum of 24 times in a 24-month period would mean that the customer has defaulted in a bill payment every month.

The following is the code for the frequency table of this variable:

```
title' Frquency table for  30-59 days past due ';
proc freq data=    cust_cred_raw_v1;
table NumberOfTime30_59DaysPastDueNotW;
run;
```

Table 7-27 lists the output of this code.

Table 7-27. Output of proc freq on cust_cred_raw_v1 (Table NumberOfTime30_59DaysPastDueNotW)

NumberOfTime30_59DaysPastDueNotW	Frequency	Percent	Cumulative Frequency	Cumulative Percent
0	211208	83.98	211208	83.98
1	26870	10.68	238078	94.66
2	7766	3.09	245844	97.75
3	2955	1.17	248799	98.92
4	1257	0.50	250056	99.42
5	547	0.22	250603	99.64
6	228	0.09	250831	99.73
7	96	0.04	250927	99.77
8	53	0.02	250980	99.79
9	22	0.01	251002	99.80
10	11	0.00	251013	99.81
11	2	0.00	251015	99.81
12	3	0.00	251018	99.81
13	1	0.00	251019	99.81
19	1	0.00	251020	99.81
96	6	0.00	251026	99.81
98	477	0.19	251503	100.00

Validation and Checkpoints in Frequencies

The following are the points that can be considered using variable frequency tables for data validation and exploration:

- Are the values as expected?

- Is the variable concentration as expected?

- Are there any missing values? What is the percentage of missing values?

- Are there any extreme values or outliers?

- Is there a possibility of creating a new variable that has a small number of distinct categories by grouping certain categories with others?

Table 7-28 shows the observations of the output of 30 to 59 DPD (days past due).

Table 7-28. Observations of the Output of 30 to 59 DPD

Number	Checkpoint	Observation	Issues?
1	Are the values as expected?	There are some unexpected values, 96 and 98.	Yes
2	Is the variable concentration as expected?	The frequency table shows that 83 percent of the customers have never defaulted more than 30 days, and 10 percent have defaulted more than 30 days once. Some customers have defaulted 10 times. But the issue here is with the values 96 and 98. How can one customer default 98 times in 24 months?	Yes
3	Are there any missing values? What is the percentage of missing values? Are there any extreme values or outliers?	No missing values.	No
4	Is there a possibility of creating a new variable with a small number of distinct categories by clubbing certain categories with others?	Yes, 96 and 98 can be clubbed as an erroneous category. But it does not affect the analysis in a major way.	No

■ **Note**   To recap, 30 to 59 DPD is the short form used for the variable defined earlier: NumberOfTime30-59DaysPastDueNotWorse. It is the number of times the borrower has been 30 to 59 DPD days past due, but no worse, in the last 2 years.

There are two more variables similar to 30 to 59 DPD: 60 to 89 DPD and 90 DPD.

■ **Note** To recap, 60 to 89 DPD is the variable NumberOfTime60-89DaysPastDueNotWorse. It is the number of times the borrower has been 60 to 89 days past due, but no worse, in the last 2 years.

90 DPD is the variable NumberOfTimes90Days. It is the number of times the borrower has been 90 days or more past due. This data is for 2 years. How many times was a customer 90 days late but not worse? One, two, three, or five times?

Consider the following code, which gives frequency tables for 60 to 89 DPD and 90 DPD:

```
title' Frquency table for  60-89, and 90+ days past due ';
proc freq data=   cust_cred_raw_v1;
table NumberOfTime60_89DaysPastDueNotW     NumberOfTimes90DaysLate;
run;
```

Table 7-29 shows the output of this code.

Table 7-29. *Frequency Tables for 60 to 89 and 90+ Days Past Due*

The FREQ Procedure

NumberOfTime60_89DaysPastDueNotW	Frequency	Percent	Cumulative Frequency	Cumulative Percent
0	238771	94.94	238771	94.94
1	9594	3.81	248365	98.75
2	1849	0.74	250214	99.49
3	534	0.21	250748	99.70
4	166	0.07	250914	99.77
5	59	0.02	250973	99.79
6	23	0.01	250996	99.80
7	16	0.01	251012	99.80
8	3	0.00	251015	99.81
9	4	0.00	251019	99.81
11	1	0.00	251020	99.81
96	6	0.00	251026	99.81
98	477	0.19	251503	100.00

(*continued*)

Table 7-29. (*continued*)

NumberOfTimes90DaysLate	Frequency	Percent	Cumulative Frequency	Cumulative Percent
0	237447	94.41	237447	94.41
1	8837	3.51	246284	97.92
2	2617	1.04	248901	98.97
3	1112	0.44	250013	99.41
4	482	0.19	250495	99.60
5	233	0.09	250728	99.69
6	124	0.05	250852	99.74
7	64	0.03	250916	99.77
8	40	0.02	250956	99.78
9	33	0.01	250989	99.80
10	10	0.00	250999	99.80
11	6	0.00	251005	99.80
12	3	0.00	251008	99.80
13	4	0.00	251012	99.80
14	2	0.00	251014	99.81
15	2	0.00	251016	99.81
16	1	0.00	251017	99.81
17	2	0.00	251019	99.81
18	1	0.00	251020	99.81
96	6	0.00	251026	99.81
98	477	0.19	251503	100.00

The issue of 96 and 98 seems to be consistent. Are these default values? The percentage will pose a problem despite it being a low value. This has to be taken note of before resolving.

The list of issues now includes the following:

1. Revolving utilization has outliers of around 5 percent.

2. The monthly income is missing in 20 percent of the cases.

3. 30 to 59 DPD, 60 to 89 DPD, and 90 DPD have issues with 96 and 98.

Similarly, frequency tables have to be created for all the variables in order to fully understand the complete data set. The following code gives the frequency tables for other variables as well:

```
Proc freq data=cust_cred_raw_v1;
tables
age
NumberOfOpenCreditLinesAndLoans
NumberRealEstateLoansOrLines
NumberOfDependents_new;
run;
```

Table 7-30 is the frequency table for NumberOfDependents_new. You will have similar frequency tables for the remaining variables, which are not shown here.

Table 7-30. *Frequency Table for NumberOfDependents_new*

NumberOfDependents_new	Frequency	Percent	Cumulative Frequency	Cumulative Percent
0	145520	59.41	145520	59.41
1	43934	17.94	189454	77.34
2	32820	13.40	222274	90.74
3	16106	6.58	238380	97.32
4	4857	1.98	243237	99.30
5	1272	0.52	244509	99.82
6	283	0.12	244792	99.93
7	101	0.04	244893	99.98
8	40	0.02	244933	99.99
9	8	0.00	244941	100.00
10	8	0.00	244949	100.00
13	1	0.00	244950	100.00
20	2	0.00	244952	100.00
43	1	0.00	244953	100.00

```
Frequency Missing = 6550
```

The variable values in the preceding output are missing 2 percent of the observations. This is an issue. As given in the following code, the missing option can be used to see the missing percentage in proc frequency:

```
title' Frequency  table for All discrete variables with missing %';
proc freq data=cust_cred_raw_v1  ;
tables
age NumberOfOpenCreditLinesAndLoans NumberRealEstateLoansOrLines
NumberOfDependents_new/missing;
run;
```

Table 7-31 is the frequency table for NumberOfDependents_new with this code. You will have similar frequency tables for the remaining variables, which are not shown here.

Table 7-31. *Frequency Table for NumberOfDependents_new with /missing Option*

NumberOfDependents_new	Frequency	Percent	Cumulative Frequency	Cumulative Percent
.	6550	2.60	6550	2.60
0	145520	57.86	152070	60.46
1	43934	17.47	196004	77.93
2	32820	13.05	228824	90.98
3	16106	6.40	244930	97.39
4	4857	1.93	249787	99.32
5	1272	0.51	251059	99.82
6	283	0.11	251342	99.94
7	101	0.04	251443	99.98
8	40	0.02	251483	99.99
9	8	0.00	251491	100.00
10	8	0.00	251499	100.00
13	1	0.00	251500	100.00
20	2	0.00	251502	100.00
43	1	0.00	251503	100.00

The list of issues so far is as follows:

1. Revolving utilization has outliers of around 5 percent.

2. The monthly income is missing in 20 percent of the cases.

3. 30 to 59 DPD, 60 to 89 DPD, and 90 DPD have an issue with 96 and 98.

4. The number of dependents is missing around 2 percent of values.

The completion of contents, print, univariate analysis, and frequency steps may be the end of data exploration and validation. Conducting all these steps on some variables brought forth the preceding four issues. Similar issues can be found with other variables as well. Make notes of them and move on to the next step of cleaning the data. This is also called *preparing* the data for analysis. Having outliers is certainly an issue, but removing them is not the solution.

After data exploration and data validation, these questions are left unanswered:

1. Some variables have missing values/outliers, but those records cannot be dropped.

2. Some variables have a high percentage of missing values/outliers. Can they be dropped?

3. Some variables have negligible missing values/outliers. How should they be treated when there are 2 percent, 40 percent, and 95 percent missing values?

4. Can the missing values be substituted with some other values?

5. Some variables are discrete and some continuous. Is the treatment (dropping or substitution) the same for discrete and continuous variables? If a continuous variable has 20 percent missing values and a discrete variable, which has just four levels, also has 20 percent missing values, can they be treated the same?

With these questions in mind, you can now proceed to the next step in the data preparation process.

Step 5: The Missing Value and Outlier Treatment

It has already been established that dropping the records because of missing values is not a good solution. Instead, a technique called *imputation* might be used. This technique involves replacing missing/erroneous values with the best possible substitutions to minimize the damage yet get accurate results. There are different types of imputations; we'll discuss two: stand-alone and those based on related variables.

Stand-Alone Imputation

The missing values need to be replaced with either the mean or median, depending on the rest of the values and what makes better business sense. The stand-alone imputation is convenient and easy to implement. The assumption here is that the missing values are not very distinct from the values that are already present.

The data in Table 7-32 shows the age of 21 athletes; one athlete's age is missing. Replace the missing value with 27, which is the average age of the rest of the values, since only one value is missing. (In general, the age of an athlete will be no more than 50 or 60.)

Table 7-32. *Age Data for Athletes*

Age	30	25	29	29	26	26	25	30	27	27	....	30	28	29	26	25	28	26	26	25	30	27

The new data will look like Table 7-33 after stand-alone imputation (using mean of the rest of the age values to replace the missing value).

Table 7-33. *Age Data for Athletes After Stand-Alone Imputation*

Age	30	25	29	29	26	26	25	30	27	27	27	30	28	29	26	25	28	26	26	25	30	27

The stand-alone imputation is a good method to use if small portions of the data are missing or fall into the category of outliers.

Imputation Based on Related Variable

In some cases, the relation between the variables might be used to impute. Consider the data given in Table 7-34, where the number of games and ages of athletes are shown as two columns.

Table 7-34. *Age vs. Appearances for Athletes*

Age	Appearances
25	201
32	265
25	206
19	154
23	193
31	257
	178
37	301
33	268
28	229
24	194
31	251
30	247
36	295
38	312
26	213
40	330
26	209
40	330
20	162
35	283
22	177

The age is missing for one athlete. If stand-alone imputation is used, an average age of 30 is derived. To determine whether the missing value can be replaced with 30, the information on the number of games played can be used. A simple sort on the number of games produces Table 7-35.

Table 7-35. *Age vs. Appeareances for Athletes After Sorting*

Age	Appearances
40	330
40	330
38	312
37	301
36	295
35	283
33	268
32	265
31	257
31	251
30	247
28	229
26	213
26	209
25	206
25	201
24	194
23	193
	178
22	177
20	162
19	154

The number of games infers that 22 is a better value than 30. Imputation based on variable relation is definitely a better method, but what are the variables to be considered while making the imputations? To answer this question, you need to have a clear idea of all variables and business problems at hand. As discussed in the following section, in all the cases of missing values, imputation may not be the best technique.

Too Many Missing Values

Imputation will not work as well if there are too many missing values (Table 7-36). For example, if 70 percent of the values in a variable are missing, they cannot be imputed based on the remaining 30 percent of values. In this case, creating an indicator variable (Table 7-37) to tell whether the parent value is missing or present is the only workable solution.

Table 7-36. *Example of Too Many Missing Values in Data*

Age	31		32		35	29	30	35

Table 7-37. *Indicator Variable for Table 7-36*

Age_ind	1	0	0	0	0	1	0	0	0	0	1	0	1	0	1	0	1

Table 7-37 shows the best that can be taken out of the variable that is listed in Table 7-36.

Hence, when there are too many missing values, the actual variable should be dropped and replaced with an indicator variable.

The Missing Value and Outlier Treatment

Earlier topics explained how to explore and validate the variables. A few issues with missing values and outliers were found while validating the data. The chart in Figure 7-9 shows the missing value and outlier treatment for both continuous and discrete variables for all levels of missing value percentages. You can first start with missing value treatment and then move on to outlier treatment. The same order applies, if a variable has both missing values and outliers.

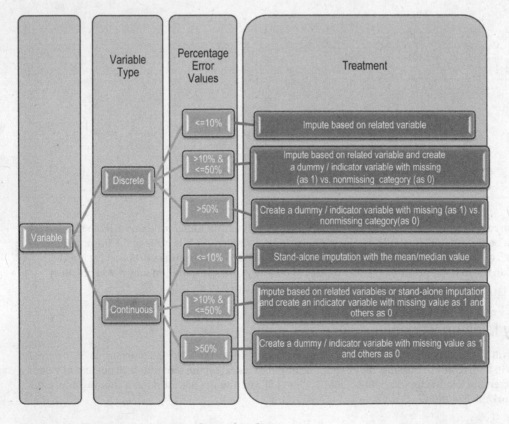

Figure 7-9. *How to treat missing value and outliers*

This chart can be used for cleaning the data. There are two types of variables: discrete and continuous. The percentage missing in each variable might be different.

Practically speaking, the following are the possible percentages of missing values in discrete and continuous variables:

- A discrete variable with less than 10 percent of missing values or outliers

- A discrete variable with 10 to 50 percent of missing values or outliers

- A discrete variable with more than 50 percent of missing values or outliers

- A continuous variable with less than 10 percent of missing values or outliers

- A continuous variable with 10 to 50 percent of missing values or outliers

- A continuous variable with more than 50 percent of missing values or outliers

The treatment for each situation follows.

For a discrete variable with less than 10 percent of missing values or outliers:

- It is best to impute based on a related variable. Since a discrete variable has levels, stand-alone imputation may not be effective.

- Consider the example Table 7-38 for the variable number of savings accounts held by customers.

Table 7-38. *Number of Savings Accounts Held by Customers*

Number of Accounts	Number of Customers
0	2,505
1	351,234
2	339,778
3	139,918
4	124,044
5	94,003
6	74,325
7	64,469
8	1,456
#N/A	34,530
Total	1,226,262

Of the total number of customers, 34,530 customers' data is missing. The number of accounts they hold is not specified. It is almost 2.8 percent of the total data. A stand-alone imputation performed on the data yields 2.7 as the average number of accounts in the population. It can simply be inferred that 34,530 customers have three accounts each, or some other related variable can be made use of.

Sound business insight is necessary to pick the best related variable to perform an imputation. This example considers one more variable called the credit card holder indicator. It takes just two values, Yes or No. Say you are trying to use this indicator to perform the imputation. How this can be done? For this, you need some more information.

Out of 1,226,262 customers, some have credit cards, and some of them don't have any. Table 7-39 still shows frequency of customers with 1,2,3,4, and 8 savings accounts. Out of 351,234 customers with one savings account, how many of them have credit cards? How many don't have any card? There are 339,778 customers with two savings accounts, so how many of them have a credit card? How many of them don't have any card? These questions will be answered using a simple cross-tab association table between the number of accounts and credit card indicator.

Table 7-39. *Frequency of Customers with Savings Accounts and Credit Cards (Data Filled in Table 7-40)*

Number of accounts	Credit Card Holder		Number of Customers
	Yes	**No**	
0			2,505
1			351,234
2			339,778
3			139,918
4			124,044
5			94,003
6			74,325
7			64,469
8			1,456
N/A			34,530
			1,226,262

The cross-tab association table shown in Table 7-40 between the numbers of saving accounts will determine what the right substitution for N/A might be.

Table 7-40. *Frequency of Customers with Savings Accounts and Credit Cards*

Number of Accounts	Credit Card Holder		Number of Customers
	Yes	**No**	
0	25	2,480	2,505
1	245,863	105,371	351,234
2	254,833	84,945	339,778
3	110,535	29,383	139,918
4	100,475	23,569	124,044
5	79,902	14,101	94,003
6	63,176	11,149	74,325
7	56,732	7,737	64,469
8	1,295	161	1,456
N/A	345	34,185	34,530
	913,181	313,081	1,226,262

Table 7-41 shows the same table in terms of percentages.

Table 7-41. *Freqeuncy of Customers with Savings Accounts and Credit Cards (% Values)*

	Credit Card Holder		
Number of Accounts	Yes	No	
0	1%	99%	2,505
1	70%	30%	351,234
2	75%	25%	339,778
3	79%	21%	139,918
4	81%	19%	124,044
5	85%	15%	94,003
6	85%	15%	74,325
7	88%	12%	64,469
8	89%	11%	1,456
N/A	1%	99%	34,530
	74%	26%	

Of the total number of customers who have one savings account each, 70 percent have at least one credit card. Similarly, 81 percent of customers with four savings accounts each have credit cards. In the group with the number of savings account as N/A, only 1 percent has credit cards, and the rest of the 99 percent don't have any credit card. The same is the case with the number of savings accounts = 0 group.

One percent of customers with 0 and N/A savings accounts are credit card holders. Hence, a good solution is replacing N/A with 0, rather than an overall mean of 3. This is imputation with respect to another related variable. There is no rule of thumb to select a related variable; it depends on the business problem that is being solved.

For a discrete variable with 10 to 50 percent missing values or outliers:

- Two tasks need to be performed in this case: imputing the missing values based on related variable and creating a dummy/indicator variable with a missing (as 1) vs. nonmissing category.

- The first task is ensuring that there are no missing values. Make sure that missing/nonmissing records are flagged in the data since a considerable portion of the data is missing. A new indicator variable that will capture the information regarding missing and nonmissing records of a variable needs to be created. This will be used in analysis later. How to create an indicator variable was already explained in the earlier sections. The second task is to impute the missing value based on a related variable.

For a discrete variable with more than 50 percent missing values or outliers:
A new indicator variable needs to be created and the original variable dropped because the minuscule percentage of the available values cannot be used.

For a continuous variable with less than 10 percent missing values or outliers:
A stand-alone imputation, is sufficient.

For a continuous variable with 10 to 50 percent missing values or outliers:
Both indicator variables need to be created and missing values should be imputed.

For a continuous variable with more than 50 percent missing values or outliers:
Create an indicator variable and drop the original variable.

Turning back to the case, the issues so far are as follows:

1. Revolving utilization, which is a continuous variable, has outliers around 5 percent.

2. The monthly income is a continuous variable and has missing values in 20 percent of the cases.

3. 30 to 59 DPD, 60 to 89 DPD, and 90 DPD are all discrete variables; values 96 and 98 have issues. The percentage of such errors is less than 10 percent.

4. The number of dependents is a discrete variable and has around 2 percent missing values.

The following sections examine what type of issue a variable has and perform the right treatment on the variable.

Cleaning Continuous Variables

In previous sections, you found outliers in variable `RevolvingUtilizationOfUnsecuredL`. Now you will try to treat them.

Let's start with the variable Revolving Utilization (`RevolvingUtilizationOfUnsecuredL`). Given in Table 7-42 is the `proc univariate` output for this variable, as taken from Table 7-21.

Table 7-42. *Quantiles Table from the Univariate Output of RevolvingUtilizationOfUnsecuredL*

Quantiles (Definition 5)	
Quantile	**Estimate**
100% Max	5070800%
99%	109%
95%	100%
90%	98%
75% Q3	56%
50% Median	15%
25% Q1	3%
10%	0%
5%	0%
1%	0%
0% Min	0%

Revolving Utilization, which is a continuous variable, has outliers of around 5 percent. Refer to Figure 7-10.

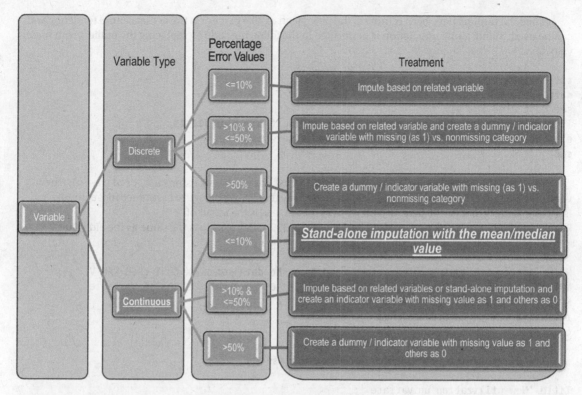

Figure 7-10. *Treating continuous variables for 5 percent outliers*

The mean value of utilization after ignoring outliers can be found by using the following code:

```
Title 'Proc Univariate on utilization less than or equal to 100% ';
Proc univariate data= cust_cred_raw_v1 plot;
Var RevolvingUtilizationOfUnsecuredL;
Where RevolvingUtilizationOfUnsecuredL<=1;
run;
```

Table 7-43 lists the output of this code.

Table 7-43. *Output of proc univariate with RevolvingUtilizationOfUnsecuredL<=1*

Basic Statistical Measures			
Location		**Variability**	
Mean	0.304325	Std Deviation	0.33852
Median	0.143964	Variance	0.11460
Mode	0.000000	Range	1.00000
		Interquartile Range	0.49351

Since the outliers have been removed, the new mean of 30.4 percent is the actual central tendency and can be used. Stand-alone imputation is performed in the following code by replacing the outliers with mean values:

```
title'Treating utilization ';
data  cust_cred_raw_v2;
set cust_cred_raw_v1;
if RevolvingUtilizationOfUnsecuredL>1then utilization_new=0.304325;
else utilization_new= RevolvingUtilizationOfUnsecuredL;
run;
```

The preceding code creates a new data set called cust_cred_raw_v2 using cust_cred_raw_v1; a new variable called utilization_new is created in such a way that if RevolvingUtilizationOfUnsecuredL is greater than 1 (that is, 100 percent), then utilization_new is the mean of RevolvingUtilizationOfUnsecuredLie, which is 0.304325. Otherwise, it is the same as the old value.

The log file for preceding code looks like this:

```
NOTE: There were 251503 observations read from the data set WORK.CUST_CRED_RAW_V1.
NOTE: The data set WORK.CUST_CRED_RAW_V2 has 251503 observations and 16 variables.
NOTE: DATA statement used (Total process time):
real time          0.84 seconds
cpu time           0.40 seconds
```

The following is the new variable's univariate analysis:

```
Title 'New utilization univariate ';
Proc univariate data= cust_cred_raw_v2 plot;
Var utilization_new;
run;
```

The output of this code is given in Table 7-44 and Figure 7-11.

Table 7-44. *Output of proc univariate on utilization_new Variable*

The UNIVARIATE Procedure Variable: utilization_new			
Moments			
N	251503	Sum Weights	251503
Mean	0.30432542	Sum Observations	76538.7562
Std Deviation	0.33479563	Variance	0.11208811
Skewness	0.9652513	Kurtosis	-0.4765405
Uncorrected SS	51483.0736	Corrected SS	28190.3844
Coeff Variation	110.012377	Std Error Mean	0.00066759

(*continued*)

Table 7-44. (*continued*)

Basic Statistical Measures

Location		Variability	
Mean	0.304325	Std Deviation	0.33480
Median	0.153575	Variance	0.11209
Mode	0.000000	Range	1.00000
		Interquartile Range	0.47923

Tests for Location: Mu0=0

Test	Statistic		p Value	
Student's t	t	455.8585	Pr> |t|	<.0001
Sign	M	116657	Pr>= |M|	<.0001
Signed Rank	S	1.361E10	Pr>= |S|	<.0001

Quantiles (Definition 5)

Quantile	Estimate
100% Max	1.00000000
99%	0.99999990
95%	0.99999990
90%	0.93255772
75% Q3	0.50920204
50% Median	0.15357490
25% Q1	0.02997462
10%	0.00299276
5%	0.00000000
1%	0.00000000
0% Min	0.00000000

Extreme Observations

Lowest		Highest	
Value	Obs	Value	Obs
0	251479	1	224166
0	251455	1	228890
0	251452	1	231077
0	251438	1	238536
0	251434	1	244591

```
            Histogram                              #        Boxplot

  1.025+*                                          29          |

      .**************                           23431          |

  0.925+***                                       4807          |

      .***                                        4364          |

  0.825+***                                       4155          |

      .***                                        4052          |

  0.725+***                                       4137          |

      .***                                        4243          |

  0.625+***                                       4523          |

      .***                                        4772          |

  0.525+****                                      5398       +-----+

      .****                                       5611       |     |

  0.425+****                                      6283       |     |

      .*****                                      6931       |     |

  0.325+********                                 12975       |  +  |

      .******                                    8744       |     |

  0.225+******                                   10008       |     |

      .********                                  12257       *-----*

  0.125+**********                               16663       |     |

      .****************                          27497       |     |

  0.025+****************************************** 80623      +-----+


      ----+----+----+----+----+----+----+----+---

      * may represent up to 1680 counts
```

Figure 7-11. *Box plot for utilization_new variable*

utilization_new will be used for analysis henceforth. The issue with the outliers is now resolved. In the next section, you will treat monthly income, which has 20 percent missing values.

Treating Monthly Income

The monthly income is a continuous variable and is missing values in 20 percent of the data. Refer to Figure 7-12.

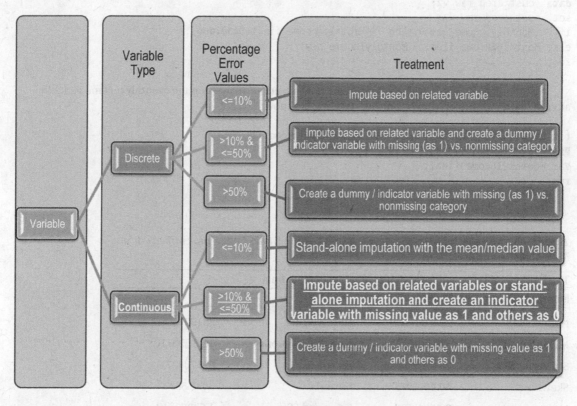

Figure 7-12. *Treating a continuous variable with 20 percent missing values*

Since the monthly income is continuous and is missing 20 percent of the values, a monthly income indicator variable needs to be created, and the missing values with the median will be imputed, as shown in the following code:

```
title'Treating Monthly income ';
data  cust_cred_raw_v2;
set  cust_cred_raw_v2;
if    MonthlyIncome_new =.then    MonthlyIncome_ind =1 ;
else MonthlyIncome_ind = 0;
run;
```

The preceding code creates a new variable called MonthlyIncome_ind, which takes value 1, when MonthlyIncome_new is . (a missing value), otherwise 0.

The following code creates a new variable called MonthlyIncome_final, which takes a value of 5400 when MonthlyIncome_new is . (a missing value). Otherwise, the new variable is the same as the old variable. This only replaces missing values with the mean.

```
data cust_cred_raw_v2;
set cust_cred_raw_v2;
if    MonthlyIncome_new =.then    MonthlyIncome_final=5400.000 ;
else MonthlyIncome_final = MonthlyIncome_new;
run;
```

Now, in the following code, you are performing univariate analysis on the monthly income variable after the treatment. Everything seems to falls into place now.

```
title'Univariate Analysis on Final  Monthly income ';
Proc univariatedata= cust_cred_raw_v2 ;
Var MonthlyIncome_final;
run;
```

Table 7-45 shows the output of this code.

Table 7-45. *Monthly Income After Resolving the Missing Values Issues (proc univariate on MonthlyIncome_final)*

The UNIVARIATE Procedure Variable: MonthlyIncome_final

Moments

N	251503	Sum Weights	251503
Mean	6383.21654	Sum Observations	1605398110
Std Deviation	7847.22102	Variance	61578877.8
Skewness	40.4062915	Kurtosis	3036.58327
Uncorrected SS	2.57348E13	Corrected SS	1.54872E13
Coeff Variation	122.935216	Std Error Mean	15.6474762

Basic Statistical Measures

Location		Variability	
Mean	6383.217	Std Deviation	7847
Median	5400.000	Variance	61578878
Mode	5400.000	Range	835040
		Interquartile Range	3485

(*continued*)

Table 7-45. (*continued*)

Tests for Location: Mu0=0

Test	Statistic		p Value	
Student's t	t	407.939	Pr> |t|	<.0001
Sign	M	124424.5	Pr>= |M|	<.0001
Signed Rank	S	1.548E10	Pr>= |S|	<.0001

Quantiles (Definition 5)

Quantile	Estimate
100% Max	835040
99%	23250
95%	13500
90%	10700
75% Q3	7385
50% Median	5400
25% Q1	3900
10%	2348
5%	1516
1%	0
0% Min	0

Extreme Observations

Lowest		Highest	
Value	Obs	Value	Obs
0	251388	702500	93565
0	251374	729033	151133
0	251359	730483	123292
0	251305	772700	174163
0	251300	835040	122544

The monthly income variable is cleaned now. We can use these two new variables for the analysis from here on.

30 to 59 Days Past Due

30 to 59 DPD, 60 to 89 DPD, and 90 DPD are all discrete variables. They have issues with the values 96 and 98. The percentage of such errors is less than 10 percent. Figure 7-13 gives the solution.

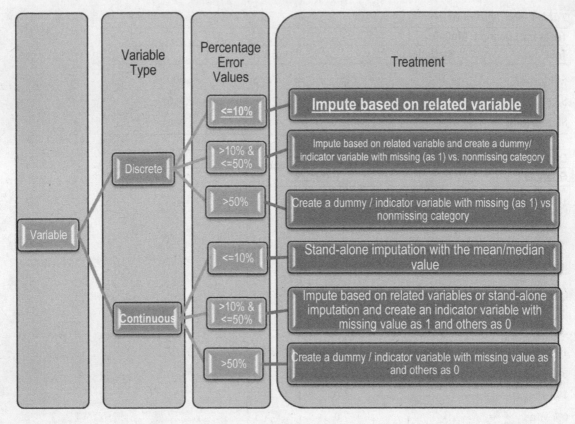

Figure 7-13. *Treating a discrete variable with less than 10 percent error values*

The treatment for this variable is imputation based on a related variable. Experience/observation shows the distribution of 30 DPD in combination with serious delinquency, which is the target variable and the most important of all. The cross-tab frequency of serious delinquency (the bad indicator), and 30 DPD is found in the following code:

```
Title 'Cross tab frequency of NumberOfTime30_59DaysPastDue1 and SeriousDlqin2yrs';
proc freq data=cust_cred_raw_v2;
tables NumberOfTime30_59DaysPastDueNotW*SeriousDlqin2yrs;
run;
```

Table 7-46 gives the output of the preceding code.

Table 7-46. *The Cross-Tab Frequency of NumberOfTime30_59DaysPastDueNotW*SeriousDlqin2yrs*

Frequency Row Pct Col Pct	Table of NumberOfTime30_59DaysPastDueNotW by SeriousDlqin2yrs		
	NumberOfTime30_59DaysPastDueNotW	SeriousDlqin2yrs	
		0	1	Total
	0	120977	5041	126018
		80.65	3.36	84.01
		96	4	
		86.43	50.28	
	1	13624	2409	16033
		9.08	1.61	10.69
		84.97	15.03	
		9.73	24.03	
	2	3379	1219	4598
		2.25	0.81	3.07
		73.49	26.51	
		2.41	12.16	
	3	1136	618	1754
		0.76	0.41	1.17
		64.77	35.23	
		0.81	6.16	
	4	429	318	747
		0.29	0.21	0.5
		57.43	42.57	
		0.31	3.17	
	5	188	154	342
		0.13	0.1	0.23
		54.97	45.03	
		0.13	1.54	
	6	66	74	140
		0.04	0.05	0.09
		47.14	52.86	
		0.05	0.74	

(*continued*)

Table 7-46. (*continued*)

Frequency Row Pct Col Pct	Table of NumberOfTime30_59DaysPastDueNotW by SeriousDlqin2yrs		
	NumberOfTime30_59DaysPastDueNotW	SeriousDlqin2yrs		
		0	1	Total
	7	26	28	54
		0.02	0.02	0.04
		48.15	51.85	
		0.02	0.28	
	8	17	8	25
		0.01	0.01	0.02
		68	32	
		0.01	0.08	
	9	8	4	12
		0.01	0	0.01
		66.67	33.33	
		0.01	0.04	
	10	1	3	4
		0	0	0
		25	75	
		0	0.03	
	11	0	1	1
		0	0	0
		0	100	
		0	0.01	
	12	1	1	2
		0	0	0
		50	50	
		0	0.01	
	13	0	1	1
		0	0	0
		0	100	
		0	0.01	

(*continued*)

Table 7-46. (*continued*)

Frequency Row Pct Col Pct	Table of NumberOfTime30_59DaysPastDueNotW by SeriousDlqin2yrs		
	NumberOfTime30_59DaysPastDueNotW	SeriousDlqin2yrs	
		0	1	Total
	19	0	0	0
		0	0	0
		.	.	
		0	0	
	96	1	4	5
		0	0	0
		20	80	
		0	0.04	
	98	121	143	264
		0.08	0.1	0.18
		45.83	54.17	
		0.09	1.43	
	Total	139974	10026	150000
		93.32	6.68	100

Frequency Missing = 101503

The cross-tab frequency shows overall frequency, overall percent, row percent, and column percent of that value. The percentage of zeros in class 98 is 45.83, and the nearest group with a percent of zeros is group 6, in other words, 47.14. So, the bad rate in group 98 is 54 percent, and the nearest group with a bad rate is 52.8 percent. The apt substitution for 98 will be 6, since there is no other group whose bad rate is similar to this group. Since group 6 is admissible and has a bad rate of 52.8 percent, 98 can be safely replaced with 6. Since there are just five observations in 96, both 96 and 98 can be replaced with 6.

The following code simply creates a new variable that takes value 6 when 30 DPD is 96 or 98. Otherwise, it is the same as 30 DPD.

```
title'Treating 30DPD';
data cust_cred_raw_v2;
set   cust_cred_raw_v2;
if    NumberOfTime30_59DaysPastDueNotW in (96, 98) then    NumberOfTime30_59DaysPastDue1= 6;
else  NumberOfTime30_59DaysPastDue1= NumberOfTime30_59DaysPastDueNotW;
run;
```

In the preceding code, a new variable called NumberOfTime30_59DaysPastDue1 is created based on NumberOfTime30_59DaysPastDueNot. The new variable takes a value of 6 whenever NumberOfTime30_59DaysPastDueNotW takes 96 or 98. Otherwise, it is same as the old variable.

The following is the log file and output:

```
NOTE: There were 251503 observations read from the data set WORK.CUST_CRED_RAW_V2.
NOTE: The data set WORK.CUST_CRED_RAW_V2 has 251503 observations and 19 variables.
NOTE: DATA statement used (Total process time):
real time          0.60 seconds
cpu time           0.49 seconds
```

The new variable is now free from errors. The following code runs PROC FREQUENCY on the variable NumberOfTime30_59DaysPastDue1:

```
Proc freq data=cust_cred_raw_v2;
tables NumberOfTime30_59DaysPastDue1;
run;
```

Table 7-47 gives the output of this code.

Table 7-47. *Output of proc freq on NumberOfTime30_59DaysPastDue1*

NumberOfTime30_59DaysPastDue1	Frequency	Percent	Cumulative Frequency	Cumulative Percent
0	211208	83.98	211208	83.98
1	26870	10.68	238078	94.66
2	7766	3.09	245844	97.75
3	2955	1.17	248799	98.92
4	1257	0.50	250056	99.42
5	547	0.22	250603	99.64
6	711	0.28	251314	99.92
7	96	0.04	251410	99.96
8	53	0.02	251463	99.98
9	22	0.01	251485	99.99
10	11	0.00	251496	100.00
11	2	0.00	251498	100.00
12	3	0.00	251501	100.00
13	1	0.00	251502	100.00
19	1	0.00	251503	100.00

Similarly, the rest of the variables can be treated using the treatment chart depicted in Figure 7-13. Once we are done with all the variables and finish all the exploration, validation, and cleaning steps, the data is ready for analysis.

■ **Note**　The previous treatment chart, represented in Figure 7-13, is suggestive only; a slightly different approach for outliers can also be used. When the outliers are 10 to 50 percent, they can be treated as a different class, or a subset can be taken for the population. This depends on the problem objectives at hand.

Conclusion

This chapter covered various ways of exploring and validating data. It also dealt with identifying issues in the data and then resolving them by using imputation techniques. As you can see, data cleaning requires a lot of time. Using junk data in analysis will only lead to useless insights. The methods provided here are guidelines rather than rules, and they arise from our work experience in the field. Using wisdom and rationale is important in preparing the data for analysis. Hence, the same amount of importance needs to be given to data cleaning as to data analysis. Preparing the data for analysis is the second step in analysis. Subsequent chapters will examine analysis and predictive modeling techniques.

CHAPTER 8

■ ■ ■

Testing of Hypothesis

You should now have an understanding of the basics of SAS and the fundamentals of statistics. You've mostly used descriptive statistics to explain the data and get some quick insights without applying any advanced techniques. One advanced technique you'll learn to apply in this chapter is how to test your hypotheses. Learning how to test a hypothesis is important for analysts because they will use the process in many situations, such as when testing correlation, testing regression coefficients, testing parameter estimates in time-series analysis, testing the goodness of fit in logistic regression, and so on. You'll learn about those topics in the coming chapters.

In this chapter, the focus is on testing of hypothesis and the important concepts involved in doing so. Specifically, you will learn about the null hypothesis, the alternate hypothesis, the process of testing a hypothesis, different types of tests to use, and finally the possibility of errors in the overall testing process.

Testing: An Analogy from Everyday Life

Let's use a simple real-life example to conduct a test. Say you want to buy a 50-pound cake for a big party. You walk into a cake shop and ask for one. The store manager says it's ready, and she shows it to you. You might get suspicious about its taste and quality. Fifty pounds is a giant cake, and obviously you don't want to take any risks, even if the store manager assures you that it's the best quality. In fact, you may want to test the cake. In other words, you would like to test the statement made by the store manager that the cake is of good quality. Obviously, you can't eat the whole cake and claim you are just testing. So, you will ask the manager to cut a small piece out of the cake give it to you for testing. You might want to cut this test sample randomly from the cake. The following are the possibilities that might result from your test:

- The test piece is awesome and tastes like the best cake you have ever had. It may be an instant buy decision.

- The test piece is contradictory to your expectations. You will definitely not buy it in that case.

- The quality is not the best, but it is still satisfactory. You may want to buy it if nothing better is available.

You had an assumption to begin with, you then took a sample to test it, and you made a conclusion based on a simple test. In statistical terms, you made an *inference* on the whole population based on testing a random sample. This process was the essence of the testing of hypothesis, in other words, the science of confirmatory data analysis.

261

Let's consider one more example. A giant e-commerce company claims that half of its customers are male and another half female. To test this statement, you take a random sample of 100 customers and count how many of them are male. Again, the following three scenarios may arise:

- Exactly 50 percent are males, and the other 50 percent are females.

- One gender dominates. For example, almost 90 percent are males, and only 10 percent are females.

- One gender is near 50 percent. For example, 52 percent are males in the sample.

In the first scenario, you agree to the statement made by the e-commerce company that the count of male and female customers is the same. In the second scenario, you simply reject the company's claim. In the third scenario, you may tend to agree with the claim. Once again, you are making an inference on the whole population based on the sample measures.

These are reasonably good examples of the process of testing a hypothesis. It is summarized as follows:

1. You start with an assumption.

    a. The whole cake is good in the first example.

    b. Overall, the gender ratio is 50 percent in the second example.

2. You take a sample that represents the population.

    c. You try a piece of cake in the first example.

    d. You look at 100 customers in the second example.

3. You do some kind of test on the sample gathered in step 2.

    e. You test the piece of cake by putting it in your mouth.

    f. You actually count the number of male and female customers in the sample.

4. You make a final interpretation and inference based on the testing of random samples.

    g. You make a decision about whether the cake is good or bad.

    h. You make an inference about whether the gender ratio is really 50 percent or not.

What Is the Process of Testing a Hypothesis?

Testing a hypothesis is a process similar to the examples discussed in the previous section. Using this process you make inferences about the overall population by conducting some statistical tests on a sample. You are making statistical inferences on the population parameter using some test statistic values from the sample. You can refer to Chapter 5 for more about population, parameters, samples, and statistics, but here is a quick recap (see also Figure 8-1):

- *Population*: Population is the totality, the complete list of observations, or the complete data about the subject under study. Examples are all the credit card users or all the employees of a company.

- *Sample*: A sample is a subset of a population or a small portion collected from a population. Examples are credit card users between 20 and 30 years old and employees who are part of the HR team.

- *Parameter*: Any measure that is calculated on the population is a parameter. Examples are the average income of all credit card users and the average income of all the employees in your company.

- *Statistic*: Any measure that is calculated on the sample is a statistic. Examples are the average income of a sample from credit card users (users between 20 and 30 years old) and the average income of HR team employees.

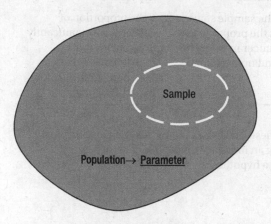

Figure 8-1. Parameter and statistic

In inferential statistics, you make an assumption about the population. That assumption is called the *hypothesis* (the *null hypothesis* to be precise). You take a sample and calculate a test statistic, and you expect this test statistic to fall within certain limits if the null hypothesis is true.

Table 8-1 contains a few more examples involving the process of testing a hypothesis.

Table 8-1. Examples of Testing a Hypothesis

Scenario	Null Hypothesis	Sample	Sample Statistic	Inference
Bank customers salary	The average income is $35,000.	You take a simple random sample of 300 customers.	The sample statistic is the average salary of 300 sampled customers.	Accept the null hypothesis if the salary of the sample falls near $35,000, or reject the null hypothesis.
Drug testing	The drug has 1.5 percent alcohol.	You take a random sample of 100 ml.	The sample statistic is the measured alcohol percentage in the sample.	Accept the null hypothesis if the sample test value is near 1.5 percent.
Product feedback	Our product customer satisfaction is 80 percent.	You take a simple random or stratified sample of users across various segments.	You conduct a survey and take the sample C-SAT score (formal customer satisfaction score).	Accept the null hypothesis if the sample C-SAT falls near 80 percent.

(*continued*)

Table 8-1. (*continued*)

Scenario	Null Hypothesis	Sample	Sample Statistic	Inference
Student training	The training has no significant effect on students.	You take a sample of students who took the training.	Students take a test before the training and a test after the training.	If there is a significant increment in the marks, then accept the null hypothesis.
Smoking causes cancer	Smoking does not cause cancer (smoking and cancer are independent).	You take a random sample from the population (contains smokers and nonsmokers).	The sample statistic is the proportion of cancer in smokers and nonsmokers.	If the proportion of cancer is not significantly different in smokers than in nonsmokers, then accept the null hypothesis.

Up to now, we've talked about framing the null hypotheses, taking a random sample, sampling the distribution, calculating the probability, and finally making an inference on the population. Figure 8-2 illustrates the main steps involved in the process of testing a hypothesis.

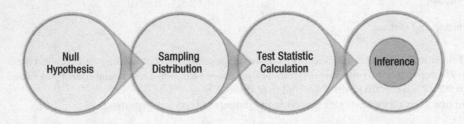

Figure 8-2. *An overview of steps involved when testing a hypothesis*

Here are the steps again:

1. State the null hypothesis on the population. Start with an assumption or hypothesis.

2. Select the sample and sampling distribution.

3. Calculate the test statistic value.

4. Make a final inference based on the test statistic results.

Here's a short business case study to help you understand the process of testing a hypothesis. A soap production company claims that the average weight of soap is 250 grams, and the population has a standard deviation of 5 grams. You want to test this by taking a sample of 50 bars of soap. You need to note that the soaps are priced based on the assumption that they weigh 250 grams. Anything more than 250 grams is a loss to the company. If the average weight is significantly less than 250 grams, then the company loses its customers; if it is greater than 250 grams, then the company is pricing its soap too low. In this case study, you can test a hypothesis to verify that the average weight of the bars of soap is 250 grams. Imagine that you take a sample of 50 soap bars and the sample average weight is 260 grams. Now the question is, should you accept or reject the null hypothesis?

The flowchart in Figure 8-3 shows the approach of testing the example hypothesis.

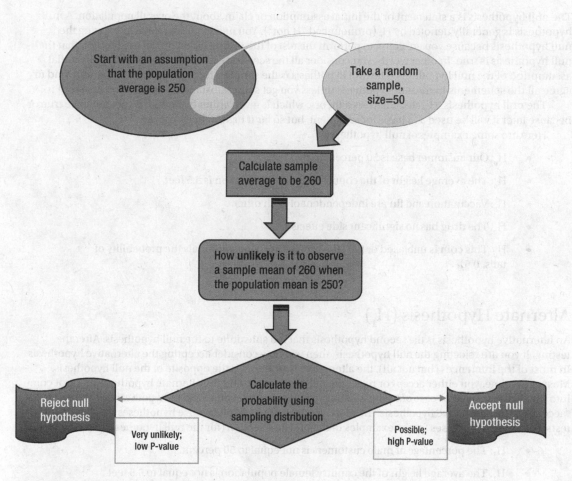

Figure 8-3. *Testing a hypothesis for the soap example*

Figure 8-4 shows all the steps involved in the process of test of hypothesis.

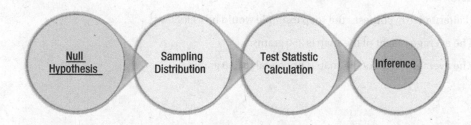

Figure 8-4. *Establishing the null hypothesis, the first step*

State the Null Hypothesis on the Population: Null Hypothesis (H₀)

The null hypothesis is a statement or the initial assumption or claim about the overall population. A null hypothesis is generally denoted by H_0 (pronounced "H not"). You need to be careful while stating the null hypothesis because you are going to perform the rest of the testing based on the assumption that the null hypothesis is true. In other words, you consider all the scenarios and measures based on the initial assumption of the null hypothesis. The null hypothesis is the simplest form of hypothesis. You will tend to agree all the statements in the null hypothesis unless you get some substantial evidence to not accept it.

The null hypothesis, H_0, characterizes a theory, which is given either because it is thought to be true or because later it will be used as a basis for argument, but so far it has not been proved.

Here are some examples of null hypotheses:

- H_0: Our customer base is 50 percent male.

- H_0: The average height of the country's female population is 5.5 feet.

- H_0: Vaccination and flu are independent of each other.

- H_0: The drug has no significant side effects.

- H_0: This coin is unbiased or fair (the probability of heads equals the probability of tails, 0.5).

Alternate Hypothesis (H₁)

An alternative hypothesis is the second hypothesis that is a substitute to the null hypothesis. After the testing, if you are rejecting the null hypothesis, then you may consider accepting the alternative hypothesis. In most of the inferences (but not all), the alternative hypothesis is the opposite of the null hypothesis. Most of the time, you either accept or reject the null hypothesis, and the alternate hypothesis may not come into the picture at all. For ease of understanding, people use "rejecting the null hypothesis" as the same as "accepting the alternative hypothesis." In short, you may not need alternative hypotheses for all types of tests. Alternate hypotheses for the examples in the previous section (for the null hypotheses) are shown here:

- H_1: The percentage of male customers is not equal to 50 percent.

- H_1: The average height of the country female population is not equal to 5.5 feet.

- H_1: Vaccination and flu are dependent on each other.

- H_1: The drug has significant side effects.

- H_1: This coin is biased (the probability of heads and the probability of tails are not equal to 0.5).

The null and alternative hypotheses the soap example would be as follows:

- H_0: The average weight of the soap is 250 grams.

- H_1: The average weight of the soap is not equal to 250 grams.

Sampling Distribution

Once you have defined the null hypothesis, you need to take a sample and calculate a statistic. Even before calculating the statistic, you should have a clear idea about what the distribution of the sample statistic will be. This is also known as the *sampling distribution* (Figure 8-5).

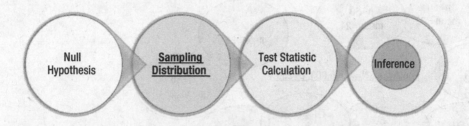

Figure 8-5. *Sample distribution, the second step*

A sampling distribution is the distribution of the statistic on simple random sampling with replacement (SRSWR). In the soap example, imagine there are a million bars of soap in the population. What will the sampling distribution of the average weight be when you take the average weight from multiple samples? You simply take a sample of 50 soaps and find the average weight; let's say it's 260 grams. What will the distribution of the values of the average weight be if you repeat this same sampling exercise?

■ **Note**   While dealing with multiple samples in SRSWR method, you take a sample and replace it in the population before taking any other sample.

Figure 8-6 shows that the overall population size is 1,000,000 and the overall average is 250 grams. Now you take a random sample of 50 soaps, find the average weight, and replace them in the population. You take another random sample of 50 soaps and again calculate the average weight. If you repeat this process of simple random sampling with replacement and find the average weight for each sample, what will the distribution of weights be? Read on to get the answers!

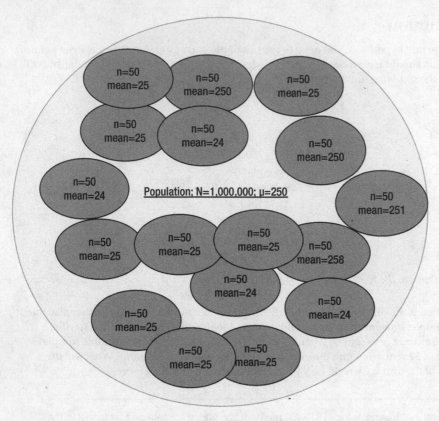

Figure 8-6. *Multiple random samples in the soap example*

If you repeat this sampling infinite times and find the average weight of each sample, the probability distribution of average weight is called the *sampling distribution*. Figure 8-7 shows an imaginary frequency distribution for this example, when you take 3,000 simple random samples with replacement. From this frequency distribution, you can get an idea of how the probability distribution will look. The probability distribution, as discussed earlier, will require an infinite number of samples to be taken.

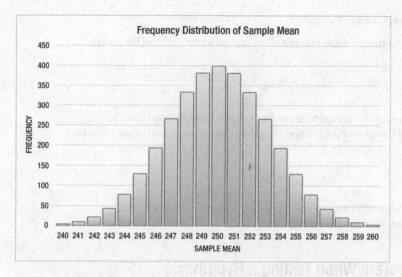

Figure 8-7. Frequency distribution of the sample mean in the soap example

Figure 8-7 probably answers all of the previous questions. Your understanding will further develop as you read about the central limit theorem in the following section.

You can see from the frequency distribution that for most of the samples you get a sample mean of 250 or a value near to it. Rarely you get a sample with a high deviation from the overall population mean.

The example discussed in Figure 8-7 is still not sufficient for you to get an understanding of the exact sampling distribution (the probability distribution of sample mean) of a sample mean. A frequency distribution of 3,000 test cases is not sufficient; you want to get the sampling distribution when you infinitely repeat the process of simple random sampling with replacement. The central limit theorem, discussed next, helps you in understanding the distribution of the sample mean.

Central Limit Theorem

To understand the philosophy behind the process of testing a hypothesis, you need to understand the central limit theorem. The following statement is a simplified form of the central limit theorem:

> *The distribution of the means of large samples tends to be normal, regardless of the distribution of the parent population.*

We'll now elaborate on this theorem. Imagine that a large number of random observations are taken from a population and an arithmetic mean of these observed values is computed. If you repeat this procedure a number of times and get a large number of such computed arithmetic means, the central limit theorem states that these computed means will be distributed as per an approximate normal distribution (the bell curve), regardless of the distribution of the underlying parent population.

The central limit theorem has three subtheorems:

- The mean of the sample means distribution is always equal to the mean of the parent population.

$$mean\left(\overline{x}_i\right) = \mu$$

- The standard deviation of the sample means distribution is always equal to the standard deviation of the parent population divided by the square root of the sample size.

$$SD\left(\overline{x}_i\right) = \sigma / \sqrt{n}$$

- The distribution of the sample means will increasingly approximate a normal distribution as the size, n, of samples increases.

Use of Central Limit Theorem When Testing a Hypothesis

You use the central limit theorem to test whether the population mean is equal to μ (the population arithmetic mean). To test the population mean, you take the sample mean, $\overline{x}$. The central limit theorem states that the sample mean follows the normal distribution. You try to quantify the probability of getting a sample mean of $\overline{x}$ or more. If that probability is much less, then you reject the null hypothesis.

In Figure 8-8, you can see the sample distribution of the mean values of the samples.

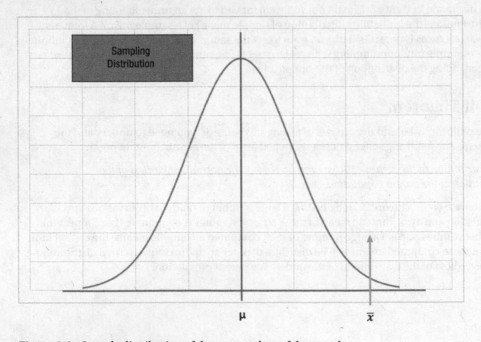

Figure 8-8. *Sample distribution of the mean values of the samples*

The mean of the sample means distribution is same as the population mean; it's in accordance with the central limit theorem. Here you try to find the probability of getting a sample mean of $\bar{x}$ or more if it is really taken from a population of mean μ. You accept or reject the null hypothesis based on the resultant probability. If the resultant probability is much less, then you reject the null hypothesis. Here, you are basically trying to test whether the sample is really coming from a population of mean μ.

In simple terms, if your calculations show that the probability of getting a specific value of sample mean (or worse) is much less and you still get that value of mean from your sample, then something is fishy. It simply means, with a reasonably accuracy, that the sample is not coming from the population for which it is claimed. In the soap example, the population mean is claimed to be 250 grams, and the sample mean is 260. Now you calculate the probability of getting a sample mean of 260 (or more) when the sample is taken from a population of mean 250. If this probability is much less (less than a predetermined threshold value), then with a *reasonable confidence* you can reject the null hypothesis that the population mean is 250.

Sampling distribution of the mean values of samples follows normal distribution. The sample mean follows normal distribution, but you can't use the same theorem for all sample statistics. Sample variance doesn't follow normal distribution. If the sample size is less than 30, then the sampling distribution is not normal. Table 8-2 contains some more examples of hypotheses, samples, and their corresponding distributions.

Table 8-2. *Examples: Samples and Their Corresponding Distributions*

Scenario	Null Hypothesis	Sample	Sampling Distribution
Bank customers Salary	The average income is $35,000.	A simple random sample of 300 customers.	Normal distribution
Product feedback	Our product customer satisfaction is 80 percent.	Simple random or stratified sample of 25 users across various segments.	T-distribution
Students training	The training has no significant effect on students.	A sample of students who took the training.	T-distribution
Smoking causes cancer	Smoking does not cause cancer (smoking and cancer are independent).	A random sample from population (contains smokers and nonsmokers).	Chi-square distribution
Variance in stock price of two stocks	The variance of two stocks is same.	Two random samples from stock 1 and stock 2.	F-distribution

In the soap example, you are also testing the population mean only, so you can apply the central limit theorem. The sampling distribution for the soap example looks like Figure 8-9.

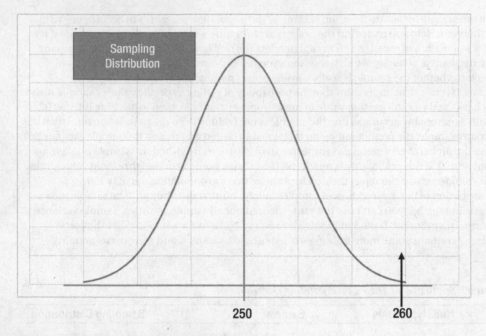

Figure 8-9. Sampling distribution for soap example

Now you need to decide whether a sample mean value of 260 is far or near to the population mean of 250. How far is really far? You need to answer the question, how likely is it to get a 260 sample mean when the sample is taken from a parent population of mean 250? You need to measure the probability of getting a sample mean of 260 or more when its parent population has a mean of 250. You need to calculate and transform the statistic to find the probability and then accept or reject the hypothesis.

Test Statistic

A test statistic (Figure 8-10) is the measure that is calculated from the sample. Sometimes the sample statistic alone might not be directly useful for finding the probability, but you can transform it and find the probability.

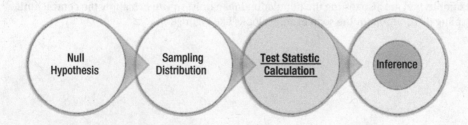

Figure 8-10. Test statistic, the third step

The test statistic selection will depend on the assumed probability distribution and the null hypothesis. Most of the time, it's simply the sample statistic value. In the soap example, the sample mean is the test statistic value. The sample mean follows the normal distribution. Sometimes, to find normal distribution probabilities, you need to transform normal distribution to a standard normal variate. Table 8-3 shows a few examples of test statistics. The test statistic here is chosen based upon the type of problem.

Table 8-3. *Examples of Test Statistic*

Scenario	Null Hypothesis	Sample	Test Statistic
Bank customers salary	The average income is $35,000.	A simple random sample of 300 customers.	z=Mean (income) or $z = \dfrac{\bar{x}_i - \mu}{s/\sqrt{n}}$
Drug testing	The drug has 1.5 percent alcohol.	A random sample of 100 ml.	p=Alcohol percentage per 100 ml
Product feedback	Our product customer satisfaction is 80 percent.	Simple random or stratified sample of users across various segments.	P=Satisfied customers/ total customers
Students training	The training has no significant effect on students.	A sample of students who took the training.	$t = \dfrac{\bar{d}}{s/\sqrt{n}}$
Smoking causes cancer	Smoking does not cause cancer (smoking and cancer are independent).	A random sample from population (contains smokers and nonsmokers).	$\chi2 = \sum \dfrac{(O-E)^2}{E}$

Note: There are some unknown test statistic values in Table 8-3.

In the soap example, the test statistic is the mean value of the sample, that is, 260 grams. To find the probability, you transform it to the standard normal distribution. Instead of calculating probabilities directly from the given sampling normal distribution, you are converting the statistic to the standard normal distribution. This way it is easy to calculate the probability, and standard normal probability tables are available in several documents and tools. $\bar{x}$ follows the normal distribution with mean 250, and the population standard deviation is given as 5 grams. The following formula is the test statistic for the sample means (when the sample size is more than 30):

$$z = \frac{\bar{x} - \mu}{s/\sqrt{n}}$$

The formula will use the following values:

$\bar{x} = 260$ is the mean value of the sample.

$\mu = 250$ is the population mean, which is the null hypothesis

S = 5 is the population standard deviation.

N = 50 is the sample size.

$$z = \frac{260 - 250}{5 / \sqrt{50}}$$

$$z = \frac{10}{1 / \sqrt{2}}$$

$$z = 14.14$$

Now that the test statistic is ready, you need to give your inference; that is, you need to accept or reject the null hypothesis.

Now you know the exact test statistic value. Is it near to the null hypothesis? Or is it far away from the null hypothesis parameter? You can accept or reject the null hypothesis based on this test statistic point and its distance from the null hypothesis parameter. The acceptance or rejection of hypothesis is called *inference*.

Inference

Inference is the last step in the process of testing a hypothesis (Figure 8-11).

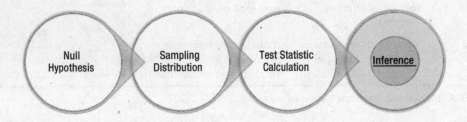

Figure 8-11. *Inference, the fourth step*

In this step, you decide whether you should accept or reject the null hypothesis. You accept the null hypothesis if the test statistic is within the permissible limits. If it is beyond the permissible limits, then you reject it. You can decide by looking at the *P-value*, or the probability value.

P-value

The P-value is the most important measure when testing a hypothesis. As discussed, you consider a sample and then calculate a test statistic. For example, if you are testing the population mean and if the sample mean is far away from the hypothesized mean, you reject the null hypothesis. But how far is really far? If the null hypothesis says the mean is 250 and your sample gives a value of 260, is it really far from 250? Are you OK with 251 as a sample mean? Is 252 fine? What are the actual boundaries you should have? To decide these boundaries, you need to understand sampling distribution.

When testing the sample, you got a sample mean of 260. Now you can actually find the probability of getting a sample mean of 260 or more when the sample is coming from a population having a mean value of 250. This is called the *P-value*.

- The P-value tells you what the probability is of observing this current test statistic value, or worse, when the sample is coming from a population that is complying with the null hypothesis.

- Imagine that your null hypothesis is true (the population mean equals 250). You can't expect every sample statistic (sample mean) to have the same value as the population parameter. It can be higher or lower than the population parameter. Fortunately, the P-value gives you the probability of such an event. For example, you are drawing a sample from a population that really has a mean of 250. Obviously, you can't expect every sample to have exactly a 250 mean. By chance if you get a mean of 260, the P-value will tell you what the probability is of getting a sample mean of 260 or more when the sample is drawn from a population of mean 250.

- What is the probability of observing the current statistic or an extreme value? The P-value is the answer to this question.

- The P-value is the measure of likelihood of obtaining a test statistic (as much or more) when H_0 is true.

Figure 8-12 illustrates the definition of a P-value.

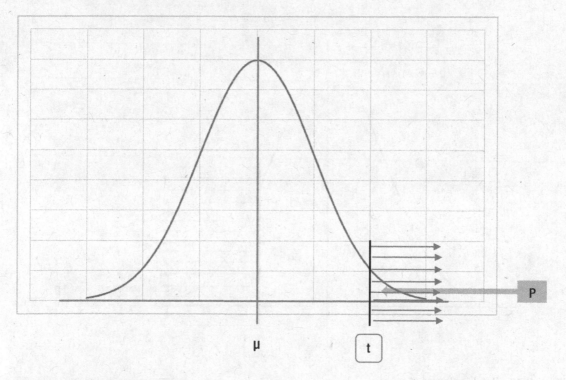

Figure 8-12. *P-value explained*

In Figure 8-12, imagine the null hypothesis states that the population has a mean of μ and the observed sample mean is t. The P-value is the area under the probability curve from t and above P(x>=t). The probability can be easily calculated by getting help from the sampling distribution properties. Here is a summary of the P-value:

1. You start with a null hypothesis.

2. You take a sample and calculate a statistic to test the null hypothesis.

3. If the null hypothesis is true, then the statistic is expected to follow certain probability distribution.

4. If it follows a particular probability distribution, then you can find the probability of getting the test statistic or beyond. This is the P-value.

After understanding the P-value, it's not too difficult to comprehend its practical applications. If the P-value is high, then t is very close to μ, which means you may accept the null hypothesis. If the P-value is low, then t is very far from μ, which means you may have to reject the null hypothesis. Generally 5 percent is taken as an industry standard for the P-value. If the P-value (the probability) is more than 5 percent, then the null hypothesis is accepted. For a P-value less than 5 percent, the null hypothesis is rejected. If the P-value is less than 5 percent, then the sample statistic, t, is said to be significantly different from the population parameter.

Figure 8-13 shows a high P-value of around 25 percent. The sample statistic is also seen close to the population parameter.

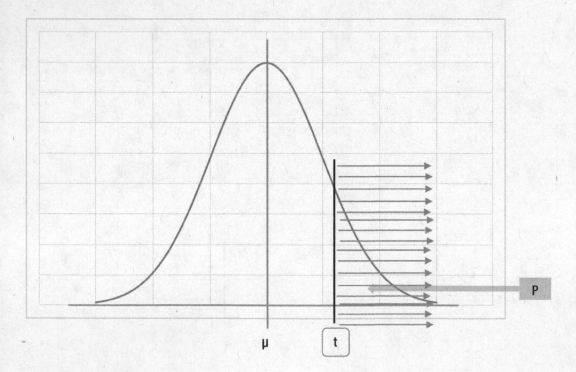

Figure 8-13. *P-value close to 25 percent*

The graph in Figure 8-14 shows a low P-value of around 1 percent.

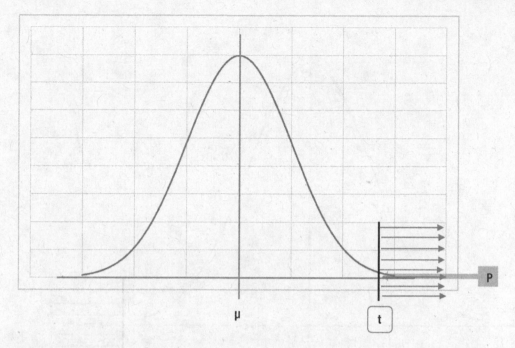

Figure 8-14. *P-value close to 1 percent*

The sample statistic is far from the population parameter, which means the statistic value is significantly different from the population parameter. If the sample is really coming from the population of mean μ, it should not be significantly different. In other words, the sample value should be near μ, which is not the case, so you reject the null hypothesis.

In the soap example, you got a sample mean of 260, when the population mean is 250. To see the probability of getting a sample mean of 260 or more (P-value), you used a standard Z-value normal probability curve that gave you the P-value.

In the "Test Statistic" section earlier in this chapter, we had calculated the z value for this example.

$$z = 14.14$$

Now, what is the probability of getting a Z-value greater than or equal to 14.14?

P(z>=14.14)=?

From the standard normal tables, you can find the previous probability value of Z>=14.14,

$P(z>=14.14)=1.54*(10^{-44})$

which means P(z>=14.14) < 0.000000000000000000000000000000000001.
Figure 8-15 shows the P-value for the soap example.

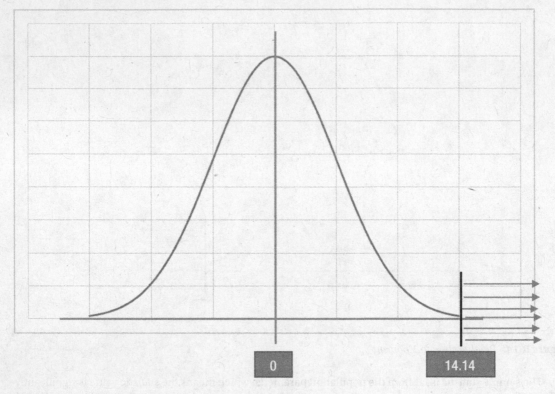

Figure 8-15. *P-value for the soap example*

Generally, if the P-value is less than 5 percent, then you reject the null hypothesis. Here, the P-value is not only less than 0.05, it is even less than 0.0000000000000001. Hence, you reject the null hypothesis. Here is another way of looking at this P-value:

- The null hypothesis states that the population mean is 250. If the sample is coming from a population of mean 250, then its sample mean should be nearly 250. In this exercise, you found the probability of getting 260. The sample doesn't seem to come from a parent population of mean 250. You got a sample mean of 260. The probability of such a chance is less than 0.000000000000001. Hence, you reject the null hypothesis. If it would have been greater than 5 percent (2.5 percent on either side), you would have accepted the null hypothesis.

- The null hypothesis states that the population mean is 250, and you got a sample mean of 260. How far is it from the hypothesized mean of 250? Given the P-value, it looks very far. If the sample statistic would have been near to the population parameter of 250, then the P-value should have been 20 percent, 25 percent, 30 percent, or at least 5 percent (the least acceptable value). But in this example, the P-value is less than 0.0000000000000001.

- Based on the P-value, you can conclude that the null hypothesis is not correct, and you should reject it. Let's consider another example. What is the probability of a football team winning 144 matches (with equal opposition) successively? You see that it is nearly equal to getting a sample mean of 260 when it is coming from a population mean of 250. The following bullets quantify these two probabilities:

  - Probability of a football team winning 144 consecutive matches = 4.48*(10-44)

  - Probability of getting a sample mean of 260 when the population mean is 250 = 1.48*(10-44)

Both these probabilities are extremely low. In fact, both these events are almost impossible; hence, you *reject* the null hypothesis.

Critical Values and Critical Region

Critical values are the boundaries between the acceptance and rejection regions. The critical region is the rejection region beyond these values (Figure 8-16). You have a 5 percent probability region where you reject the null hypothesis that is the critical region. Before the start of test, you decide this critical region for 5 percent or 10 percent or 1 percent.

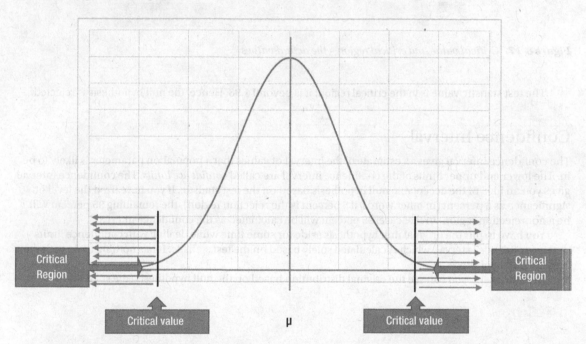

Figure 8-16. *Critical value and critical region*

If a test statistic value falls in the critical region, then the null hypothesis will be rejected.

A 5 percent tolerance means 2.5 percent on the either side of the null hypothesis value. Similarly, 10 percent means 5 percent on both sides, and so on. Generally, the P-value is sufficient to make a decision on accepting or rejecting the null hypothesis.

In the soap example, you got a Z-statistic value of 14.14. It falls within the rejection region or critical region. For a 5 percent critical region, on either side of null hypothesis mean, the critical values (based on standard normal probability tables) are -1.96 and 1.96 (Figure 8-17).

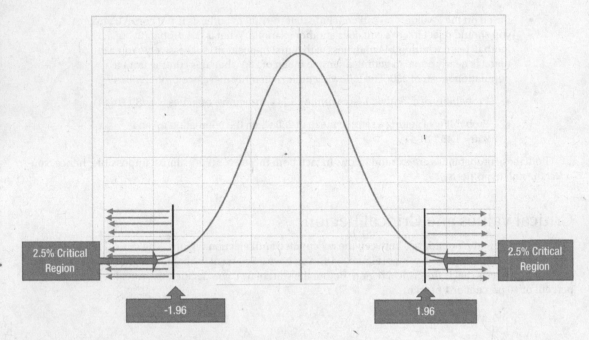

Figure 8-17. *Critical value and critical region – the actual values*

The test statistic value is in the critical region; it is beyond 1.96. Hence, the null hypothesis is rejected.

Confidence Interval

The confidence interval gives an estimate of the interval of values that a population parameter is likely to be in. The lower and upper limits of the confidence interval are called *confidence limits*. The confidence interval gives you an idea of the acceptable null hypothesis based on the test statistic. If you have fixed the level of significance as 5 percent (in other words, if 5 percent is the rejection region), the remaining 95 percent will be a nonrejection region. In that case, 95 percent will be calculated as the confidence interval.

You have to put the original null hypothesis aside for some time while dealing with confidence limits or the confidence interval, which is calculated solely based on the test statistic. It has nothing to do with the original null hypothesis.

In Figure 8-18, you can see the original distribution based on the null hypothesis (the solid line).

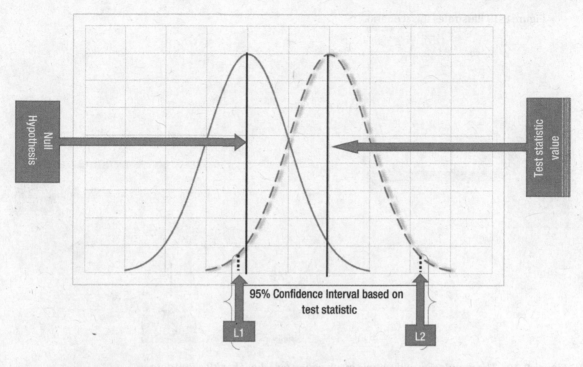

Figure 8-18. Confidence intervals

The dotted line shows a distribution based on the calculated sample statistic. The confidence interval in this example includes the null hypothesis. So, there is a higher chance to accept the null hypothesis.

If you get a test statistic of Z, then the confidence interval will be given by two limits of Z-l, and Z+l. The number l represents the confidence limits on both sides of Z. These are the two limits within which any null hypotheses will be accepted.

Let's look at another example of working with confidence intervals. Say a health program in a preschool involves checking the weight of all the students. The school claims that the average weight of the students is 30 pounds. You need to test this claim and will use the following hypotheses:

The null hypothesis: $H_0=30$ (the average weight of students is 30 pounds)

Alternative hypothesis $H_1 \neq 30$ (the average weight of students is not 30 pounds)

Since you don't know the population standard deviation and you are testing the sample mean, you will use a T-test. (See Table 8-8 later in this chapter for more details.) The choice of test depends upon the null hypothesis and the sampling distribution of the test statistic.

We will explain the details of the SAS code and testing procedure later in the chapter. For now just consider the confidence interval for the output (Table 8-4).

Table 8-4. Confidence Interval for the School's Health Check Data

Mean	95% CL Mean	
40.3796	38.9624	41.7967

Table 8-4 shows that the sample mean is 40.37. And the 95 percent confidence limits around the mean are 38.96 and 41.79. This mean is based on the sample statistic.

Figure 8-19 illustrates this scenario.

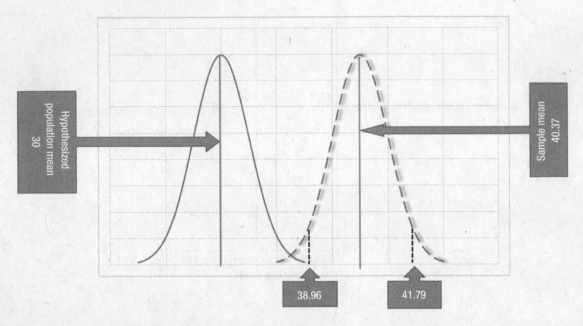

Figure 8-19. *The population and sample mean values for school health check data*

You can expect the population mean to be between 38.96 and 41.76 almost 95 percent of the time. If the null hypothesis is beyond these limits or beyond this interval, you can reject it. So, the null hypothesis of $H_0 = 30$ will be rejected in this case.

Table 8-5 summarizes some facts about the process of testing a hypothesis. We are using the example discussed earlier about a company claiming that there are 50 percent males in its customer population.

Table 8-5. *Summary of Facts for the Process of Testing a Hypothesis Using an Example*

Sr No	Testing of Hypothesis Facts	Example Claim: 50 Percent Male Population in the Customer Base
1	If the null hypothesis is true, then you expect the sample statistic to be within certain limits.	If the overall population has a 50 percent gender ratio, then you expect the sample also to have a gender ratio of 50 percent.
2	By chance if the sample statistic falls beyond the expected limits, then you calculate the probability of that chance or how likely that event is, called the P-value.	If the proportion is not 50 percent in the sample and if you get, say, 70 percent males by chance in the sample, then you use sampling distribution to calculate the probability of observing 70 percent or worse.
3	If it is very unlikely to witness what you observed in the sample, then you reject the null hypothesis or if the probability of that rare event is less than a predefined threshold value, then you reject the null hypothesis.	If you have a sample with a 70 percent male population (the event) when the sample is taken from a population where there are 50 percent male and if you see that the probability of such an event is extremely low or lower than 5 percent, then you reject the null hypothesis.

Tests

The type of test depends upon the sample statistic and the distribution of the sample statistic. If your sample statistic is a sample mean for a large sample, then you use a Z-test. If you are dealing with the sample mean of a small sample or if you don't know the standard deviation of the population, then you use a T-test. Let's look at a T-test for mean and some other example tests.

T-test for Mean

You use a T-test for the mean to test the null hypothesis, which may be something like the mean of a population being equal to a certain value. You use a T-test when the sample size is small. When the sample size is large, the sampling distribution is normal. But when the sample size is small, then the sample mean does not follow a normal distribution. The sample mean for small samples (less than 30) follows T-distribution with n-1 degrees of freedom (where n is the sample size). The T-distribution is similar to a normal distribution. As the sample size increases, the T-distribution tends toward a normal distribution. You use a T-test when the sample size is small or when the population standard deviation is not available.

■ **Note** *Degree of freedom*, in simple terms, is the number of observations that can be chosen independently to form one particular T-distribution. You will further deal with degree of freedom in the following case study. The test statistic for a T-test is also the same as the test statistic for the Z-test, except that you don't know this standard deviation of the population in a T-test.

$$t = \frac{\bar{x} - \mu}{s / \sqrt{n}}$$

Let's look at a brief case study to apply the concepts discussed so far.

Case Study: Testing for the Mean in SAS

Imagine a business scenario in which the city marketing manager of a smart TV company wants to devise a marketing strategy for its new TV model. She wants to have an effective strategy to reach the customers. For this, she needs to get an idea of the average monthly expenses of customer households in her city. An average estimate of expenses given by a local market research company is $4,500. The manager is not really confident about the accuracy of this figure. She wants to validate this figure before using it to devise her marketing strategy. She advises one of her teammates to collect some sample data on the expenses of a few randomly chosen families. Table 8-6 contains the sample data for 27 families.

Table 8-6. *Monthly Expense Data for a Sample of Households*

Sr No	Expense
1	4650
2	4248
3	4961
4	3200
5	4438
6	4993
7	4620
8	1100
9	4237
10	4991
11	4533
12	4457
13	4663
14	4613
15	4484
16	4997
17	5027
18	4807
19	4435
20	4503
21	9000
22	3950
23	4972
24	4983
25	8300
26	4516
27	4426

You are trying to test whether the population really has $4,500 as their average monthly expenses.
You are testing the population mean, and the sample size is small, so you use a T-test. Here are the steps that you follow:

1. **State the null hypothesis.**

   The null hypothesis (H_0): The average expense is equal to $4,500.

   The alternative hypothesis (H_1): The average expense is not equal to $4,500.

2. **Decide the sampling distribution and level of significance.**

   You are testing the sample mean for a small sample; hence, you will use T-distribution. And the level of significance is the same as the industry standard, in other words, 5 percent. So, the sampling distribution is a T-distribution with 26 degrees of freedom. T-distribution is not the same for all types of sample sizes. The T-distribution needs to be mentioned along with the degrees of freedom, which are equal to the sample size-1 (n-1). To reiterate, *degree of freedom* for a T-distribution is the number of observations that can be chosen independently to form one particular T-distribution.

3. **Test the statistic.**

$$t = \frac{\bar{x} - \mu}{s / \sqrt{n}}$$

$$t = \frac{4744.59 - 4500}{1367.55 / \sqrt{27}}$$

$$t = 0.93$$

4. **Make an inference.**

   To make an inference on the overall population, you need to calculate the P-value using T-distribution with 26 degrees of freedom.

$$P = Probability\ of\ (t > 0.93)$$

You can use distribution tables or a T-distribution probability value in any tool to get the previously mentioned probability. The value of P is as follows:

$$P = 0.361$$

The final inference is that the P-value is much more than 0.05; hence, you don't have much evidence to reject the null hypothesis. You agree with the null hypothesis, in other words, that the value of the average expenses given by the market research team is correct or that there is not much evidence to prove that the values given by the market research team are wrong.

You can also use the following SAS code to perform the same test:

```
proc ttest data =Expenses H0 = 4500;
var expense;
run;
```

- The H0 option in the SAS code refers to the population hypothesized mean.

- The variable name in the data set is expense.

- The data set name is Expenses.

Table 8-7 shows the output of this code.

Table 8-7. *Output of proc ttest (Data =Expenses)(the T-test Procedure; Variable: expense)*

N	Mean	Std Dev	Std Err	Minimum	Maximum
27	4744.6	1367.5	263.2	1100.0	9000.0

Mean	95% CL	Mean	Std Dev	95% CL	Std Dev
4744.6	4203.6	5285.6	1367.5	1077.0	1874.1

DF	t Value	Pr > |t|
26	0.93	0.3613

The output gives the sample size, sample mean, sample standard deviation, and minimum and maximum values of the variable expenses.

The third table in Table 8-7 is the most important one for the inference. It shows the T-statistic value. The T-statistic value is calculated from the test statistic equation. The P-value of the previous test is 0.3613. Based on the P-value, you accept the null hypothesis.

In the second table of the output tables in in Table 8-7, 95 percent confidence limits for the mean are given. The interval (4203.6 to 5285.6) has a 95 percent chance to contain the population mean.

Other Test Examples

Table 8-8 contains examples of other types of tests, including a Z-test, Chi-square test, and more.

Table 8-8. *Some More Test Examples*

Scenario	Null Hypothesis	Test
Testing sample mean with large sample	Population mean = X1.	Z-test
Testing sample mean with small sample	Population mean = X2.	T-test
Testing the independence of two variables	The two variables are independent.	Chi-square test
Testing correlation between two variables	Correlation The correlation between two variables is zero.	Person correlation test
Testing the significance of an independent variable on a dependent variable	Coefficient The coefficient of independent variable is zero in the regression model.	Test for Beta
Testing the difference in variance of two populations	There is no difference in the variance between two populations.	F-test
Testing the proportion of a particular class in a variable in population	The population proportion = P1.	Z- test for proportion

Two-Tailed and Single-Tailed Tests

In some cases of testing the hypothesis, in the critical region, the opposite of the null hypothesis defines the alternative hypothesis and is called the *nondirectional alternative hypothesis*. Up until now we have discussed only the nondirectional alternative hypothesis. There is a *directional* alternative hypothesis as well. Let's look at the details in more depth.

Two-Tailed Tests

The types of tests done with the nondirectional hypothesis are called *two-tailed tests* because the decision-making critical regions are on the two extreme tails of the distribution. In the case of a nondirectional alternate hypothesis, the critical region can be on the either side of the central tendency (Figure 8-20). For example, in soap weight example, the null hypothesis is that the average weight of soap is 250 grams; the alternative hypotheses says that the soap weight is not equal to 250 grams. You draw a sample of 50 bars of soap to test this hypothesis. You may want to reject the null hypothesis if you get a sample average much higher than 250 or two much less than 250. You are making a decision independent of the direction of the value. Figure 8-20 is an example for the nondirectional alternative hypothesis.

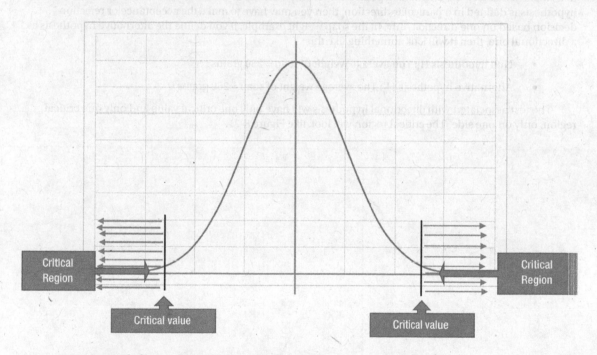

Figure 8-20. *Example for nondirectional alternative hypothesis*

The following are some familiar examples of nondirectional hypotheses:

- General hypothesis
  - Null hypothesis H_0: The population mean = μ
  - Alternative hypothesis H_1: The population mean $\neq \mu$

- Soap weight example
  - Null hypothesis H_0: The average weight of soap = 250 grams
  - Alternative hypothesis H_1: The average weight of soap ≠ 250 grams
- Gender ratio example
  - Null hypothesis H_0: The preparation of male = 50 percent
  - Alternative hypothesis H_1: The preparation of male ≠ 50 percent

There will be two critical values on either side of null hypothesis; the critical region will be split into two parts. If the test statistic falls in any of the two critical regions, you will reject the null hypothesis. If you are deciding the minimum significance level for accepting the null hypothesis as 5 percent, then it will be split into two parts of 2.5 percent, each on both sides of null hypothesis.

Single-Tailed Tests

In a single-tailed test, the alternate hypothesis is on only one side of the null hypothesis. If the alternative hypothesis is defined in a particular direction, then you may have to make the acceptance or rejection decision based on one direction only. In the soap weight example, if you define the alternative hypothesis as a directional one, then it will look something like this:

- Null hypothesis H_0: The average weight of soap = 250 grams
- Alternative hypothesis H_1: The average weight of soap > 250 grams

The test associated with directional hypothesis will have only one critical value and only one critical region, only on one side. The critical region will look like Figure 8-21.

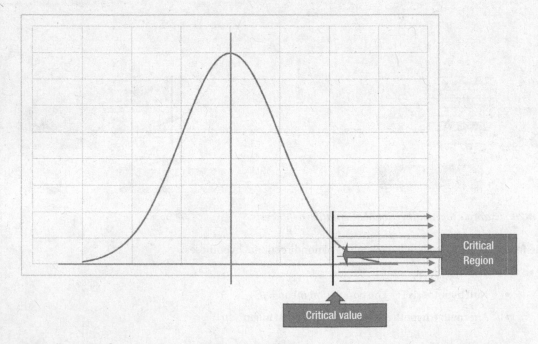

Figure 8-21. *Critical value and critical region for directional alternate hypothesis (right-tailed test)*

The minimum significance level probability (the generally accepted value is 5 percent) will be considered on only one side of the distribution. If the test statistic falls in that 5 percent of the critical region, you reject the null hypothesis. The single-tail test depicted in Figure 8-21 is a right-tailed test because the critical region falls on the right side. Similarly, you can have a single-tail test in which the critical region falls on the left side (Figure 8-22). In the same soap example, the following alternative hypothesis will result in a left-tailed test:

- Null hypothesis H_0: The average weight of soap = 250 grams

- Alternative hypothesis H_1: The average weight of soap < 250 grams

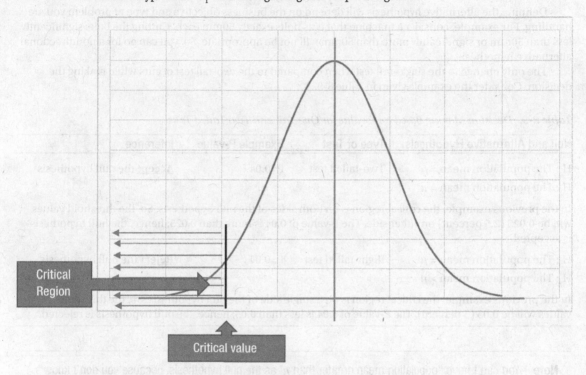

Figure 8-22. *Critical value and critical region for directional alternate hypothesis (left-tailed test)*

Here are more examples of a directional hypothesis:

- General hypothesis (right-tailed test)
  - Null hypothesis H_0: The population mean = μ
  - Alternative hypothesis H_1: The population mean > μ
- General hypothesis (left-tailed test)
  - Null hypothesis H_0: The population mean = μ
  - Alternative hypothesis H_1: The population mean < μ

- Gender ratio example (right-tailed test)
  - Null hypothesis H_0: The proportion of male = 50 percent
  - Alternative hypothesis H_1: The proportion of male > 50 percent
- Gender ratio example (left-tailed test)
  - Null hypothesis H_0: The proportion of male = 50 percent
  - Alternative hypothesis H_1: The proportion of male < 50 percent

Defining the alternative hypothesis will depend on the business objective and type of problem you are handling. For example, consider a machine that cuts bolts exactly 50mm each. Cutting the bolts significantly less than 50mm or significantly more than 50mm will not be appropriate. So, you can go for a nondirectional alternative hypothesis.

The only change in the single-tail test when compared to the two-tail test occurs while making the decision. Consider the example given in Table 8-9.

Table 8-9. *Decision Making Based on P-value for One-Tail and Two-Tailed Tests*

Null and Alternative Hypothesis	Type of Test	Example P-value	Inference
H_0: The population mean $= \mu$ H_1: The population mean $\neq \mu$	Two-tailed test	P=0.04	**Accept** the null hypothesis
In the previous example, the critical region is on both sides of the null hypothesis. So, the threshold values will be 0.025 (2.5 percent) on either side. The P-value of 0.04 is more than 0.025; hence, the null hypothesis is accepted.			
H_0: The population mean $= \mu$ H_1: The population mean $> \mu$	Right-tailed test	P=0.04	**Reject** the null hypothesis
In the previous example, the critical region is on a single side of the null hypothesis. So, the threshold values will be 0.05 (5 percent). The P-value of 0.04 is less than 0.05; hence, the null hypothesis is rejected.			

■ **Note** You can't have "population mean greater than μ" as the null hypothesis, because you don't know what the distribution of the sample mean will be when the null hypothesis is "population mean greater than μ." The central limit theorem works only for the null hypothesis - the population mean is equal to μ.

Errors in Testing

When testing a hypothesis, you are always taking a sample and making an inference about the population. There is always an element of error in doing so. If the population size is really large, then it may not be possible to get 100 percent accurate results. In other words, it's not always easy to infer something about a huge population just by taking a sample from it. Consider the following two examples:

- In the cake example, you are making a decision on the whole cake by tasting a small sample. It might happen that only the piece you tried is not good and the rest of the cake is good. In that case, you will make a wrong inference on the whole cake. Or it can happen the other way around. Only the sample piece of cake is good, but the rest is not good. You may go ahead and buy the cake, which will be a wrong decision anyway.

- In the soap weight example, it might so happen that the population average weight is not equal to 250, but you will end up accepting the null hypothesis if your sample shows 250 as the average weight by chance. It might happen the other way around as well. You might end up rejecting the null hypothesis if you get a sample with an average weight far away from what is assumed in the null hypothesis.

Yes, there is a chance to wrongly reject the null hypothesis, and also there is a chance to wrongly accept the null hypothesis. Refer to Table 8-10 for more details.

Table 8-10. *Errors in Accepting or Rejecting the Null Hypothesis*

	NULL HYPOTHESIS IS TRUE	NULL HYPOTHESIS IS FALSE
Reject the null hypothesis	**Type 1 error**	Right inference
Accept the null hypothesis	Right inference	**Type 2 Error**

Type 1 Error: Rejecting the Null Hypothesis When It Is True

Sometimes your sample might deceive you. By looking at the test statistic that is calculated from the sample, you may end up rejecting the null hypothesis wrongly. If you reject a null hypothesis when it needed to be accepted, then it is called a Type 1 error. As you will appreciate in the following text, a Type 1 error is decided by the analysts well before the start of testing.

What is the maximum value of P that you want to reject? The probability of a Type 1 error, also known as the level of significance, is denoted by α. Yes! You may actually reject the null hypothesis wrongly, in fact, generally 5 percent of the time. It is similar to a lottery example. There is a positive probability to win a countrywide lottery four years consecutively, but that probability is very small. It's probably less than 0.00000000000000000001; hence, you reject the null hypothesis. So, you will not a buy a lottery ticket if you don't see at least 5 percent significance. Therefore, for all tests, the level of significance is fixed at 5 percent at the beginning. Sometimes it can be 1 percent or 10 percent depending on the type of variable.

Figure 8-23 shows the distribution of the test statistic. The distribution shows that there can be values beyond C, but the probability of getting C or more is less than 5 percent. Hence, you reject the null hypothesis if a value falls beyond C, even when the null hypothesis is true.

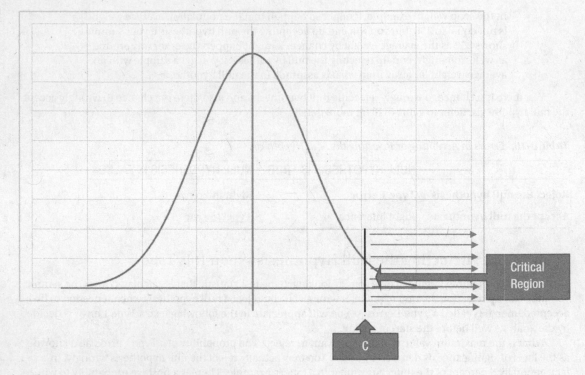

Figure 8-23. *Critical region for Type 1 error*

The following bullets further clarify the level of significance. The level of significance is

- The probability of Type 1 error.
- Also known as the probability of rejection error.
- Also known as the size of the test.
- Predefined probability to wrongly reject the null hypothesis.
- Minimum P-value required for accepting the null hypothesis.

Every test will begin with a level of significance, generally fixed as 5 percent, 10 percent, or 1 percent.

Type 2 Error: Accepting the Null Hypothesis When It Is False

A Type 2 error is when you wrongly accept the null hypothesis. It is also known as an *acceptance error*. The probability of a Type 2 error is denoted by β. For example, falling for a typical sales trap is an example of a Type 2 error. Say a salesman tries to sell you a defective product. He says that the product is really good (say the null hypothesis). If you go ahead and buy that item, then you are accepting a wrong null hypothesis. This is a Type 2 error. The probability of a Type 2 error, or β, is used for calculating the power of the test.

The following bullets clarify further the power of the test:

- $1 - \beta$
- $1 - P$ (accepting H_0 wrongly)
- $1 - P$ (accepting H_0 when H_0 is false)
- P (rejecting H_0 when H_0 is false)

Which Error Is Worse: Type 1 or Type 2?

We simply can't state, in general, which error is worse. It depends on the null hypothesis and the problem under consideration. Consider the two scenarios demonstrated in Table 8-11. In the first scenario, the Type 1 error is dangerous, and in the second scenario, the Type 2 error is dangerous.

Table 8-11. *Which Error Is Bad? Example Scenarios*

Scenario and Null Hypothesis	Type 1 Error	Type 2 Error
Testing whether a drug is poison or not H_0: **Drug is poisonous**	The Type 1 error is rejecting the null hypothesis (saying that the drug is not poisonous) when it is really poisonous. Stating that the drug is safe might yield serious consequences.	The Type 2 error is accepting the null hypothesis (saying that the drug is poisonous) when it is really not poisonous. Stating that the good drug is unsafe might not cause a big damage.
Testing whether a customer is good or bad before giving a huge personal loan H_0: **Customer is good**	The Type 1 error is rejecting the null hypothesis (saying the customer is bad) when the customer is really good. You might lose one good customer and the profits associated with that customer.	The Type 2 error is accepting the null hypothesis (saying the customer is good) when the customer is bad. If you approve one bad loan, that one bad customer could eat up 100 good customers' profits by running away with the loan amount.

■ **Note**   Though we are using the phrase "accepting the null hypothesis," when really testing, you don't really *accept* a null hypothesis. You either reject a null hypothesis or fail to reject the null hypothesis. We used "accepting the null hypothesis" to make it easier to understand.

Conclusion

This chapter discussed one of the most applicable statistical concepts called testing a hypothesis. This concept is used even in court trials. The central limit theorem and similar theorems that explain the sampling distribution about the sampling statistic are the keys when testing a hypothesis. The P-value helps in final inference. Significance tests are used in many other statistical concepts such as regression, correlation, testing independence, testing the impact of independent variable on the dependent variable, and so on.

So far, you've read chapters about basic descriptive statistics, data cleaning techniques, and testing the statistical significance. In the coming chapters, you'll learn some predictive modeling techniques such as linear regression, logistic regression, time-series forecasting, and so on.

■ ■ ■

## CHAPTER 9

Correlation and Linear Regression

In the previous chapter, we covered how to prepare data for model building. In this chapter, we discuss model building, which may be the most important step in a data analytics project. We will discuss the most popular technique of model building—linear regression. We will first discuss correlation in detail and then discuss the differences between correlation and regression. Finally, we will show a detailed example of modeling using linear regression. We will explain the concepts using real-world scenarios wherever possible.

What Is Correlation?

Do the following questions sound familiar to you?

- Is there any association between hours of study and grades?

- Is there any association between the number of church buildings in a city and the number of murders?

- What happens to sweater sales with an increase in temperature? Are the two very strongly related?

- What happens to ice-cream sales when the weather becomes hot?

- Is there any association between health and the gross domestic product (GDP) of a country?

- Can you quantify the association between asthma cases in a city and the level of air pollution?

- Is there any association between average income and fuel consumption?

In all these scenarios, the question is about whether two factors are related. And if they are associated, what is the strength of association? You probably know that there is an association between asthma cases and air pollution; you also know that there is an association between study hours and grades. But how strong are these associations? How do you quantify this association? Is there any measure to quantify the association between two variables? The answer is yes. *Correlation* is a measure that quantifies the association between two variables.

For example, say a newly opened e-commerce site spends a considerable amount of money on billboard and online advertising. It runs a rigorous advertisement campaign in two phases. The first phase is billboard advertisement, and the second one is an online campaign. Both billboard and online advertising fetch a great response. Now the question is, is there any association between the money spent on advertising and the number of responses? If there is an association, how strong is it? Is the association different for online and billboard marketing campaigns?

The following two scatterplots show the association between the budget and the responses for the two marketing campaigns. Figure 9-1 shows the association between budget and responses in the online segment. You can see a clear strong relation because as the budget increased, the responses increased. Figure 9-2 shows the billboard campaign, where there is no such clear trend.

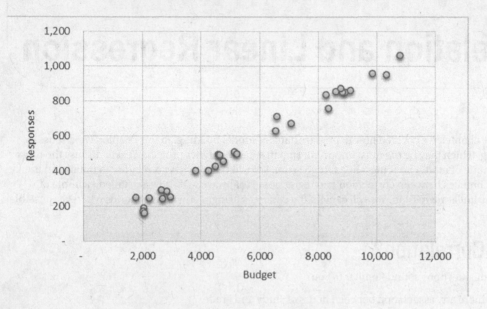

Figure 9-1. *Online campaign: number of responses by budget*

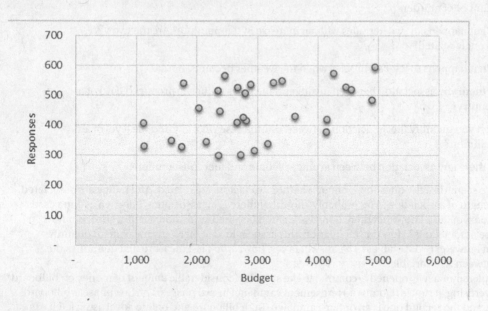

Figure 9-2. *Billboards: number of responses by budget*

As discussed, correlation is a technique that can be used to quantify the association. It assigns a number to the association, which tells you to what extent the given variables are dependent on each other. The correlation technique translates itself into a measure, which is called a *correlation coefficient*. The correlation coefficient is represented by *r*.

Pearson's Correlation Coefficient (r)

Pearson's correlation coefficient is defined based on the concept of variances. It is a measure of the linear relation or dependence between two variables, x and y. Its value varies between +1 and -1 (both inclusive). A value of 1 indicates a total positive correlation, 0 indicates an existence of no correlation, and -1 indicates a total negative correlation. This coefficient is widely used in statistics and analytics as a measure of degree of linear dependence between two variables.

Variance and Covariance

As discussed in Chapter 6, the variance gives you an idea about the dispersion within a given variable. If there are two variables x and y, the dispersion within x and dispersion within y are denoted by their variance.

Variance in x is as follows:

$$\frac{1}{N}\sum_{i=1}^{N}(x_i - \overline{x})^2$$

Variance in y is as follows:

$$\frac{1}{N}\sum_{i=1}^{N}(y_i - \overline{y})^2$$

If you want to see the dispersion in y with respect to the corresponding dispersion in x, then you use *covariance*. So, covariance always has two factors, and it can be understood as combined variance or parallel variance.

The covariance between x and y is as follows:

$$\frac{1}{N}\sum_{i=1}^{N}(x_i - \overline{x})(y_i - \overline{y})$$

As you can see, the formula of covariance looks similar to correlation except that it's considering the preparation of the variance of both x and y.

$$correlation(x,y) = \frac{\frac{1}{N}\sum_{i=1}^{N}(x_i - \overline{x})(y_i - \overline{y})}{\sqrt{\frac{1}{N}\sum_{i=1}^{N}(x_i - \overline{x})^2 * \frac{1}{N}\sum_{i=1}^{N}(y_i - \overline{y})^2}}$$

So, Pearson's correlation is nothing but the proportion of covariance in the product of individual variances. If the two variables are independent of each other (that is, the deviation in y doesn't depend on the corresponding deviation in x), then their covariance will be near to zero, which results in the correlation being zero.

Correlation Matrix

If you have several variables, say, x1, x1, x3...xn, and you are interested in all the combination of correlations, say, x1 versus x3, x2 versus x4, x1 versus x4, and so on, then you represent all the correlation combinations in the form of the following matrix. Correlation can be found between a pair of variables only. Correlation doesn't give you an idea of the association of x1 with x2 and x3 together.

	X_1	X_2	X_3
X_1	r_{11}	r_{12}	r_{13}
X_2	r_{21}	r_{22}	r_{23}
X_3	r_{31}	r_{21}	r_{33}

	X_1	X_2	X_3
X_1	Correlation between x_1 and x_1	Correlation between x_1 and x_2	Correlation between x_1 and x_3
X_2	Correlation between x_2 and x_1	Correlation between x_2 and x_2	Correlation between x_2 and x_3
X_3	Correlation between x_3 and x_1	Correlation between x_3 and x_2	Correlation between x_2 and x_3

Calculating Correlation Coefficient Using SAS

In the e-commerce example given earlier, let's say you load the required data into SAS and find out the correlation between budget and responses, for both online and billboard marketing cases. Here is the code:

```
PROC IMPORT OUT= WORK.add_budget
            DATAFILE= "C:\Users\VENKAT\Google Drive\Training\Books\Content\
Regression Analysis\Add_budget_data.xls"
            DBMS=EXCEL REPLACE;
     RANGE="budget$";
     GETNAMES=YES;
     MIXED=NO;
     SCANTEXT=YES;
     USEDATE=YES;
     SCANTIME=YES;
RUN;
proc contents data= add_budget varnum;
run;
```

Table 9-1 shows the output for the previous code.

Table 9-1. *Output of PROC CONTENTS Procedure on add_budget Data Set*

Variables in Creation Order				
#	Variable	Type	Len	Label
1	Day	Num	8	Day
2	Online_Budget	Num	8	Online Budget
3	Responses_online	Num	8	Responses_online
4	Live_Budget	Num	8	Live Budget
5	Responses_live	Num	8	Responses_live

The procedure used for finding correlation is PROC CORR. In the code, we need to mention the dataset name and the variable names that we want to use for finding the correlation. The following is the SAS code using PROC CORR:

```
proc corr data=add_budget ;
var Online_Budget Responses_online ;
run;
```

The previous code generates the output shown in Table 9-2.

Table 9-2. *Output of PROC CORR Procedure on add_budget Data Set*

The CORR Procedure

2 Variables:	Online_Budget Responses_online

Simple Statistics							
Variable	N	Mean	Std Dev	Sum	Minimum	Maximum	Label
Online_Budget	30	5672	2894	170156	1773	10801	Online Budget
Responses_online	30	556.13333	278.06384	16684	163.00000	1063	Responses_online

Pearson Correlation Coefficients, N = 30 Prob > |r| under H0: Rho=0		
	Online_Budget	Responses_online
Online_Budget	1.00000	0.99337
Online Budget		<.0001
Responses_online	0.99337	1.00000
Responses_online	<.0001	

The correlation matrix in the Pearson correlation coefficient table shows the correlation and p-value of correlation. The correlation between budget and response is 99.3 percent (.99337), which indicates a strong association. Any number close to 1(or -) is a strong association.

■ **Note** The p-value in the same table is shown as <.0001. This p-value is the result of testing the null hypothesis that the correlation coefficient of the current sample is not significant. If the p-value is less than 5%, we reject that null hypothesis, which means that the correlation coefficient is significant.

Similarly, you can find the correlation between the billboard advertising budget and corresponding responses by using following code:

```
proc corr data=add_budget ;
var Live_Budget Responses_live ;
run;
```

The previous code generates the output given in Table 9-3.

Table 9-3. Output of PROC CORR Procedure on add_budget Data Set

The CORR Procedure

| 2 Variables: | | | | | Live_Budget Responses_live | | |

Simple Statistics

Variable	N	Mean	Std Dev	Sum	Minimum	Maximum	Label
Live_Budget	30	2940	1056	88205	1118	4946	Live Budget
Responses_live	30	446.16667	92.03900	13385	300.00000	594.00000	Responses_live

Pearson Correlation Coefficients, N = 30 Prob > |r| under H0: Rho=0

	Live_Budget	Responses_live
Live_Budget	1.00000	0.43573
Live Budget		0.0161
Responses_live	0.43573	1.00000
Responses_live	0.0161	

The correlation between the budget and the response is 43.6 percent, which indicates a weak association.

From the correlation analysis done earlier, you can conclude that it may be a good idea to increase the budget for online marketing campaigns because they produce more responses. There is a strong correlation between responses and the advertising money spent in the online campaign. You saw a week correlation between the budget spent and responses in the case of the billboard campaign, so additional spending on this kind of campaign may not be advisable.

Correlation Limits and Strength of Association

What is the correlation between x and x? You can find it by using the correlation equations discussed earlier.

$$correlation(x,x) = \frac{\frac{1}{N}\sum_{i=1}^{N}(x_i - \overline{x})(x_i - \overline{x})}{\sqrt{\frac{1}{N}\sum_{i=1}^{N}(x_i - \overline{x})^2 * \frac{1}{N}\sum_{i=1}^{N}(x_i - \overline{x})^2}}$$

$$correlation(x,x) = \frac{\frac{1}{N}\sum_{i=1}^{N}(x_i - \overline{x})^2}{\sqrt{\frac{1}{N}\sum_{i=1}^{N}(x_i - \overline{x})^2}} = 1$$

So, the correlation of x with x, or the maximum value of the correlation, is 1. Similarly, you can show that the correlation between x and –x (the opposite of x) is -1. As a conclusion, the extreme values of correlation are -1 and +1.

A perfect positive correlation between a pair of variables will stand at +1, while a perfect negative correlation will have the correlation coefficient value as -1.

The following can be taken as rules of thumb:

- If the correlation is 0, then there is no linear association between two variables.

- If the correlation is between 0 and 0.25, then there is a negligible positive association.

- If the correlation is between 0.25 and 0.5, then there is a weak positive association.

- If the correlation is between 0.5 and 0.75, then there is a moderate positive association.

- If the correlation is between 0.75 and 1, then there is a strong positive association.

Figure 9-3 shows the scatterplot examples between two variables.

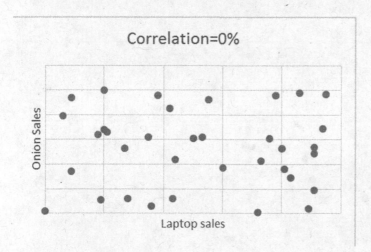

Figure 9-3. *Scatterplot examples with different values of correlation percentage*

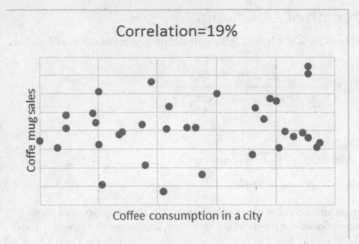

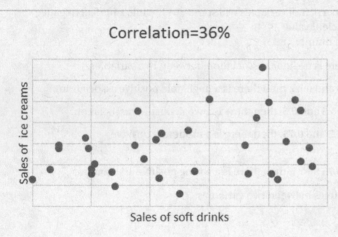

Figure 9-3. (*continued*)

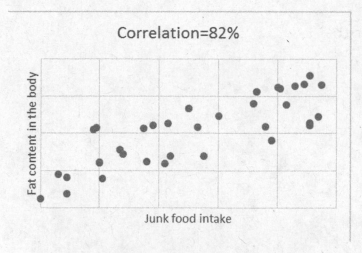

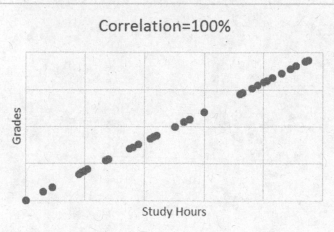

Figure 9-3. (*continued*)

The following are the rules of thumb for negative correlation coefficient values:

- If the correlation is 0, then there is no linear association between two variables.

- If the correlation is between 0 and -0.25, then there is an insignificant negative association.

- If the correlation is between -0.25 and -0.5, then there is a weak negative association.

- If the correlation is between -0.5 and -0.75, then there is a moderate negative association.

- If the correlation is between -0.75 and -1, then there is a strong negative association.

Figure 9-4 shows the scatterplots between two variables.

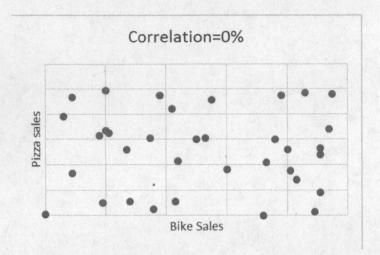

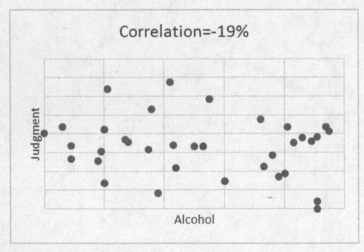

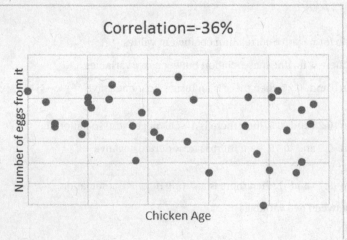

Figure 9-4. *Scatterplot examples with different values of correlation percentage, for different set of variables*

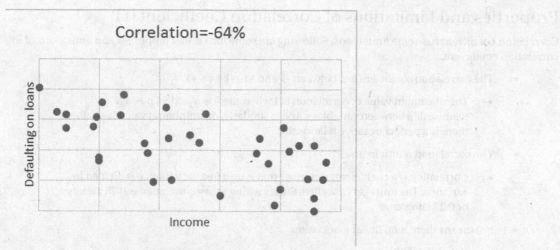

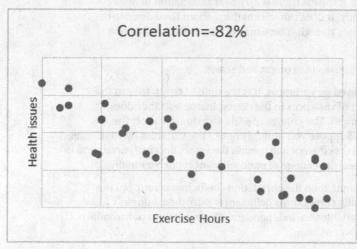

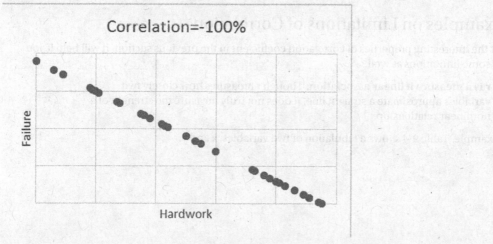

Figure 9-4. (*continued*)

Properties and Limitations of Correlation Coefficient (r)

Correlation coefficient has some limitations. Following are some interesting properties and limitations of the correlation coefficient.

- The correlation coefficient lies between -1 and +1 ; $-1 \leq r \leq +1$.

  - The maximum value of correlation is 1 when there is a perfect positive relationship between variables x and y; similarly, the minimum value is -1 when there is a perfect negative relationship.

- The correlation is unit free.

  - Correlation is a coefficient, a number that is unit free. Its variance is divided by variance. The units get cancelled. So, it's wrong to say correlation is 0.75 meters or 0.23 kilograms.

- r=0 means there is no linear association.

  - If the correlation is zero, it means there is no linear association between two variables under study. It does not tell anything about the existence of any nonlinear association, though. There may or may not be a nonlinear association.

- Correlation is independent of change of origin and scales.

  - Correlation is purely based on variances. It is the ratio of covariance in the numerator and product of variances in the denominator. Variance doesn't depend on change of origin. The change of scale also doesn't affect the correlation coefficient. Suppose you multiply one of the variables by 10 (change of scale); the correlation coefficient will remain the same. Because correlation is the ratio of two variances, the change of scale will cancel out eventually.

- If two variables are independent, then the correlation coefficient is zero, but the opposite may not be true. If the correlation coefficient is zero, then it doesn't necessarily imply that two variables are independent. There may be a polynomial (nonlinear) dependency between them.

Some Examples on Limitations of Correlation

Having seen the interesting properties of correlation coefficient in the previous section, it will help if you know about some limitations as well.

1. **r is a measure if linear association:** Though r measures how closely two variables approximate a straight line, it does not truly measure the strength of a nonlinear relationship.

As an example, Table 9-4 shows a tabulation of two variables, x and y.

Table 9-4. *Tabulation of x and y*

x	y
-31	900
-25	625
-24	576
-19	361
-13	169
-6	36
-1	1
3	9
10	100
11	121
14	196
15	225
24	576
24	576
29	841

The correlation coefficient for this example is -0.12, which is negligible. But a simple scan of the table reveals that x and y are directly related by the equation $y=x^2$, which is a nonlinear equation.

Figures 9-5 to 9-9 show some example plots representing a nonlinear relation between two variables. In all these cases, just measuring correlation coefficient is not good enough to predict the existence of an association.

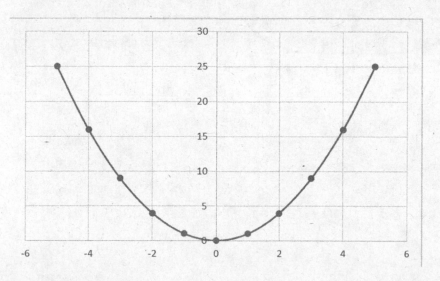

Figure 9-5. *$y=x^2$ curve*

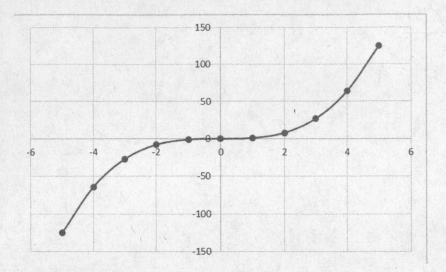

Figure 9-6. *y=x³ curve*

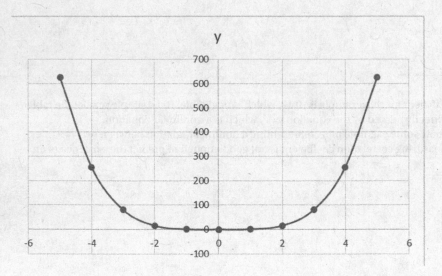

Figure 9-7. *y=X⁴ curve*

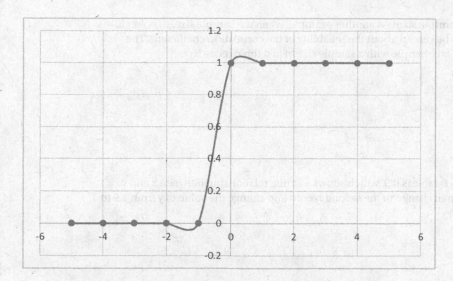

Figure 9-8. *A plot for the nonlinear function y=0 if x<0; y=1 if x>0*

Another nonlinear sine wave looks like the plot in Figure 9-9.

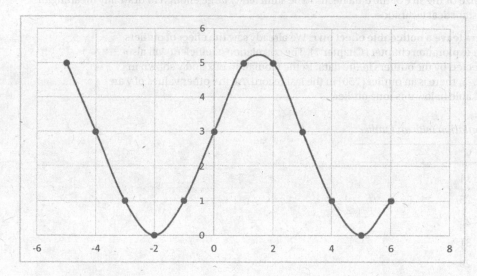

Figure 9-9. *A sine curve*

2. **Sufficient sample size:** As another example, when the sample size, n, is small, you need to be careful about the reliability of the correlation coefficient. The following is an example with a sample size of just three records:

x	y
1	2
2	2.9
3	3

The r for the previous table is 0.9, which shows a strong relationship between x and y. Now you make a small change to the second record and change the value of y from 2.9 to 4.

x	y
1	2
2	4
3	3

The r for new table suddenly becomes 0.5, a huge change from the previous value of 0.9. As a matter of fact the sample size or the size of the data needs to be sufficiently large enough to draw any meaningful insights using any statistical technique.

3. **Outliers leave a noticeable effect on r**: We already saw the effect of outliers in data exploration chapter (Chapter 7). The correlation coefficient will also get affected by the outliers in the data. In the tabulation of x and y shown in Table 9-5, there is an outlier (750) in the last record. All the other values of y are remarkable below this lone outlier.

Table 9-5. *x-y Tabulation with an Outlier*

x	y
10	14
17	25
22	23
21	31
24	29
34	60
25	19
31	35
45	45
33	38
60	50

(*continued*)

Table 9-5. (*continued*)

x	y
46	56
47	45
48	70
50	750

The correlation coefficient (r) for this relationship between x and y is 0.36, which shows a weak positive association. After removing the outlier record, r suddenly becomes 0.81, which is a strong positive association. That's a remarkable change from the previous value!

The following are two plots for this example (see Figure 9-10 and 9-11). One is with the outlier value of y, and another is without it. The extensive difference can be easily noticed. From an analyst point of view, the second plot without outliers will obviously be more meaningful.

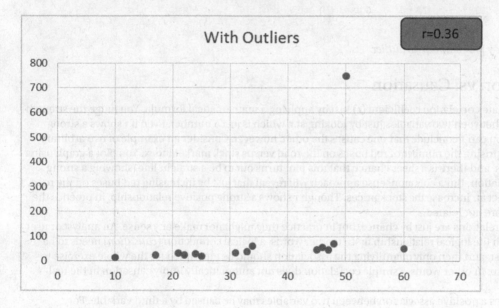

Figure 9-10. *x-y plot with an outlier*

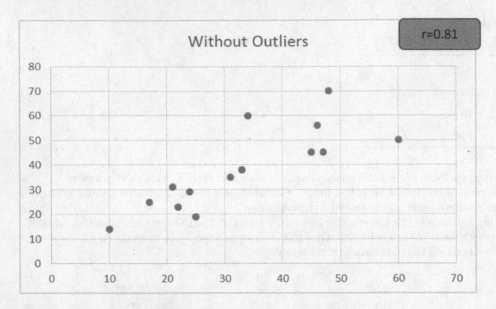

Figure 9-11. *x-y plot without an outlier*

Correlation vs. Causation

You can calculate correlation coefficient (r) just by applying a mathematical formula. You judge the strength of association between two variables just by looking at r, which is just a number. Even if r shows a strong association, you can't conclude that one causes the other, however. Consider an example of two arbitrary variables comprising the number of red buses on the road versus stock market index. You plot a graph using these variables, and there is a sheer chance that this plot turns out to be a straight line, showing a strong positive association. But a commonsense approach will reveal that just by increasing red buses on the road, you can't expect an increase the stock prices. Though r shows a strong positive relationship, in practice the two variables are not related.

Some correlations are just by chance, but in practice they might not make any sense. An analyst or user has to establish the logical relationship first. In other words, a logical connection (*causation*) needs to be established first, and then only quantifying the association should be attempted for the whole exercise to make any sense. In other words, a simple correlation does not automatically imply causation in the real-world sense.

Sometimes a positive association between two variables may be caused by a third variable. For example, say it's noticed in the real-world data that there is generally a high correlation between the number of fire accidents in a city and the sale of ice cream. Obviously, the two are not logically related. But this phenomenon is usually observed in the summer, when the atmospheric temperatures are high. And it's the temperature, the third variable, which is driving the fire accidents and ice-cream sales in this city.

The following are the types of relationships that generally exist:

- Direct cause and effect; that is, x causes y. For example, smoking causes cancer.

- Both cause and effect. Sometimes x causes y, and y might also cause x. For example, imagine KFC and McDonald's on the same block. An increase in sales of KFC causes a decrease in sales of McDonald's, while an increase in sales of McDonald's causes a decrease in sales of KFC.

- Relationships between y and x caused by a third variable, the fire accidents and ice-cream sales case discussed earlier is a good example for this case.

- Coincidental relationships or spurious correlations. The correlation between tooth brush sales and crime rates can exist only by chance. And if, by chance, you get an r value of 0.95 (95 percent correlation) between the daily milk sales in a city versus the daily sales of smartphones, it can happen only by chance.

As a conclusion, the existence of correlation doesn't assure you of any causation or logical relationship between any two variables. Correlation just quantifies the causation, which is defined or established by an analyst.

Correlation Example

Now we will discuss the case of a telecom service provider that conducts a customer satisfaction survey. The satisfaction of a customer (c_sat) depends upon many factors. The major ones are service quality, issue resolution ability, call center response, and price plans. The survey was taken with more than 1,000 randomly selected customers. The company wants to study the association of customer satisfaction with all other variables in order to streamline its future investment plans, which are intended to further improve the customer satisfaction levels.

Here is the snapshot of the survey data:

```
PROC IMPORT OUT= WORK.telecom
        DATAFILE= "C:\Users\VENKAT\Google Drive\Training\Books\Conte
nt\Regression Analysis\Telecom_Csat_correlation.csv"
        DBMS=CSV REPLACE;
    GETNAMES=YES;
    DATAROW=2;
RUN;

proc print data=telecom(obs=10);
run;
```

Table 9-6 shows the output of this code.

Table 9-6. *Output of PROC PRINT on Telecom Dataset (obs=10)*

Obs	C_sat	service	issue_resolution	call_center	price_plans
1	24	37	31	51	34
2	38	20	36	0	0
3	10	10	16	24	1
4	51	24	49	0	0
5	20	35	32	0	0
6	52	52	51	15	11
7	30	25	44	20	1
8	26	40	36	55	41
9	22	25	31	21	16
10	44	33	57	31	19

There are five variables in all: c_sat, the overall customer satisfaction score; service score; issue resolution score; call center quality score; and finally price plans score. The company tries to determine the correlation coefficient (r) of c_sat with the remaining four variables. The first four scatterplots are drawn to visualize the relation between c_sat and the other variables.

```
/*Scatterplots between c_sat and other variables*/

proc gplot data= telecom;
plot  c_sat*service;
run;

proc gplot data= telecom;
plot  c_sat*issue_resolution;
run;

proc gplot data= telecom;
plot  c_sat*call_center;
run;

proc gplot data= telecom;
plot  c_sat*price_plans;
run;
```

The previous code lines generate the output in Figures 9-12 to 9-15.

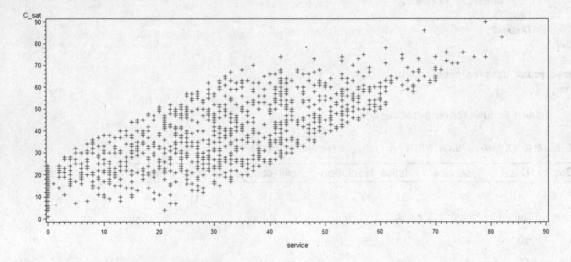

Figure 9-12. *A plot of c_sat*service*

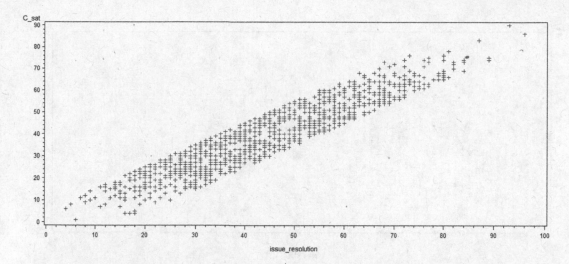

Figure 9-13. *A plot of c_sat*issue_resolution*

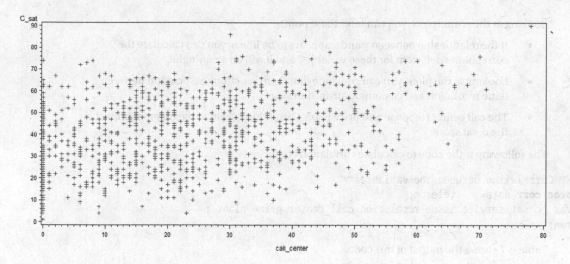

Figure 9-14. *A plot of c_sat*call_center*

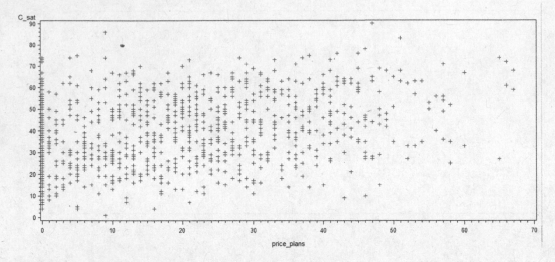

Figure 9-15. *A plot of c_sat*price_plans*

Looking at the scatterplots, you can infer these points:

- If the relationship between y and x appears to be linear, you can calculate the correlation coefficient for these variables, and it will be meaningful.

- Looking at the plots, you can conclude that there is a strong association between issue resolution and customer satisfaction scores.

- The call center response and price plans scores don't seem to have a great impact on the c_sat score.

The following is the code to calculate correlation coefficients:

```
/* Correlation between the variables*/
proc corr data=   telecom;
var  c_sat service issue_resolution call_center price_plans ;
run;
```

Table 9-7 shows the output of this code.

Table 9-7. *Output of PROC CORR on Telecom Data Set with Five Variables*

5 Variables:	c_sat service issue_resolution call_center price_plans					

Simple Statistics

Variable	N	Mean	Std Dev	Sum	Minimum	Maximum
c_sat	1009	39.42121	16.11116	39776	1.00000	90.00000
service	1009	31.81863	17.22263	32105	0	82.00000
issue_resolution	1009	44.37958	17.27128	44779	4.00000	96.00000
call_center	1009	21.15659	16.61155	21347	0	78.00000
price_plans	1009	18.46184	15.89009	18628	0	67.00000

(*continued*)

Table 9-7. (*continued*)

Pearson Correlation Coefficients, N = 1009 Prob > |r| under H0: Rho=0					
	c_sat	service	issue_resolution	call_center	price_plans
C_sat	1.00000	0.73398	0.93624	0.38258	0.33521
		<.0001	<.0001	<.0001	<.0001
service	0.73398	1.00000	0.68022	0.57126	0.46948
	<.0001		<.0001	<.0001	<.0001
issue_resolution	0.93624	0.68022	1.00000	0.34705	0.30864
	<.0001	<.0001		<.0001	<.0001
call_center	0.38258	0.57126	0.34705	1.00000	0.76341
	<.0001	<.0001	<.0001		<.0001
price_plans	0.33521	0.46948	0.30864	0.76341	1.00000
	<.0001	<.0001	<.0001	<.0001	

As expected, the output (Table 9-7) shows a high correlation (93.6%) between issue resolution score and customer satisfaction, followed by service quality score and c_sat (73.4%). There is a weak correlation between call center score and c-sat, as with price plans and c-sat. The following table takes out the correlation part separately just for the reading convenience. It is a part of Table 9-7.

	c_sat
C_sat	1.00000
service	0.73398
	<.0001
issue_resolution	0.93624
	<.0001
call_center	0.38258
	<.0001
price_plans	0.33521
	<.0001

With these results in hand, the customer care executive of the company can make a strong case to management to increase the investments in the area of issue resolution capabilities and to increase overall service quality, if the company wants to improve the customer satisfaction levels.

Correlation Summary

We have discussed correlation coefficient, its derivation, and it properties. You learned about the types of relationships and quantified the strength of association. You studied the difference between correlation and causation. You also learned about how to create a correlation matrix using SAS. You need to keep in mind that correlation is a measure of linear relationship only. There are some other measures of association such as odds ratio, Kruskal's Lambda, chi-square, and so on. They are meant to quantify some specific type association between variables. In the sections that follow, we will cover more about the quantification of the relationship between the variables.

Linear Regression

Regression is one of the most commonly used analytics techniques to study the relation between variables. At this stage, it's important to understand the difference between correlation and regression, especially when correlation is also used to study the relation between two variables. We will show an example of smartphone sales to demonstrate this difference.

Smartphone sales depend upon a number of factors such as cost, features offered, and the review ratings by critics and users. Current stock market indicators can also affect the sales because they indicate the overall state of economy and indicate the amount of spare money available to people. If the stock market is doing well, there is a chance that spare money is available to people, which can be diverted to the purchase of new smartphones. The sale of a new phone model is also dependent upon the marketing money spent by the phone manufacturer.

Table 9-8 shows a snapshot of the data collected on sales of some cell phone models and some of the variables that affect the sales.

Table 9-8. *Data for the Sales of Cell Phone Models with Variables Affecting Sales*

Model_Id	Ratings	Price	Num_new_ features	Stock_market_ind	Market_promo_ budget	Sales
M001	5	80	4	14077	96943	12737092
M002	2	412	3	13441	68123	6895692
M003	6	397	9	13540	15208	9747432
M004	3	781	6	9102	54911	7460614
M005	6	72	2	12898	8767	3670827
M006	6	55	6	9945	35473	9292936
M007	6	949	4	11315	74243	12614243
M008	6	376	4	9734	11592	3742912
M009	4	138	8	9270	6833	5700263
M010	2	667	7	8810	100020	11990734

The following is the SAS code for importing the data from a CSV file:

```
/* Importing the data into SAS*/
PROC IMPORT OUT= WORK.mobiles
           DATAFILE= "C:\Users\VENKAT\Google Drive\Training\Books\Conte
nt\Regression Analysis\Regression_mobile_phones.csv"
           DBMS=CSV REPLACE;
    GETNAMES=YES;
    DATAROW=2;
RUN;

/*Printing the contents of the data*/
proc contents data=    mobiles varnum;
run;
```

Table 9-9 shows the output for the previous code.

Table 9-9. *Output of PROC CONTENTS on Cell Phone Data Set*

Data Set Name	WORK.MOBILES	Observations	58
Member Type	DATA	Variables	7
Engine	V9	Indexes	0
Created	Tuesday, July 01, 2008 12:04:33 AM	Observation Length	56
Last Modified	Tuesday, July 01, 2008 12:04:33 AM	Deleted Observations	0
Protection		Compressed	NO
Data Set Type		Sorted	NO
Label			
Data Representation	WINDOWS_32		
Encoding	wlatin1 Western (Windows)		

Engine/Host Dependent Information	
Data Set Page Size	8192
Number of Data Set Pages	1
First Data Page	1
Max Obs per Page	145
Obs in First Data Page	58
Number of Data Set Repairs	0
Filename	C:\Users\VENKAT\AppData\Local\Temp\ SAS Temporary Files_TD7248\mobiles.sas7bdat
Release Created	9.0201M0
Host Created	W32_VSPRO

Variables in Creation Order

#	Variable	Type	Len	Format	Informat
1	Model_Id	Char	4	$4.	$4.
2	Ratings	Num	8	BEST12.	BEST32.
3	Price	Num	8	BEST12.	BEST32.
4	Num_new_features	Num	8	BEST12.	BEST32.
5	Stock_market_ind	Num	8	BEST12.	BEST32.
6	Market_promo_budget	Num	8	BEST12.	BEST32.
7	Sales	Num	8	BEST12.	BEST32.

You can see that there is data for 58 smartphone models. This is data available from the first table of the output, where the observations are shown as 58. A detailed interpretation of PROC CONTENTS output was explained in Chapter 4.

Correlation to Regression

As discussed earlier, correlation quantifies the relation between two variables. Correlation can tell up to what extent the two variables are related. It is usually determined in terms of a strong association, a weak association, or no association. If there is a strong association between two variables A and B, then it is possible to accurately predict the value of A given the value of B. In the same way, the variations in A can be predicted accurately if one knows the variations in B. In the smartphone example discussed earlier, it may be natural to conclude that smartphone sales will be dependent upon the price. Now to quantify this association, that is, to determine how closely or strongly these two variables are related, you can use correlation. The following is the SAS code to determine this relationship:

```
/* Correlation */
proc corr data=    mobiles;
var price sales;
run;
```

Table 9-10 shows the output of the previous code.

Table 9-10. Output of PROC CORR on the Cell Phone Data Set

2 Variables:			Price Sales			

Simple Statistics

Variable	N	Mean	Std Dev	Sum	Minimum	Maximum
Price	58	2121	2806	123030	51.00000	9840
Sales	58	9394688	3930243	544891907	777363	18414201

(continued)

Table 9-10. (continued)

Pearson Correlation Coefficients, N = 58 Prob > |r| under H0: Rho=0		
	Price	Sales
Price	1.00000	-0.21985
		0.0973
Sales	-0.21985	1.00000
	0.0973	

It shows that there is a weak negative association between price and mobile phone sales (r = -0.219885). Given this weak relationship, an analyst might be further interested to quantify how many sales dollars will be produced when the price is $800 or $1,500. Just correlation is not sufficient here; the analyst needs a model or an equation that can give her the sales dollars when she gives price as the input. That's where regression comes into the picture. A regression model (Figure 9-16) helps determine the value of the dependent variable (sales in this case) given the value of independent variables such as price.

Figure 9-16. *A pictorial representation of the modeling process*

The previous example shows that correlation is just a coefficient, which indicates whether the relationship between variables is weak or strong. Regression, on the other hand, is a modeling technique, which gives the actual relationship between two variables in the form of an equation. This equation can be used to get an estimated or predicted value of a dependent variable, given the values of independent variable. The dependent variable is denoted by Y, while the independent variables, which can be more than one, are denoted by X_i (i=1,2,3,4 ....).

Table 9-11 shows different substitutes given to dependent and independent variables.

Table 9-11. *Substitutes for x and y in regression modelling*

x	y
Independent	Dependent
Input	Output
Predictor	Response
Input reading	Labels
Cause	Effect
Explanatory variable	Explained variable
Regressor variable	Regressand
Controlled variable	Measured variable
Manipulated variable	Criterion variable
Feature	Experimental variable
Exposure variable	Outcome variable
RHS (Right Hand Side) variable	LHS (Left Hand Side) variable

Estimation Example

Take the example of a fast-food shop that is situated in a busy downtown area. The last 30 days of data tells you that the numbers of burgers sold on any given day are directly proportional to the number of visitors to the shop. The shop manager is interested in predicting the number of burgers sold when a given number (say 4,500) visitors come into the shop.

The following are SAS code snippets to read the historical data for the shop and find a correlation between the number of visitors and the number of burgers sold:

```
/* Importing burger sales data*/
PROC IMPORT OUT= WORK.burgers
            DATAFILE= "C:\Users\VENKAT\Google Drive\Training\Books\Content\
Regression Analysis\Burger_sales.csv"
            DBMS=CSV REPLACE;
    GETNAMES=YES;
    DATAROW=2;
RUN;
/* Correlation between visitors and burger sales*/
proc corr data=    burgers;
var visitors burgers;
run;
```

Table 9-12 shows the result of the correlation.

Table 9-12. *Output of PROC CORR on Burgers Data Set*

2 Variables: **Visitors Burgers**

Simple Statistics

Variable	N	Mean	Std Dev	Sum	Minimum	Maximum
Visitors	35	3630	1405	127066	1645	6664
Burgers	35	723.45714	298.45931	25321	293.00000	1455

Pearson Correlation Coefficients, N = 35 Prob > |r| under H0: Rho=0

	Visitors	Burgers
Visitors	1.00000	0.87226
		<.0001
Burgers	0.87226	1.00000
	<.0001	

The Pearson correlation table in Table 9-12 shows that the burgers to visitor coefficient is approximately 0.87. That shows a strong correlation. But how do you estimate the number of burgers given the number of visitors? You draw the graph shown in Figure 9-17 for the historical data.

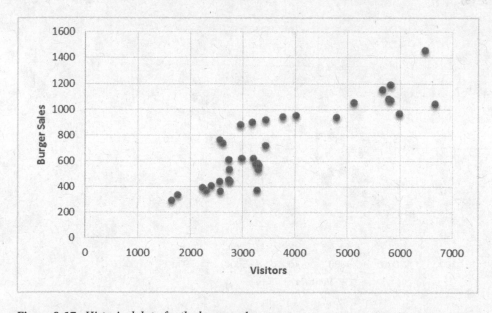

Figure 9-17. *Historical data for the burger sales*

The graph shows a strong relationship between the numbers of burgers sold and the visitor to the shop. The number of burgers sold is showing a clear increase with the increase in the number of visitors. To know the number of burgers sold when the visitors for the day are 4,500, the exercise is simple. The rough estimate comes out as 800 (Figure 9-18).

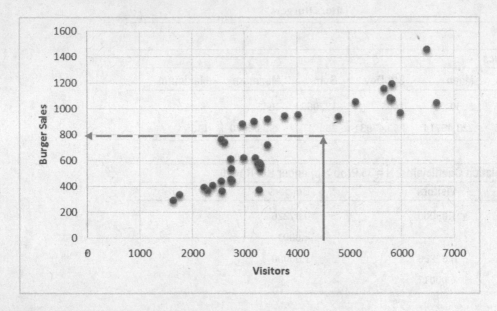

Figure 9-18. *Burger sales for number of visitor = 4500*

Just imagine a straight line passing through the core data and taking the y-axis value for an x-axis value of 4,500 (Figure 9-19).

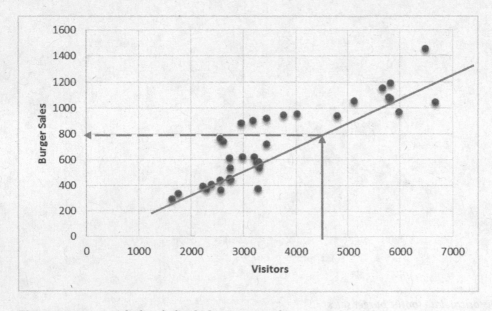

Figure 9-19. *A straight-line fit for the burger example*

The trick here is to fit the best curve, which gives the best representation of historical data. This is required to keep the estimation or prediction errors to a bare minimum. This curve (a straight line in this example) can be extended to get the estimates beyond the historical data. Regression also follows this procedure.

Simple Linear Regression

Depending upon the relationship between the dependent variable y and the independent variable x, you need to estimate the best possible fit. It can be a simple straight line or a curvilinear line, whichever best suits the data. In the burger example given in the earlier section, a simple straight-line fit solves the purpose because it appears to define the relationship with a fair amount of accuracy.

These linear regressions, where you use only one independent variable for estimations, are known as *simple linear regressions*.

As it's generally defined in high-school math, the equation of a straight line is

$$y = mx + c,$$

where *m* is the slope and *c* is the intercept.

The notations used in the regression analysis are slightly different.

$$y = \beta_0 + \beta_1 x$$

where

- β_1 is the slope.

- β_0 is the intercept.

- x = Independent variable (we provide this)

- y = Dependent variable (we estimate this)

The effort to fit a straight line to the data just takes finding the values of β_0 and β_1. Once you have these values, you can simply estimate the values of y given the values of x. For example, consider the following regression line:

$$y = 2 + 0.5x.$$

What is the value of y when x stands at 10? Simple! And this equation is called the *linear regression model*. You can have a similar regression model for the burger example also.

Regression Line Fitting Using Least Squares

In regression analysis, you try to fit a straight line that best represents the data. To do that, the values of β_0 and β_1 need be determined so that the error in the representation is at its minimum. The intercept β_0 and slope β_1 are also known as *regression coefficients*.

The following are the steps to find the regression coefficients:

1. Imagine a straight line through the data.

2. Since it can't go through all the points, you will make sure that it will be close to almost all the points.

3. In Figure 9-20, you can see the actual values (represented as circles) and the predicted values, which are along the straight line. The errors are represented as dotted lines.

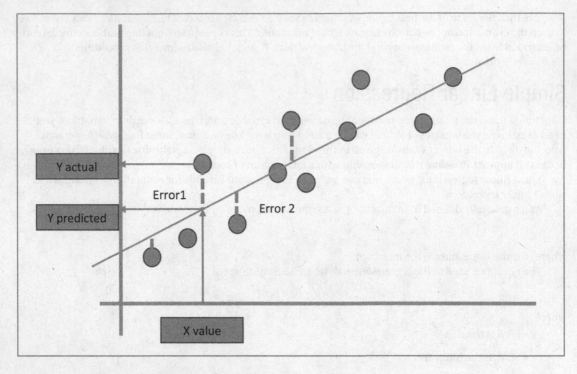

Figure 9-20. *Actual values, predicted values, and errors*

4. The best fit line will be the one with minimum overall error. Some errors are positive, while some are on the opposite side, so it may be a good idea to square the errors and sum them to find the aggregate. The best fit line will have the minimum aggregate value.

5. You need to find the regression coefficients in such a way that they will minimize the sum of squares of errors.

$$\sum e^2 = \sum (y - \hat{y})$$

$$\sum e^2 = \sum (Acutal - Predicted)^2$$

$$\sum e^2 = \sum (y - (\beta_0 + \beta_1 x))^2$$

6. You can use calculus and find the values of β_1, the slope, and β_0, the intercept that will minimize the above error function. This method is called the least square estimation method, and the reason is obvious. There's nothing to worry about since usually software like SAS will do this estimation job for you.

The Beta Coefficients: Example 1

For the burger example, you have the independent variable as the number of visitors and the dependent variable as the number of burgers sold. So, fitting a line to this data doesn't require anything other than finding the regression coefficients in the following line:

$$y = \beta_0 + \beta_1 x$$

$$Burgers = \beta_0 + \beta_1 \; visitors$$

Given the data set, you have to run the least squares algorithm and get the regression coefficient values. SAS already has built-in libraries that will run this algorithm and give the results. The following is the code:

```
/*Fitting a regression line*/
proc reg data= burgers;
model burgers=visitors;
run;
```

Here is what it means:

- proc reg is for calling regression procedure.

- The data set name is burgers.

- You need to mention the dependent and independent variables in a model statement. For example, model y=x.

Table 9-13 shows the output of this code.

Table 9-13. Output of PROC REG on Burgers Data Set

Number of Observations Read	35
Number of Observations Used	35

Analysis of Variance

Source	DF	Sum of Squares	Mean Square	F Value	Pr > F
Model	1	2304330	2304330	104.99	<.0001
Error	33	724321	21949		
Corrected Total	34	3028651			

Root MSE	148.15238	R-Square	0.7608
Dependent Mean	723.45714	Adj R-Sq	0.7536
Coeff Var	20.47839		

Parameter Estimates

Variable	DF	Parameter Estimate	Standard Error	t Value	Pr > |t|
Intercept	1	50.85364	70.25853	0.72	0.4743
Visitors	1	0.18527	0.01808	10.25	<.0001

There are so many measures and values in Table 9-13 that we will focus on the numbers that matter the most. You are looking for estimates of intercept β_0 and slope β_1. From the last table of parameter estimates, you get the regression parameters as follows:

- Intercept β_0: 50.854

- Slope, which is coefficient of x (that is, coefficient of visitors in the model equation β_1): 0.185

$$Burgers = 50.8534 + 0.185*visitors$$

You are only looking for the regression equation or regression line. So, for a day when the numbers of visitors are 4,500, the number of burgers can be estimated as follows:

$$Burgers = 50.854 + 0.185*4500$$

$$Burgers = 883.35$$

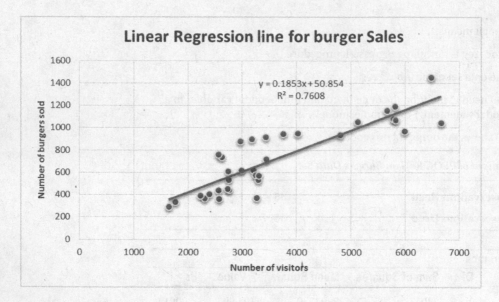

Figure 9-21. *Regression Line for the Burgers Sales Example*

We built a regression line and gave our predictions using the model. But how do we ensure that our regression model is correct and the estimates are trustworthy? How do we define goodness of fit in a model? The following section will answer this question.

How Good Is My Model?

The regression line that you just built for the burger example will give you reasonably good estimates of burgers sold, given the number of visitors. But it can still be challenged. Someone might come up with another model with the claims that their estimates are better. She might propose a new line with intercepts of 23 and a slope of 0.5. Under such conditions, how should the current model be defended? In other words, how can the accuracy of a given model be quantified? How can the errors in the estimation be calculated? We will show an example to explain this.

Table 9-14 shows some observations in the data.

Table 9-14. *Snapshot of Actual Numbers of Visitors vs. Burgers Sold*

Visitors	Burgers
3268	373
2299	371
2566	363
1759	335
3440	720

The burger numbers in Table 9-14 are the actual number of burgers sold based upon the historical data. Now you calculate the same values using the regression model, built in the earlier section, and list the values in the Burgers (Predicted) column. A good model should give estimates near to the actual values (see Table 9-15).

Table 9-15. *Predicted and Actual Values of the Burgers Sold*

Visitors	Burgers (Actual)	Burgers (Predicted)
3268	373	656
2299	371	477
2566	363	526
1759	335	377
3440	720	688

Now in the following sections, you will go on to estimate the error between predicted and actual values.

Sum of Squares of Error (SSE)

A perfect model will make sure that the estimated or predicted values are almost the same as the actual values. If they are not the same, at least they should be very close. So, the difference between the actual values denoted by y and predicted values denoted by ŷ should be insignificant at any point. Since these errors can be either positive or negative in nature, you can square and sum them to get an aggregate value. And the aggregate value (in other words, the sum of squares of deviations between actual and predicted values) should be as small as possible for a model to be categorized as good.

$$Sum\ of\ Squares\ of\ Error\ (SSE) = \sum (y_i - \hat{y}_i)^2$$

y_i is the actual value of y, and $\hat{y}_i$ is the predicted or estimated value of y.

If you want to compare two models built on the same data set, then the one with less SSE will obviously be considered better.

There no special code to produce the SSE value; by default the regression code will display SSE in the output.

```
/*Fitting a regression line*/
proc reg data= burgers;
model  burgers=visitors;
run;
```

In the output (shown Table 9-16), you can read the sum of squares of error when analyzing the variance table.

Table 9-16. *Output of PROC REG on Burgers Data Set*

Number of Observations Read	35
Number of Observations Used	35

Analysis of Variance

Source	DF	Sum of Squares	Mean Square	F Value	Pr > F
Model	1	2304330	2304330	104.99	<.0001
Error	33	724321	21949		
Corrected Total	34	3028651			

Root MSE	148.15238	R-Square	0.7608
Dependent Mean	723.45714	Adj R-Sq	0.7536
Coeff Var	20.47839		

Parameter Estimates

Variable	DF	Parameter Estimate	Standard Error	t Value	Pr > |t|
Intercept	1	50.85364	70.25853	0.72	0.4743
Visitors	1	0.18527	0.01808	10.25	<.0001

For the model that you built, the error sum of squares is 724, 321.

Total Sum of Squares

Just the error sum of squares is not sufficient to decide the goodness of the fit of a model. There is a small challenge in observing SSE in isolation. What are the limitations of SSE? You know it should be close to zero. What if SSE is greater than zero? How big is too big? If the value of dependent variables in a model is in the thousands, SSE can run even in the tens of thousands. If the values of a dependent variable are in fractions (less than 1), then SSE will naturally be less. Under these conditions, based on SSE, how do you interpret the fitness of a model? You need to consider SSE with respect to total variation within y, to determine the exact error percentage. The measure of variance in y or the total sum of squares of y is as follows:

$$Total\ sum\ of\ squares\ (SST) = \sum(y_i - \bar{y})^2$$

Where $\bar{y}$ is the mean of y, the variance in y (SST) is nothing but the squared deviation of y from the mean of y. So SSE/SST is a good measure of accuracy of the model. We can call it as the proportion of unexplained variance in y. What is the proportion of explained variation in y then? It is explained in the next section.

Sum of Squares of Regression

So, we are trying to explain the total variance in dependent variable(y) using the independent variable. Since we could not fit the perfect model or since there is no perfect fit for the data, we settled for a line with some error. So, the best model that can fit the given information will have a small error percentage. This section gives another way of looking at it.

The total variance in y is the sum of the error sum of squares, that is, the variance unexplained and regression model sum of squares (the variance explained successfully).

Total sum of squares = Sum of Squares of Error + Sum of Squares of Regression
SST = SSE + SSR
Total variance = unexplained variance + Explained variance

$$\sum (y_i - \bar{y})^2 = \sum (y_i - \hat{y}_i)^2 + \sum (\bar{y} - \hat{y}_i)$$

SST, SSE, and SSR are shown pictorially in Figure 9-22.

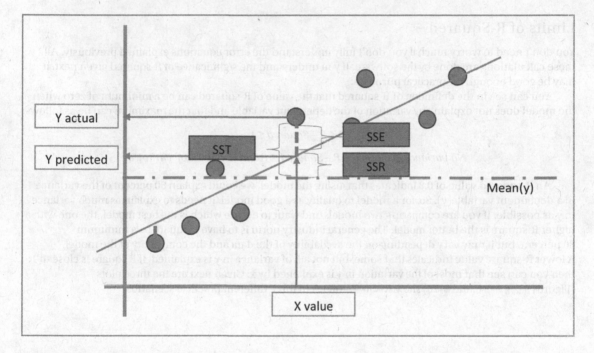

Figure 9-22. *Plot to represent SSE, SST, and SSR*

So, to have greater accuracy in the estimates, you need to have less error or you need to have a greater regression sum of squares or you need to have an error sum of squares near zero and a regression sum of squares near the total sum of squares.

So, for a best model, the value of SSE/SST should be close to zero, and the value of SSR/SST should be near one. The ratio SSR/SST is known as *coefficient of determination* or R-squared, denoted by R^2.

The Coefficient of Determination R-Square

R-square is a goodness of fit or accuracy measure. The higher the R-square, the better the model. So, the coefficient of determination is nothing, but the ratio of the variation explained to the total variation (of the dependent variable).

$$R - Squared = \frac{Sum\ of\ Squares\ of\ Regression}{Total\ Sum\ of\ Squares}$$

$$R - Squared = \frac{SSR}{SST}$$

$$R - Squared = \frac{Variance\ Explained}{Total\ Variance}$$

$$R - Squared = 1 - \frac{SSE}{SST}$$

Limits of R-Squared

You don't need to worry much if you don't fully understand the error equations explained previously. All these calculations are done by the software. If you understand the significance of R-squared given next, it may be good enough for practical purposes.

You can see in the definition of R-squared that the value of R-squared can be a minimum of zero when the model does not explain any variation of the dependent variable and that the maximum can be as follows:

$$0 \le R - Squared \le 1$$

$$No\ variance\ explained \le R - Squared \le Explained\ 100\%\ of\ variance$$

An R-squared value of 0.8 indicates that, using the model, we could explain 80 percent of the variance in a dependent variable(y). So, for a model to qualify as a good model, it needs to explain as much variance in y as possible. If you are comparing two models and want to decide which is the best model, the one with a higher R-square is the better model. The general industry norm is to have R-square be a minimum 80 percent, but it may vary depending on the availability of the data and the complexity of the model. A lower R-square value indicates that some, but not all, of variance in y is explained. If R square is close to 1, then you can see that most of the variation in y is explained by x. Given next are the three plots (Figures 9-23, 9-24, and 9-25) show R-square values in three different practical scenarios.

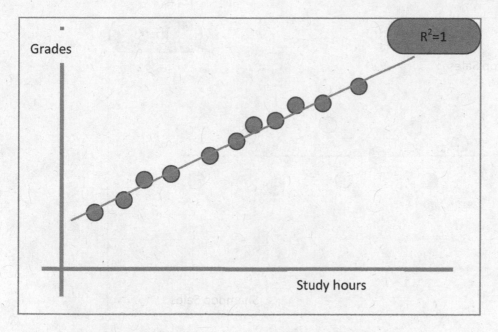

Figure 9-23. *R-squared for grades versus study hours regression model*

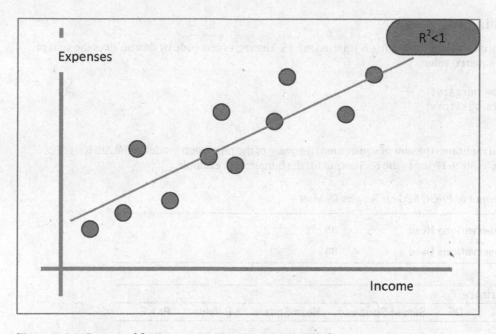

Figure 9-24. *R-squared for expenses versus income regression line*

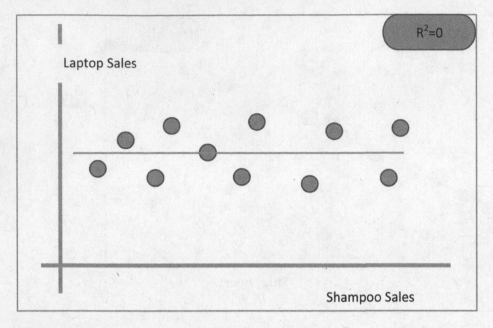

Figure 9-25. R-squared for laptop computer sales versus shampoo sales regression line

R-Squared in SAS

There is no special code for calculating R-squared in SAS. The regression code by default gives the sum of squares and R-squared value.

```
proc reg data= burgers;
model  burgers=visitors;
run;
```

SAS output mentions the sum of squares and R-square of the regression model in the Analysis of Variance Table. Table 9-17 shows the SAS output for the burger sales example.

Table 9-17. Output of PROC REG on Burgers Dataset

Number of Observations Read	35
Number of Observations Used	35

Analysis of Variance					
Source	DF	Sum of Squares	Mean Square	F Value	Pr > F
Model	1	2304330	2304330	104.99	<.0001
Error	33	724321	21949		
Corrected Total	34	3028651			

(*continued*)

Table 9-17. (*continued*)

Root MSE	148.15238	R-Square	0.7608
Dependent Mean	723.45714	Adj R-Sq	0.7536
Coeff Var	20.47839		

Parameter Estimates

Variable	DF	Parameter Estimate	Standard Error	t Value	Pr > |t|
Intercept	1	50.85364	70.25853	0.72	0.4743
Visitors	1	0.18527	0.01808	10.25	<.0001

Following are the sum of squares from the Anova table (analysis of Variance) of the output Table 9-17.
Sum of squares of regression model (SSR) =2,304,330
Sum of squares of error (SSE) =724,321
Total sum of squares (SST) =3,028,651
R-squared =0.7608, this can be verified by the R-squared actual formula

$$R - Squared = \frac{SSR}{SST}$$

The R-squared value indicates that the model's accuracy is only 76.08 percent; in other words, the predications of y (number of burgers sold) based on the input values of x (number of visitors) are done with only 76.08 percent accuracy. Is that good enough? Maybe not because an analyst looks for an R-squared value of greater than or equal to 80 percent.

Regression Assumptions

In the burger example, you assumed a linear relationship between the number of burgers sold and the number of visitors in the shop. But in actual practice this may not be the case. If the relationship is not linear in the true sense, a linear regression model should not be attempted. There are other nonlinear techniques available to do the job. The following are some assumptions under which R-squared value is valid. In some cases, you have also given the techniques to test them.

Linearity Assumption

R-squared value is calculated under the assumption that the relationship between the independent variable(x) and the dependent variable (y) is linear.

Consider the plot shown in Figure 9-26. It more or less represents the y=e^x relationship. If it's assumed that the fit here is a straight line, it will never pass through the maximum of the data points. R-squared has no meaning here.

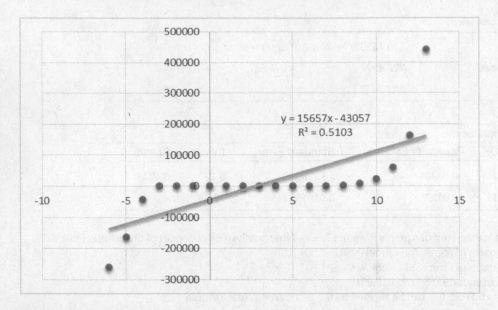

Figure 9-26. *An exponential relationship between y and x. R-squared has no meaning here*

Detection of Linearity and Fixing the Violation of Linearity

Detecting a linear relationship is fairly simple. In most cases, linearity is clear from the scatterplot. So, to find a linear fit, you can draw a scatterplot for x and y using an analytics software like SAS. If the relation between x and y is not linear, transformations such as square of x, log(x), ex, sqrt(x), and so on, can be tried. Transformation of data can be found in any standard statistics book.

We discussed creating scatterplots in correlation. The following is the code to create a scatterplot in SAS. The following code is for a sample scatterplot customer satisfaction and service in telecom data.

```
/*Scatterplots between CSAT and service score*/
proc gplot data= telecom;
plot  C_sat*service;
run;
```

Normality

R-squared is calculated under the assumption that the dependent variable y is distributed normally for each value of the independent variable x.

An easy way to understand this assumption is to look at the opposite of it. In other words, what if y is not distributed normally at each point of x? What if y is a constant for some observations of x and for rest of the data points y is distributed normally? The graph in this case might look like Figure 9-27.

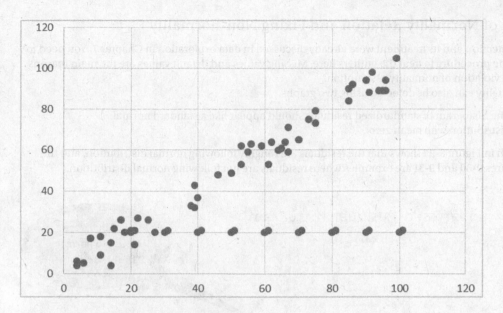

Figure 9-27. *y versus x, showing two trends*

The graph in Figure 9-27 shows that Y behaves differently when X is near multiples of 10, assuming all values are behaving normally. You can clearly see two trends in the data. There is no way you can fit one straight line that goes through the maximum points of the data. Say if you fit a linear regression line, the errors (the deviation between actual and predicted values) should follow normal distribution with the mean close to zero. Normality is a basic assumption while calculating regression coefficients. Generally, outliers are the main reason for the violation of this assumption.

The graph in Figure 9-28 shows some outliers; a model fit on this data will produce errors that are non-normal. If you try to fit a regression line on this data, it will not be valid.

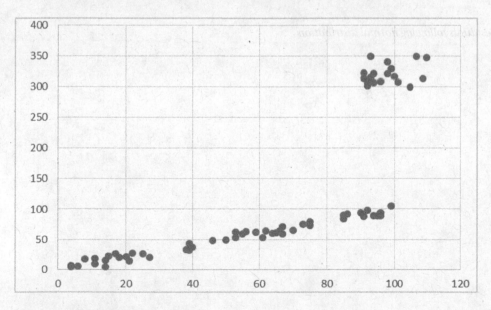

Figure 9-28. *y versus x with outliers*

Detection of Normality Relation and Fixing Non-normality

The outlier detection and its treatment were already discussed in data exploration in Chapter 7. You need to follow the same procedure to treat the outliers here. Missing values and default values are the main reasons that cause the violation of normality assumption.

The normality can also be detected using two graphs.

1. The histogram of standardized residuals should appear like a standard normal distribution with mean zero.

The graph in Figure 9-29 shows that the residuals are roughly following normal distribution, and the graphs in Figures 9-30 and 9-31 are examples where residuals are not following normal distribution.

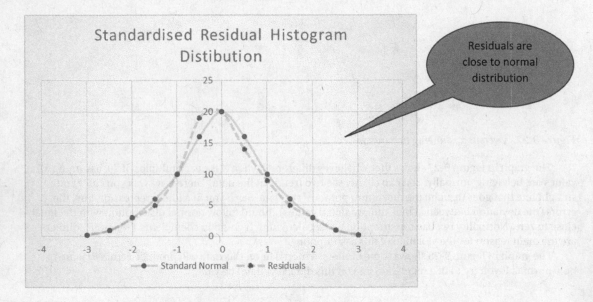

Figure 9-29. *Residuals following normal distribution*

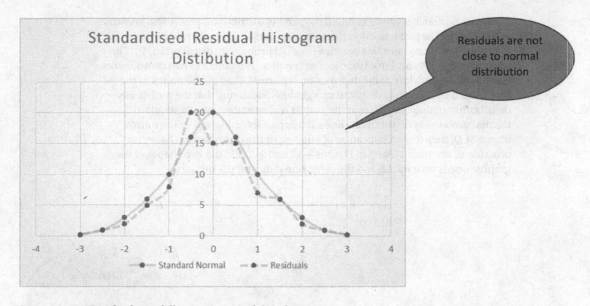

Figure 9-30. *Residuals not following a normal distribution*

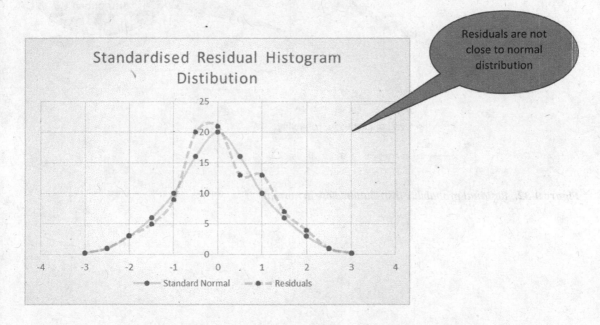

Figure 9-31. *Another example pf residuals not following a normal distribution*

2. A normal probability versus residual probability distribution plot is also known as a P-P plot. The plot basically tries to draw a perfect normal probability distribution (P) and compare it with a residual probability distribution (P). The normal probability and residual distribution plot (P-P plot) has two components, a straight line and dots. If the distribution is normal, then all the points in the P-P plot should fall close to a diagonal straight line, indicating that the probability distribution of residuals (dots in the graph) is almost the same as standard normal probability distribution (line). If this plot looks like a *D* or the mirror image of *D*, then it is an indication of outliers, or the errors are too heavy on one side of the mean. Given in Figures 9-32 and 9-33 are the two shapes of the graphs: one is acceptable, and the other one indicates the outliers.

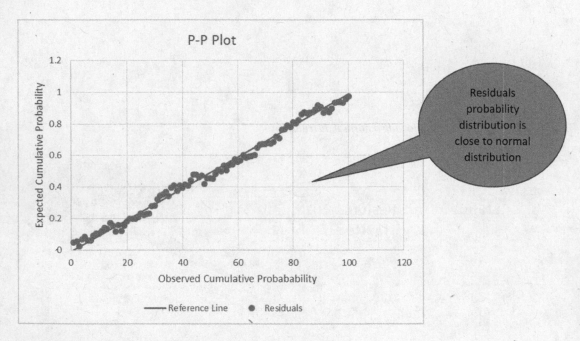

Figure 9-32. *Residual probability distribution close to normal*

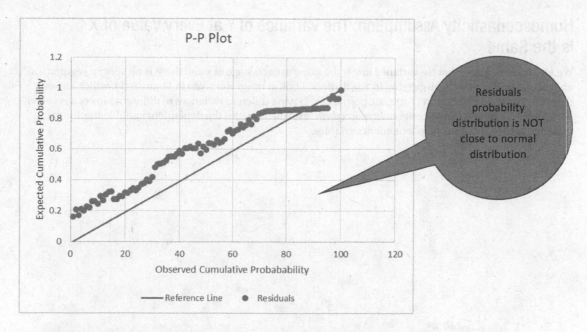

Figure 9-33. *Residual probability distribution not following a normal distribution*

To fix this normality violation, you need apply the outlier treatment or transform the variables. The outlier treatment and missing value treatment is discussed in Data exploration Chapter.

Independence Assumption: The Observations Are Independent

Independence of y is one of the most common assumptions while attempting to fit a linear regression. You can expect the values of y to depend on independent variables but not on its own previous values. For example, while analyzing sales data for one year, if the current month's sale is dependent upon the previous month's sale, then R-squared doesn't make any sense.

We are trying to predict y using x. If y dependents on its own previous y values then we are not building a robust model. The variable x will never be able to explain y variable's self-dependency.

Detection and Fixing Independence Violation

The violation of this assumption is observed mostly in time series data. Autocorrelation function (ACF) graphs can give you a clear picture about the dependencies. ACF graphs are explained in Time series analysis chapter. If there is a seasonal dependency, then you might get seasonal indices, and the data can be adjusted accordingly. If the data shows a clear trend, then some smoothing techniques might be used to correct or adjust the data. In such cases, even variable transformation using various mathematical functions can be considered.

Homoscedasticity Assumption: The Variance of Y at Every Value of X Is the Same

We are also assuming that the variance in y is the same at each stage of x and there is no special segment or an interval in X where the dispersion in Y is distinct. Look at the graph given in Figure 9-34, which shows different variance patterns in y for some points of x. Having different variances in different bins of x is called *hetroscedasticity*. It is the opposite of *homoscedasticity*, which means the dependent variable has the same variance across all points of independent variables.

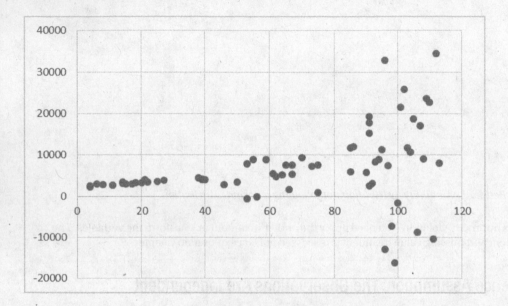

Figure 9-34. Note the different variance patterns in y for different ranges of x

In this graph, the variance in y seems to be slightly less when x is less than 40. In the range of x varying from 40 to 80, you see a slightly different variance pattern. Again, after x crossing 80, the pattern is different. In this case, it is hard to fit one regression line and expect it to pass through the maximum points of the data. As shown in the graph in Figure 9-35, the straight line fit can't be considered the best fit for the entire set of data.

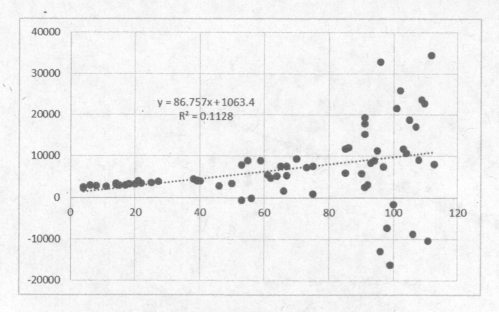

Figure 9-35. *Unsuccessfully trying to fit a straight line between y and x*

Detection of Homoscedasticity Violation and Fixing It

Again, a scatterplot between x and y gives you a fair idea of homoscedasticity. A more clever idea is to draw the residual versus predicted values (not the actual values). This gives you a picture of whether y is increasing or decreasing or whether residual variation is growing or shrinking or whether it's random around the predicted line.

The plots in Figures 9-36 and 9-37 show some examples of homoscedasticity and heteroscedasticity.

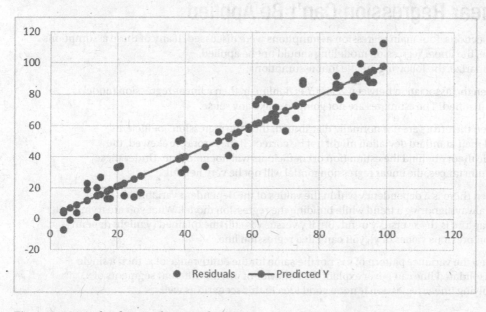

Figure 9-36. *A plot showing homoscedasticity*

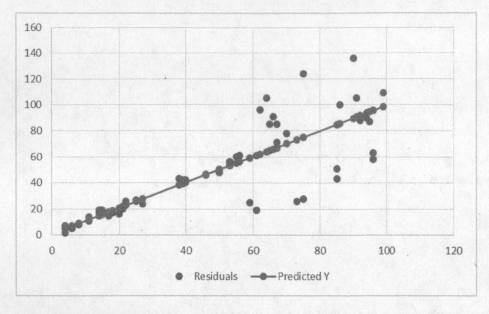

Figure 9-37. *A plot showing heteroscedasticity*

Residuals or the errors here are the difference between actual and predicted values.

The simple way to deal with this problem is to segment out the data and build different regression lines for different intervals. In general, if the first three assumptions are satisfied, then heteroscedasticity might not even exist. As a rule of thumb, first three assumptions need to be fixed before attempting to fix heteroscedasticity.

When Linear Regression Can't Be Applied

In the previous sections, four main regression assumptions were discussed. If any of these assumptions doesn't hold true, the linear regression modelling should not be applied.

Just to summarize, the following are the four assumptions:

- When the association between X and Y is nonlinear, then a linear regression model is not justified. The estimates are not going to make any sense.

- When the errors are not normally distributed, then the regression coefficients and their standard deviation might not be correct. If the errors are skewed, the basic theory behind the estimation of coefficients will not hold true. Under these circumstances, the linear regression model will not be very helpful.

- When there is a dependency within the values of the dependent variable, then you are always ignoring a trend while building the regression model. What you are trying to capture is the x versus y trend. But if y versus y is still unexplained (y might depend upon previous values of y), you can't fit a regression line.

- When the variance pattern of y is not the same for the entire range of x, then a single consolidated line can never explain the behavior of y in the different segments of x. Applying linear regression is not a good idea in this scenario as well.

Simple Regression: Example

In the beginning of the regression topic, we had discussed about the cell phone sales data. We will use regression technique to answer the questions like

- What happens to cell phone sales when the price increases or decreases?

- At a given price tag, what will be the estimated sales of a given phone model?

The following is the code for predicting sales using price.

```
proc reg data= mobiles;
model sales=price;
run;
```

Table 9-18 shows the result of this code.

Table 9-18. *PROC REG Output for Cell Phone Data Set (model sales=price)*

Number of Observations Read	58
Number of Observations Used	58

Analysis of Variance

Source	DF	Sum of Squares	Mean Square	F Value	Pr > F
Model	1	4.255764E13	4.255764E13	2.84	0.0973
Error	56	8.379103E14	1.496268E13		
Corrected Total	57	8.80468E14			

Root MSE	3868163	R-Square	0.0483
Dependent Mean	9394688	Adj R-Sq	0.0313
Coeff Var	41.17394		

Parameter Estimates

Variable	DF	Parameter Estimate	Standard Error	t Value	Pr > |t|
Intercept	1	10047931	638755	15.73	<.0001
Price	1	-307.95807	182.60286	-1.69	0.0973

Looking at the R-square, you can tell that a mere 4.8 percent of variation in sales is determined by the price of the smartphone. So, the model is not good enough for estimation purposes.

345

Now you will try to use the rest of the variables individually and see whether they will yield a good regression line. Here is the first example:

The following is the code for predicting sales using ratings.

```
proc reg data= mobiles;
model sales=Ratings  ;
run;
```

Table 9-19 shows the result of this code.

Table 9-19. *PROC REG Output for Cell Phone Data Set (model sales=ratings)*

Analysis of Variance

Source	DF	Sum of Squares	Mean Square	F Value	Pr > F
Model	1	7.974544E13	7.974544E13	5.58	0.0217
Error	56	8.007225E14	1.429862E13		
Corrected Total	57	8.80468E14			

Root MSE		3781351	R-Square	0.0906
Dependent Mean		9394688	Adj R-Sq	0.0743
Coeff Var		40.24989		

Parameter Estimates

Variable	DF	Parameter Estimate	Standard Error	t Value	Pr > |t|
Intercept	1	6850723	1186143	5.78	<.0001
Ratings	1	441766	187063	2.36	0.0217

R-square in this case is only 9.06 percent. So again, the model is not a good fit to predict the sales. The following is the code for predicting sales using number of new features.

```
proc reg data= mobiles;
model sales= Num_new_features  ;
run;
```

Table 9-20 shows the result of this code.

Table 9-20. *PROC REG output for Cell Phone Data Set (model sales= num_new_features)*

Analysis of Variance

Source	DF	Sum of Squares	Mean Square	F Value	Pr > F
Model	1	1.835206E13	1.835206E13	1.19	0.2796
Error	56	8.621159E14	1.539493E13		
Corrected Total	57	8.80468E14			

(continued)

Table 9-20. (*continued*)

Root MSE	3923637	R-Square	0.0208	
Dependent Mean	9394688	Adj R-Sq	0.0034	
Coeff Var	41.76442			

Parameter Estimates

Variable	DF	Parameter Estimate	Standard Error	t Value	Pr > |t|
Intercept	1	8163039	1240143	6.58	<.0001
Num_new_features	1	182234	166907	1.09	0.2796

R-square in this case is only 2.08 percent. Again, this is not a good fit.
The following is the code for predicting sales using stock market index.

```
proc reg data= mobiles;
model sales= Stock_market_ind   ;
run;
```

Table 9-21 shows the result of this code.

Table 9-21. *PROC REG Output for Cell Phone Data Set (model sales= stock_market_ind)*

Analysis of Variance

Source	DF	Sum of Squares	Mean Square	F Value	Pr > F
Model	1	4.726778E13	4.726778E13	3.18	0.0801
Error	56	8.332002E14	1.487857E13		
Corrected Total	57	8.80468E14			

Root MSE	3857276	R-Square	0.0537	
Dependent Mean	9394688	Adj R-Sq	0.0368	
Coeff Var	41.05805			

Parameter Estimates

Variable	DF	Parameter Estimate	Standard Error	t Value	Pr > |t|
Intercept	1	14382036	2843599	5.06	<.0001
Stock_market_ind	1	-428.98038	240.67750	-1.78	0.0801

R-square in this case is only 5.3 percent. That's not a good fit again.
The following is the code for predicting the sales using marketing budget:

```
proc reg data= mobiles;
model sales= Market_promo_budget ;
run;
```

Table 9-22 shows the result of this code.

Table 9-22. *PROC REG Output for Cell Phone Data Set (model sales= Market_promo_budget)*

Analysis of Variance

Source	DF	Sum of Squares	Mean Square	F Value	Pr > F
Model	1	1.167034E14	1.167034E14	8.56	0.0050
Error	56	7.637646E14	1.363865E13		
Corrected Total	57	8.80468E14			

Root MSE	3693055	R-Square	0.1325
Dependent Mean	9394688	Adj R-Sq	0.1171
Coeff Var	39.31003		

Parameter Estimates

Variable	DF	Parameter Estimate	Standard Error	t Value	Pr > |t|
Intercept	1	6866361	991064	6.93	<.0001
Market_promo_budget	1	38.86021	13.28462	2.93	0.0050

R-square in this case is only 13.25 percent. This is slightly better, but again it's not good enough for predictions.

Table 9-23 shows the consolidated table of R-squared values.

Table 9-23. *Table of R-squares Values for Independent Variables*

Independent Variable	R-Squared	Variance Explained	Variance Unexplained
Price	0.0483	5%	95%
Ratings	0.0906	9%	91%
Num_new_features	0.0208	2%	98%
Stock_market_ind	0.0537	5%	95%
Market_promo_budget	0.1325	13%	87%

From Table 9-23 it is evident that none of the variables could explain much of the variation in y. None of these models is sufficient enough to get a good estimate on the sales. Is there any other way to estimate sales? Yes, instead of using these variables individually, you may want to try using them together. All these independent variables together might explain the variation in y. You might want to try a regression line using several independent variables. Regression is called *simple* when there is a single independent variable, but if there are multiple independent variables, then it is called *multiple linear regression*, which is the topic of the next chapter.

Conclusion

In this chapter you learned some relatively complicated analysis measures and techniques. Until now we've used only the simple descriptive statistics, such as mean, variance, and so on. You learned that the variable association is quantified using correlation. You also learned that the prediction of a target variable using associated variable(s) is done by using simple linear regression. Throughout the section "Simple Linear Regression," we elaborated on the conceptual part.

Real-world scenarios are not simple. Several predictor variables impact one target variable so we need to use multiple liner regression there. The next chapter covers multiple linear regressions. In upcoming chapters, you will see some other non-linear regression analysis techniques, such as logistic regression.

CHAPTER 10

■ ■ ■

Multiple Regression Analysis

In Chapter 9, we discussed correlation, which is used for quantifying the relation between a pair of variables. We also discussed simple regression, which helps predict the dependent variable when an independent variable is given. In simple regression, you use a single independent variable to predict the dependent variable. This is a simplistic approach, and in practice some dependent variables may require more than one independent variable for accurate predictions. For example, can you predict the gross domestic product (GDP) of a nation by looking just at exports? The obvious answer is that it can't be done. Predicting the GDP may need several other variables, such as per-capita income, value of natural resources, national debt, and so on. Likewise, the health of an individual depends upon many variables, such as smoking or drinking habits, eating habits, job pressure, daily workouts, genetics, sleeping habits, and more.

In real-life scenarios, you can't expect a single variable to explain all the variations in a dependent variable; several independent variables are needed to predict most dependent variables in real life. That's where multiple linear regression analysis comes into the picture. One more classic example of multiple linear regression is the credit risk analysis done by banks, which we have been discussing throughout this book.

Multiple Linear Regression

As discussed, real-world problems are multivariate; in other words, most of the target variables in real life are dependent on multiple independent variables. The overall salary of an employee may depend upon her educational qualifications, years of experience, type and complexity of work, company policies, and so on. Figure 10-1 shows a few more examples.

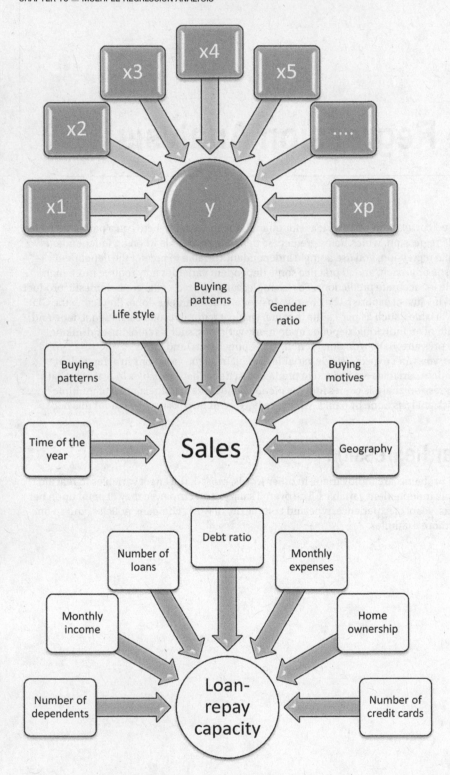

Figure 10-1. *Examples of multiple linear regression in real life*

What are the factors that a company should consider for predicting the sales of its product?

What are the factors that a bank should consider to predict the loan-repaying capabilities of its customers?

Most of the economic models to predict profits, return on investments, and so on, involve multiple variables. The multiple regression technique is not very different from that of simple regression models except that multiple variables are involved. The basic assumptions, interpretation of the regression coefficients, and R-square remain the same.

Multiple Regression Line

The simple regression line equation is $y = \beta_0 + \beta_1 x$. The multiple regression line equation is as follows:

$$y = \beta_0 + \beta_1 x_1 + \beta_2 x_2 + \beta_3 x_3 + \beta_4 x_4 + \beta_5 x_5 + \beta_6 x_6 \ldots\ldots + \beta_p x_p$$

where

- $\beta_1, \beta_2 \ldots\ldots \beta_p$, are the coefficient of $x_1 \ x_2 \ldots\ldots x_p$.
- β_0 is the intercept.

Here, in multiple regression, your goal is to fit a regression line between all the independent variables and the dependent variable. Consider the example of smartphone sales. Say you want to predict the sales of smartphones using independent variables such as ratings of the phone, price band of the phone, market promotions, and so on.

In a simple regression, you try to fit a regression line between y and x. Since there are only two variables involved in the process, the regression line is a simple straight line on a two-dimensional plot. A straight line involving a regression equation like $y = \beta_0 + \beta_1 x_1 + \beta_2 x_2$ will be in three dimensions, as shown in 10-2. It's obviously harder to imagine than a two-dimensional line involved in a simple linear regression.

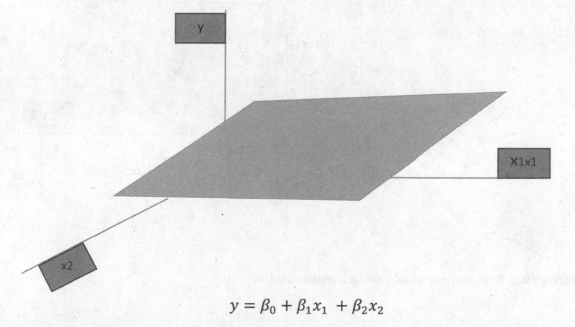

$$y = \beta_0 + \beta_1 x_1 + \beta_2 x_2$$

Figure 10-2. *Plot for Multiple Regression with Two Independent Variables*

In the smartphone sales example, the regression line will look like this:

$$sales = \beta_0 + \beta_1 price + \beta_2 ratings + \beta_3 new\ fetures + \beta_4 stock\ market + \beta_5 MarketPromo$$

Once you have the values of beta coefficients, you can create the regression line equation and use it for predicting the sales. Finding these beta coefficients is the topic of the following sections.

Multiple Regression Line Fitting Using Least Squares

The process of fitting a multiple regression line is the same as the one you studied in Simple regression chapter. The only difference is the number of independent variables and their coefficients. Here you have a number of additional independent variables ($x_1\ x_2\ ....x_p$) and you are trying to find multiple coefficients ($\beta_1, \beta_2 .... \beta_p$).

The following is what you are trying to do:

- Fitting a plane (multidimensional line) that best represents your data

- Fitting a regression line that goes through the most of the points of the data

- Representing the scattered data in the form a multidimensional line with minimal errors

Finding the values of $\beta_1, \beta_2 .... \beta_p$ and β_0, the intercept in the equation $y = \beta_0 + \beta_1 x_1 + \beta_2 x_2 .. + \beta_p x_p$. Refer to Figure 10-3.

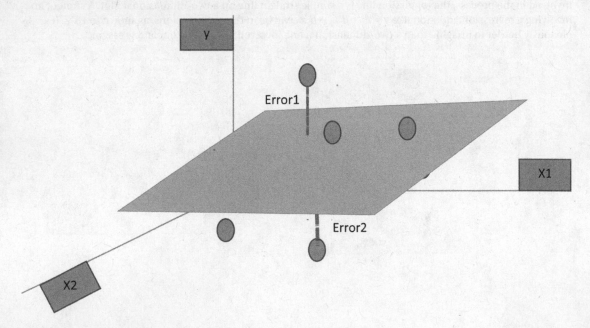

Figure 10-3. *Estimated and actual values in regression analysis*

Figure 10-3 shows the estimated regression line and the actual values seen in a three-dimensional space. The error is nothing but the distance between the two points. The estimated points on the regression plane and the actual point are denoted as circles. The dotted line is the error. As expected, you always want the difference between the actual value and estimated value to be zero.

Try to minimize the square of the errors to find the beta coefficients.

$$\text{minimize } \sum e^2 = \sum \left(y_i - \hat{y}_i \right)^2$$

$$\sum e^2 = \sum \left(y_i - (\beta_0 + \beta_1 x_1 + \beta_2 x_2 + \ldots + \beta_p x_p) \right)^2$$

Now you can use optimization techniques to find the values of $\beta_1, \beta_2, \ldots \ldots \beta_p$ and β_0 that will minimize the previous function. A line that will result from this procedure will be the best regression line for this data. This is because it makes sure that the sum of squares of errors is kept to the minimum possible value while optimizing.

Multiple Linear Regression in SAS

In SAS you need to mention the data set name and the dependent independent variable list in the model statement. SAS will do the least squares optimization and provide the beta coefficients as the result.

Example: Smartphone Sales Estimation

In the smartphone sales example, discussed earlier in Chapter simple regression chapter, you are estimating the sales using independent variables such as Price, Ratings, Num_new_features, Stock_market_ind, and Market_promo_budget. When using variables one at a time, you could not get a good model. The following (Table 10-1) is the R-squared table that shows the variables and variation that they explain in the dependent variable. Please refer to simple regression for mobile phone sales example in the previous chapter.

Table 10-1. *R-square table*

Independent Variable	R-Squared	Variance Explained	Variance Unexplained
Price	0.0483	5%	95%
Ratings	0.0906	9%	91%
Num_new_features	0.0208	2%	98%
Stock_market_ind	0.0537	5%	95%
Market_promo_budget	0.1325	13%	87%

You now fit a multiple regression line to this problem. The following is the SAS code to do this:

```
/* Multiple Regression line for Smartphone sales example*/
proc reg data= mobiles;
model sales= Ratings Price Num_new_features Stock_market_ind Market_promo_budget;
run;
```

Here is the code explanation:

- `PROC REG` is for calling regression procedure.

- The data set name is `mobiles`.

- You need to mention the dependent and independent variables in the model statement; `sales` is the dependent variable.

Table 10-2 shows the SAS output of this code.

Table 10-2. *Output of PROC REG on Mobiles Data Set*

Number of Observations Read	58
Number of Observations Used	58

Analysis of Variance

Source	DF	Sum of Squares	Mean Square	F Value	Pr > F
Model	5	7.408043E14	1.481609E14	55.16	<.0001
Error	52	1.396636E14	2.685839E12		
Corrected Total	57	8.80468E14			

Root MSE	1638853	R-Square	0.8414
Dependent Mean	9394688	Adj R-Sq	0.8261
Coeff Var	17.44446		

Parameter Estimates

Variable	DF	Parameter Estimate	Standard Error	t Value	Pr > |t|
Intercept	1	493778	1520007	0.32	0.7466
Ratings	1	651375	90775	7.18	<.0001
Price	1	-1833.67081	127.24677	-14.41	<.0001
Num_new_features	1	547167	85234	6.42	<.0001
Stock_market_ind	1	-105.38867	106.01603	-0.99	0.3248
Market_promo_budget	1	100.92891	8.38744	12.03	<.0001

The output has all the regression coefficient estimates. The coefficient of all the independent variables is mentioned against them in the final Parameter Estimates table. The intercept is 493,778. The regression coefficient for ratings is 651,375. For price, it is -1,833.67, and for a number of new features, it stands at 547,167.

The final multiple regression line equation is as follows:

$$Sales = 493778 + 651375 * Ratings - 1833.67 * Price + 547167 * Num new features$$
$$- 105.39 * Stock market ind + 100.93 * Market\_promo\_budget$$

From the previous coefficients and their signs, you can infer that the mobile phone sale number is directly proportional to customer ratings, number of new features added, and market promotion efforts. You can also observe that the sale is inversely proportional to price band and stock market indices (negative coefficient in the regression equation). The relation of stock market index with sales appears slightly against the intuition, but that is what has been happening historically. The stock market index has a negative effect on smartphone sales. If the stock market increases, smartphone sales show a decline, and vice versa for a bearish market.

With this equation, if you have the values of sample ratings, price band, number of new features, stock market indicator, and estimated promotion level of the product, you can get a fair idea about the sales. But unfortunately that's not all. There are some questions to be answered.

How good is the value of predicted sales? Is this line a good fit to this data? Can you use this regression line for predictions? What is the error or accuracy that you can guarantee using this regression line? How much of the variance in sales is explained by all the variables? How is the goodness of fit measured? Can you use R-squared here as well?

Goodness of Fit

The *goodness of fit*, or the validation measure, here is R-squared. The R-square value is nothing but the variance explained in the dependent variable by all the variables put together. Multiple regression analysis of variance (ANOVA) tables give the values of the sum of the squares of errors.

Just for your convenience, we have reproduced the analysis of variance tables from the mobile phone sales example in Table 10-3.

Table 10-3. *Analysis of Variance Tables from the Mobile Phone Sales Example*

Analysis of Variance (ANOVA)

Source	DF	Sum of Squares	Mean Square	F Value	Pr > F
Model	5	7.408043E14	1.481609E14	55.16	<.0001
Error	52	1.396636E14	2.685839E12		
Corrected Total	57	8.80468E14			

Root MSE		1638853	R-Square	0.8414	
Dependent Mean		9394688	Adj R-Sq	0.8261	
Coeff Var		17.44446			

Since the sales numbers are in the millions, the variance (sum of squares) is in multiples of millions. Since all the values are on the same scale, it is easy to interpret the ANOVA table. The total sum of squares or the measure of overall variance in Y is around 8.8 units, whereas the error sum of squares is 1.4 units. The rest is the sum of squares, which is 7.4 units. The R-squared value is 84.14 percent.

Please refer to Chapter 9 on simple regression to learn more about R-square and the sum of squares.

This regression line looks like a fairly accurate model. All the variables together are explaining almost 84 percent of the variation in y (Figure 10-4), though individually none of the variables could explain more than 10 percent of the variation in y. You can confidently go ahead and use this model for predictions.

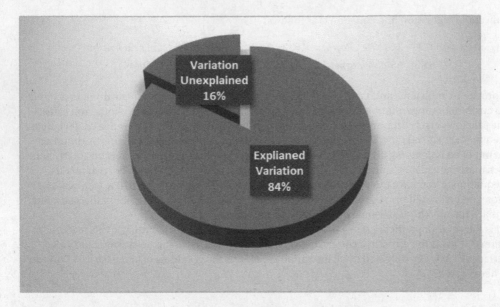

Figure 10-4. *Explained and unexplained variations in mobile phone example*

Three Main Measures from Regression Output

In the SAS output, you have some other measures also mentioned apart from just the regression coefficients, ANOVA table, and R-square. You may not need to know all of them, but the following are the statistical measures that you must know:

1. The F-test, F-value, P-value of F-test

2. The T-test, T-value, P-value of T-test

3. The R-squared and adjusted R-square

The F-test, F-value, P-value of F-test

You perform the F-test to see whether the model is relevant enough in the current context. This is the test to see the overall fitness of the model. The following questions are answered using this test:

*   Is the model relevant at all?

*   Is there at least one variable that explains some variation in the dependent variable (y)?

As an example, consider building an analytics model for smartphone sales. Some trivial independent variables such as number of buses in the city, average number of pizzas ordered, and average fuel consumption will definitely not help in predicting the values of smartphone sales. If you use more and more variables like this, the whole model will be irrelevant. The obvious reason is that the independent variables in the model are not able to explain the variation in the dependent variable y. The F-test determines this. It tells whether there is at least one variable in the model that has a significant impact on the dependent variable.

While discussing the multiple regression line earlier in this chapter, we talked about beta coefficients. The F-test uses them to test the explanatory or prediction power of a model. If at least one beta coefficient is not equal to zero, you can infer the presence of one independent variable in the model that has a significant effect on the variation of the dependent variable y. If more than one beta is nonzero, it indicates the presence of multiple independent variables that can explain the variations in y.

H_0:

- $\beta_1 = \beta_2 = \beta_3 = \beta_4 = \beta_5 = \beta_6 \ldots\ldots = \beta_p = 0$, which is equivalent to
  $\beta_1 = 0 \ and \ \beta_2 = 0 \ and \ \beta_3 = 0 \ and \ \beta_4 = 0 \ldots\ldots and \ \beta_p = 0.$

- If all betas are zero, it means that the model is insignificant and it has no explanatory prediction power.

H_1:

- $\beta_1 \neq 0 \ or \ \beta_2 \neq 0 \ or \ \beta_3 \neq 0 \ or \ \beta_4 \neq 0 \ or \ \beta_5 \neq 0 \ or \ \beta_6 \neq 0 \ or \ldots \beta_p \neq 0$, which is equivalent to saying that at least one $\beta \neq 0$.

For an explanation of H_0 and H_1, please refer to Testing of Hypothesis in Chapter 8.

As discussed earlier, even if one beta is positive, you can infer that the model has some explanatory power.

To test the previous hypothesis, you rely on the F-statistic and calculate the F-value and corresponding P-value. Based on the P-value of F-test, you accept or reject the null hypothesis. The P-value of the F-test will finally decide whether you should consider or reject the model. The F-value is calculated using the explained sum of squares and regression sum of squares. Refer to Testing of Hypothesis in Chapter 8 for more about P-values and F-tests.

If the P-value of the F-test is less than 5 percent, then you reject the null hypothesis H_0. In other words, you reject the hypothesis that the model has no explanatory power, and that means the model can be used for useful predictions. If the F-test has a P-value greater than 5 percent, then you don't have sufficient evidence to reject H_0, which may force you to accept the null hypothesis. In simple terms, look at the P-value of the F-test. If it is greater than 5 percent, then your model is in trouble; otherwise, there is no reason to worry.

Example: F-test for Overall Model Testing

In the smartphone sales example, if you want to see what the overall model fit is, you have to look at the P-value of the F-test in the output.

There will be no change in the following code:

```
/* Multiple Regression line for Smartphone sales example*/
proc reg data= mobiles;
model sales= Ratings Price Num_new_features Stock_market_ind Market_promo_budget;
run;
```

Table 10-4 gives the output of this code. In the ANOVA table, you can see the F-value and P-value of the F-test.

Table 10-4. Output of PROC REG on Mobiles Data Set

Number of Observations Read	58
Number of Observations Used	58

Analysis of Variance

Source	DF	Sum of Squares	Mean Square	F Value	Pr > F
Model	5	7.408043E14	1.481609E14	55.16	<.0001
Error	52	1.396636E14	2.685839E12		
Corrected Total	57	8.80468E14			

Root MSE	1638853	R-Square	0.8414
Dependent Mean	9394688	Adj R-Sq	0.8261
Coeff Var	17.44446		

Parameter Estimates

Variable	DF	Parameter Estimate	Standard Error	t Value	Pr > |t|
Intercept	1	493778	1520007	0.32	0.7466
Ratings	1	651375	90775	7.18	<.0001
Price	1	-1833.67081	127.24677	-14.41	<.0001
Num_new_features	1	547167	85234	6.42	<.0001
Stock_market_ind	1	-105.38867	106.01603	-0.99	0.3248
Market_promo_budget	1	100.92891	8.38744	12.03	<.0001

From the Analysis of Variance table, you can see that the F-value is 55.16, and the P-value of the F-test is less than 0.0001, which is way below the magic number of 5 percent. So, you can reject the null hypothesis. Now it's safe to say that the model is meaningful for any further analysis. Generally, when R-squared is high, you see that the model is significant. In some cases where there are too many junk or insignificant variables in the model, R-squared and F-test behave differently. Please refer to the "R-squared and Adjusted R-Square (Adj R-sq)" section later in this chapter.

F-test: Additional Example

Let's look at the simple regression example, price versus sales regression line.

```
proc reg data= mobiles;
model sales=price;
run;
```

Table 10-5 shows the result of the previous code.

Table 10-5. *Output of PROC REG on Mobiles Data Set (model sales=price)*

Number of Observations Read	58
Number of Observations Used	58

Analysis of Variance

Source	DF	Sum of Squares	Mean Square	F Value	Pr > F
Model	1	4.255764E13	4.255764E13	2.84	0.0973
Error	56	8.379103E14	1.496268E13		
Corrected Total	57	8.80468E14			

Root MSE	3868163	R-Square	0.0483
Dependent Mean	9394688	Adj R-Sq	0.0313
Coeff Var	41.17394		

Parameter Estimates

Variable	DF	Parameter Estimate	Standard Error	t Value	Pr > |t|
Intercept	1	10047931	638755	15.73	<.0001
Price	1	-307.95807	182.60286	-1.69	0.0973

The F-value is 2.84, and the P-value of the F-test is 9.73 percent (0.0973). Since the P-value is greater than 5 percent, you don't have enough evidence to reject H_0, and you need to accept that the model is insignificant. In such a case, there is no point in looking further at the R-squared value. How much variance in y is explained by the model.

The T-test, T-value, and P-value of T-test

The T-test in regression is used to test the impact of individual variables. As discussed earlier, the F-test is used to determine the overall significance of a model. For example, in a regression model with ten independent variables, all may not be significant. Even if some insignificant independent variables are removed, there will not be any substantial effect on the predicting power of the model. The explanatory power of the model will largely remain the same after removal of less impacting variables. In mathematical terms, the beta coefficients of insignificant independent variables can be considered as zero.

For example, take the model for predicting smartphone sales. Do you need to keep all the independent variables? Are all relevant or significant? What if a junk variable such as the number of movie tickets sold is also included in the model? This model with a junk variable included might pass the F-test because there is at least one variable in the model that is significant. Consider the following questions, which are faced by every analyst while working on regression models:

- How do you identify the variables that have no impact on the model outcome (the dependent variable)?

- How do you test the impact of each individual variable?

- How do you test the effect of dropping or adding a variable in the model?

- Is there any way you can filter all the insignificant variables and have only significant variables in the model?

A T-test on regression coefficients will answer all these questions.
The null hypothesis on the T-test on regression coefficients is as follows:
H_0:

- $\beta_i = 0$, which is equivalent to saying that coefficient of the independent variable (x_i) is equal to zero. It also means that the variable x_i has no impact on dependent variable y, and you can drop it from the model. The statistical inferences would be that the R-squared value will not get affected by dropping x_i, and there will be no corresponding change in y for in every unit change in x_i.

H_1:

- $\beta_i \neq 0$, which would mean that the variable x_i has some significant impact on the dependent variable and dropping x_i would badly effect your model. In statistical terms, the R-squared value will drop significantly by dropping x_i, and there will be some corresponding change in y for every unit change in x_i.

To test the previous hypothesis, you calculate the T-statistic, which is also known as the T-value. Based on the T-value, you accept or reject the null hypothesis. In other words, the T-value will finally decide whether you should consider or reject a variable in the model. The T-value is calculated using the beta coefficients against a normal distribution of beta coefficients with a zero mean. Like the F-test, here also you look at the P-value of the T-statistic to determine the impact of a variable on the model.

If the P-value of a T-test is less than 5 percent, then you reject null hypothesis H_0. In other words, you reject the premise that the variable has no impact. That means the variable is useful, and it has some explanatory power or some minimal impact on the dependent variable. If the T-test has a P-value greater than 5 percent, then you don't have sufficient evidence to reject H_0, which may force you to accept the null hypothesis (and eventually remove that variable from the model). In simple terms, look at the P-value of the T-test; if it is greater than 5 percent, then your variable is in trouble and you may need drop it. If the P-value less than 5 percent, there is no reason to worry, and you can keep the variable in your model and proceed with further analysis.

Example: T-test to Determine the Impact of Independent Variables

In the smartphone sales example, if you want to determine the impact of each independent variable, you need to look at the P-value of the T-tests in the output. Because there are five independent variables, you will have five T-tests, five T-values, and five P-values for the T-tests. There will be one more P-value for intercept in the Parameters Estimate table, which is not given much importance.

There will be no change in the following code:

```
/* Multiple Regression line for Smartphone sales example*/

proc reg data= mobiles;
model sales= Ratings Price Num_new_features Stock_market_ind Market_promo_budget;
run;
```

Table 10-6 shows the output of this code.

Table 10-6. *Output of PROC REG on Mobiles Data Set*

Number of Observations Read	58
Number of Observations Used	58

Analysis of Variance

Source	DF	Sum of Squares	Mean Square	F Value	Pr > F
Model	5	7.408043E14	1.481609E14	55.16	<.0001
Error	52	1.396636E14	2.685839E12		
Corrected Total	57	8.80468E14			

Root MSE	1638853	R-Square	0.8414
Dependent Mean	9394688	Adj R-Sq	0.8261
Coeff Var	17.44446		

Parameter Estimates

Variable	DF	Parameter Estimate	Standard Error	t Value	Pr > |t|
Intercept	1	493778	1520007	0.32	0.7466
Ratings	1	651375	90775	7.18	<.0001
Price	1	-1833.67081	127.24677	-14.41	<.0001
Num_new_features	1	547167	85234	6.42	<.0001
Stock_market_ind	1	-105.38867	106.01603	-0.99	0.3248
Market_promo_budget	1	100.92891	8.38744	12.03	<.0001

The P-value of the T-test for all the independent variables except stock market are less than 5 percent. So, the null hypothesis will be rejected in the case of Ratings, Price, Num_new_features, and Market_promo_budget. It would mean that these variables have some impact on the dependent variable, in other words, smartphone sales, whereas stock market (Stock_market_ind) has no impact on sales of smartphones. You come to this conclusion by looking at the P-value of the T-test for the stock market variable, which is greater than 5 percent (0.05), and there is not enough evidence to reject H_0. You may have to accept the hypothesis that the stock market indicator has no impact on the sales.

Verifying the Impact of the Individual Variable

In the previous example, you made two inferences based on T-tests.

- The stock market has no impact on the dependent variable (y, the smartphone sales). This variable is not explaining a significant portion of variations in y.

- The rest of the independent variables, Ratings, Price, Num_new_features, and Market_promo_budget, have significant impact on the sale of smartphones.

Let's validate the first inference. You will drop the variable Stock_market_ind and rebuild the model. The R-squared value including this variable is 84.14 percent. Let's see the R-squared value of the model excluding this variable.

```
/* Multiple Regression model without Stock_market_ind */

proc reg data= mobiles;
model sales= Ratings Price Num_new_features  Market_promo_budget;
run;
```

Table 10-7 shows the output of this code.

Table 10-7. *Output of PROC REG on Mobiles Data Set (Excluding stock_market_ind Variable)*

Number of Observations Read	58
Number of Observations Used	58

Analysis of Variance

Source	DF	Sum of Squares	Mean Square	F Value	Pr > F
Model	4	7.381502E14	1.845375E14	68.72	<.0001
Error	53	1.423178E14	2.685241E12		
Corrected Total	57	8.80468E14			

Root MSE	1638671	R-Square	0.8384
Dependent Mean	9394688	Adj R-Sq	0.8262
Coeff Var	17.44252		

Parameter Estimates

Variable	DF	Parameter Estimate	Standard Error	t Value	Pr > |t|
Intercept	1	-804177	778126	-1.03	0.3061
Ratings	1	648559	90721	7.15	<.0001
Price	1	-1855.97681	125.23876	-14.82	<.0001
Num_new_features	1	546713	85224	6.42	<.0001
Market_promo_budget	1	103.06985	8.10532	12.72	<.0001

The important metric to note here is the R-squared value, which is 83.84 percent. The change in R-square is insignificant (earlier 84.14 percent). You can now safely conclude that the stock market index has nothing to do with smartphone sales.

Let's validate the second inference. You will drop the variable ratings and rebuild the model. The R-squared value including this variable is 83.84 percent. Let's see the R-squared value excluding this variable.

```
/* Multiple Regression model without Ratings */
proc reg data= mobiles;
model sales= Price Num_new_features  Market_promo_budget  ;
run;
```

Table 10-8 shows the output for this code.

Table 10-8. *Output of PROC REG on mobiles Variable with Price, Num_new_features, and Market_promo_budget*

Number of Observations Read	58
Number of Observations Used	58

Analysis of Variance

Source	DF	Sum of Squares	Mean Square	F Value	Pr > F
Model	3	6.009139E14	2.003046E14	38.69	<.0001
Error	54	2.795541E14	5.176928E12		
Corrected Total	57	8.80468E14			

Root MSE	2275286	R-Square	0.6825
Dependent Mean	9394688	Adj R-Sq	0.6649
Coeff Var	24.21886		

Parameter Estimates

Variable	DF	Parameter Estimate	Standard Error	t Value	Pr > |t|
Intercept	1	2072232	924776	2.24	0.0292
Price	1	-1623.99248	167.95479	-9.67	<.0001
Num_new_features	1	570435	118243	4.82	<.0001
Market_promo_budget	1	106.23586	11.23739	9.45	<.0001

The new R-square value is 68.25 percent, which has dropped significantly from 83.84 percent. This forces you to accept the fact that the ratings variable has significant impact on the model outcome. By removing this variable, you are losing a considerable amount of explanatory power of the model.

In fact, the R-square value will change significantly for all the other high-impact variables. In the following text, we have repeated the regression model with different combinations of independent variables. Tables 10-9 through 10-11 list the R-square for these models.

The following is the SAS code to build a regression model with a dependent variable sales and independent variables as ratings, num_new_features, and market_promo_budget.

```
proc reg data= mobiles;
model sales= Ratings Num_new_features Market_promo_budget ;
run;
```

Table 10-9 lists the output of this code.

Table 10-9. R-square with Ratings, Num_new_features, and Market_promo_budget

Root MSE	3681899	R-Square	0.1686
Dependent Mean	9394688	Adj R-Sq	0.1224
Coeff Var	39.19129		

The following is the SAS code to build a regression model with a dependent variable sales and independent variables as ratings, price, and market_promo_budget.

```
proc reg data= mobiles;
model sales= Ratings Price Market_promo_budget ;
run;
```

Table 10-10 lists the output of this code.

Table 10-10. R-square with Ratings, price and Market_promo_budget

Root MSE	2163769	R-Square	0.7129
Dependent Mean	9394688	Adj R-Sq	0.6969
Coeff Var	23.03184		

The following is the SAS code to build a regression model with a dependent variable sales and independent variables as ratings, price, and num_new_features.

```
proc reg data= mobiles;
model sales= Ratings Price Num_new_features ;
run;
```

Table 10-11 lists the output of this code.

Table 10-11. R-square with Ratings, Price and Num_new_features

Root MSE	3267501	R-Square	0.3452
Dependent Mean	9394688	Adj R-Sq	0.3088
Coeff Var	34.78030		

Figure 10-5 shows the change in the R-squared value when the variable is in the model and the R-squared value when the variable is dropped from the model.

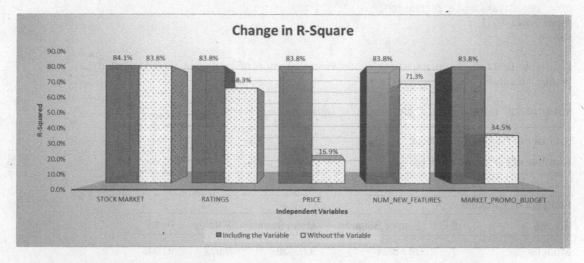

Figure 10-5. *Change in R-square values*

Looking at the previous results, you should have no hesitation in believing the T-test results about the impact of independent variables.

The R-squared and Adjusted R-square (Adj R-sq)

In the regression output you might have already observed the adjusted R-squared value near the R-squared. It is known as *adjusted R-square*. To understand the importance of the adjusted R-square, let's look at an example.

Import the sample regression data (sample_regression.csv); it is a simple simulated data set with some independent variables along with a dependent variable. Build three models on this data and note the R-square and adjusted R-square values for the three models. The following are the specifications for building the models:

- First model with independent variables: x1, x2, x3

- Second model with independent variables: x1, x2..x6

- Third model with all the variables: x1, x2...x8

```
/* Importing Sample Regression Data Set*/

PROC IMPORT OUT= WORK.sample_regression
        DATAFILE= "C:\Users\VENKAT\Google Drive\Training\Books\Content\
Multiple and Logistic Regression\sample_regression.csv"
        DBMS=CSV REPLACE;
    GETNAMES=YES;
    DATAROW=2;
RUN;
```

Model 1: Y vs. x_1, x_2, x_3

```
/* Regression on Sample Regression Data set*/
proc reg data=sample_regression;
model y=x1 x2 x3;
run;
```

Table 10-12 shows the result of this code. Please take note of the R-squared and adjusted R-squared values.

Table 10-12. *Output of PROC REG on simple_regression Data Set (model y=x1 x2 x3)*

Number of Observations Read	13
Number of Observations Used	13

Analysis of Variance

Source	DF	Sum of Squares	Mean Square	F Value	Pr > F
Model	3	10.81884	3.60628	9.02	0.0045
Error	9	3.59765	0.39974		
Corrected Total	12	14.41649			

Root MSE	0.63225	R-Square	0.7504
Dependent Mean	1.90077	Adj R-Sq	0.6673
Coeff Var	33.26281		

Parameter Estimates

Variable	DF	Parameter Estimate	Standard Error	t Value	Pr > |t|
Intercept	1	-2.28399	0.94357	-2.42	0.0386
x1	1	-0.20151	0.31685	-0.64	0.5406
x2	1	0.00276	0.00094042	2.93	0.0167
x3	1	0.37819	0.15635	2.42	0.0387

Model 2: Y vs. $x_1, x_2, \ldots x_6$

```
proc reg data=sample_regression;
model y=x1 x2 x3 x4 x5 x6;
run;
```

Table 10-13 shows the result of this code. Please take note of the R-squared and adjusted R-squared values.

Table 10-13. *Output of PROC REG on simple_regression Data Set (model y=x1 x2 x3 x4 x5 x6)*

Number of Observations Read	13
Number of Observations Used	13

Analysis of Variance

Source	DF	Sum of Squares	Mean Square	F Value	Pr > F
Model	6	10.97854	1.82976	3.19	0.0917
Error	6	3.43796	0.57299		
Corrected Total	12	14.41649			

Root MSE	0.75696	R-Square	0.7615	
Dependent Mean	1.90077	Adj R-Sq	0.5231	
Coeff Var	39.82402			

Parameter Estimates

Variable	DF	Parameter Estimate	Standard Error	t Value	Pr > |t|
Intercept	1	-3.75324	3.13791	-1.20	0.2768
x1	1	-0.25185	0.40396	-0.62	0.5559
x2	1	0.00277	0.00114	2.42	0.0521
x3	1	0.39208	0.20288	1.93	0.1015
x4	1	0.01814	0.06899	0.26	0.8013
x5	1	0.03867	0.12932	0.30	0.7750
x6	1	0.02680	0.07431	0.36	0.7307

Model 3: Y vs. $x_1, x_2, \ldots x_8$

```
proc reg data=sample_regression;
model y=x1 x2 x3  x4 x5 x6 x7 x8;
run;
```

Table 10-14 shows the result of the previous code. Please take note of the R-squared and adjusted R-squared values.

Table 10-14. *Output of PROC REG on simple_regression Data Set (model y=x1 x2 x3 x4 x5 x6 x7 x8)*

Number of Observations Read	13
Number of Observations Used	13

Analysis of Variance

Source	DF	Sum of Squares	Mean Square	F Value	Pr > F
Model	8	11.78372	1.47296	2.24	0.2276
Error	4	2.63278	0.65819		
Corrected Total	12	14.41649			

Root MSE	0.81129	R-Square	0.8174
Dependent Mean	1.90077	Adj R-Sq	0.4521
Coeff Var	42.68228		

Parameter Estimates

Variable	DF	Parameter Estimate	Standard Error	t Value	Pr > |t|
Intercept	1	12.10783	15.28092	0.79	0.4725
x1	1	0.09820	0.54286	0.18	0.8652
x2	1	0.00138	0.00185	0.75	0.4975
x3	1	0.39448	0.24749	1.59	0.1862
x4	1	0.05691	0.09191	0.62	0.5693
x5	1	0.12081	0.15736	0.77	0.4855
x6	1	-0.11731	0.17154	-0.68	0.5316
x7	1	-0.06604	0.12435	-0.53	0.6235
x8	1	-0.16166	0.17645	-0.92	0.4114

Analysis on Three Models

Figure 10-6 shows the R-squared and adjusted R-squared results from the previous three models.

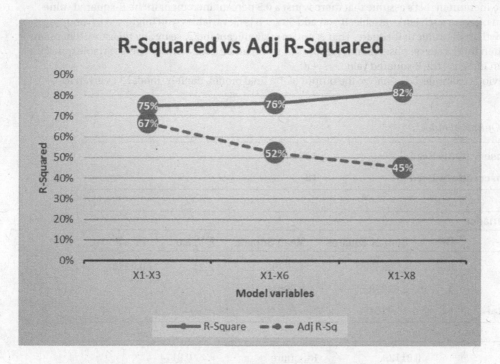

Figure 10-6. *R-squared and adjusted R-squared values for the three models*

When you build the model with just the three variables x1, x2, and x3, the R-square and adjusted R-squared values are very close. As you keep adding more and more variables, the gap between the R-square and adjusted R-squared values widens. Though R-square is a good measure to explain the variation in y, there is a small challenge in its formula.

Limitations of R-Squared

The following are the limitations:

- The R-squared will either increase or remain the same when adding a new independent variable to the model. It will never decrease unless you remove a variable.

- Once R-square reaches a maximum point with a set of variables, then it will never come down by adding another independent variable to the set. There may be some minute upward improvements in the R-square value. Even if the newly added variable has no impact on the model outcome, there can be some marginal improvements in the R-squared value.

- Here is an easy way to understand this feature: R-squared is the total amount of variation explained by the list of independent variables in the model. If you add any new junk independent variable, a variable that has no impact or relation with dependent variable, the R-square still might increase slightly, but it will never decrease.

It is favorable (to an analyst) that R-squared increases when you have a decent or high-impact on dependent variable. But what happens to the R-squared value if a junk or trivial variable is added to the model? R-squared will not show any significant increase when you add junk variables. But still there will be a small positive increment. Let's assume that there is just a 0.5 percent increment in the R-squared value for the addition of every such junk variable. If you add 50 such junk variables, you might see a whopping 25 percent growth in the value of R-square. That is not an insignificant increase by any measure. It happens particularly when there is fewer observations (records or rows) in the data. Adding a new variable quickly impacts, and an increase the R-squared value is seen.

In the previous example, let's analyze the output of the final model, namely, model 3 as given in Table 10-15.

Table 10-15. *Output of model 3*

Number of Observations Read	13
Number of Observations Used	13

Analysis of Variance

Source	DF	Sum of Squares	Mean Square	F Value	Pr > F
Model	8	11.78372	1.47296	2.24	0.2276
Error	4	2.63278	0.65819		
Corrected Total	12	14.41649			

Root MSE	0.81129	R-Square	0.8174
Dependent Mean	1.90077	Adj R-Sq	0.4521
Coeff Var	42.68228		

Parameter Estimates

Variable	DF	Parameter Estimate	Standard Error	t Value	Pr > |t|
Intercept	1	12.10783	15.28092	0.79	0.4725
x1	1	0.09820	0.54286	0.18	0.8652
x2	1	0.00138	0.00185	0.75	0.4975
x3	1	0.39448	0.24749	1.59	0.1862
x4	1	0.05691	0.09191	0.62	0.5693
x5	1	0.12081	0.15736	0.77	0.4855
x6	1	-0.11731	0.17154	-0.68	0.5316
x7	1	-0.06604	0.12435	-0.53	0.6235
x8	1	-0.16166	0.17645	-0.92	0.4114

Here are some observations from the previous result:

- There are just 13 records in the data. It's a small sample, so there are few observations to analyze.

- Most of the variables have absolutely no impact on the dependent variable as per the T-test. The P-value for all the variables in the T-test is greater than 5 percent.

- Even the F-test tells you that the overall model is insignificant because the P-value of the F-test is 22.76 percent, which is greater than the magic number of 0.5 percent.

- Still the R-square is 81.74 percent.

This is the point we want to make here. The R-square value from model 1 to model 2 jumped from 75 percent to 76 percent, and then finally for model 3 it went on to become 82 percent. The R-squared will further increase if you add some more variables to model 3. So, you need to be careful while inferring anything based on R-squared values.

A combination of fewer observations and many independent variables is a highly vulnerable situation in regression analysis. It is like cheating ourselves by adding junk independent variables and feeling thrilled about increments in R-square. But on the ground, the whole model itself may be junk.

This behavior of R-squared establishes the need for a different measure that can give a reliable measure of the predicting power of any given model. Adjusted R-squared does exactly that.

Adjusted R-squared

Adjusted R-squared is derived from R-squared only. What are the expectations from this new measure? The adjusted R-squared is expected to be as follows:

- A measure that will give an idea about the explained variations in a model.

- A measure that will penalize the model when a junk variable is added.

- A measure that will take into account both the number of observations and the number of independent variables in a model.

- A measure that will increase only when a significant or impactful independent variable is added. It will decrease for the addition of any junk or trivial variable to the model.

Adjusted R-squared meets all these expectations. The following is the formula for adjusted R-squared:

$$\bar{R}^2 = R^2 - \frac{k-1}{n-k}(1-R^2)$$

where

- R^2 is the usual R-squared.

- n is the number of records.

- k is the number of independent variables.

Adjusted R-squared adds a penalty to R-squared for every new junk variable added. It shows an increase only for the addition of meaningful or high-impact variables to the model.

Let's revisit the R-squared and adjusted R-squared comparison chart; see Figure 10-7.

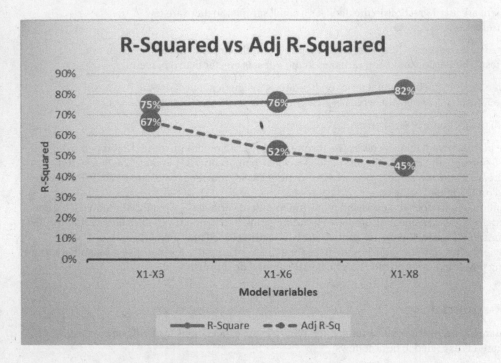

Figure 10-7. *R-squared and adjusted R-squared for the three models*

Looking at the adjusted R-squared values, you can conclude that only three variables (x1 to x3) are sufficient for the model to realize its maximum potential. As you go on adding new variables (x4 to x8), adjusted R-squared is showing a decrease. It indicates that all the incoming variables from x4 until x8 are junk variables, and they have no impact on y.

Adjusted R-square: Additional Example

For the smartphone sales example, you will first observe the full model results; in other words, the regression model will use all five independent variables.

```
/* Smartphone sales R-Squared and Adj-R Squared*/
proc reg data= mobiles;
model sales= Ratings Price Num_new_features Stock_market_ind Market_promo_budget;
run;
```

Table 10-16 gives the result of this code.

Table 10-16. *Output of PROC REG on Mobiles Data Set*

Number of Observations Read	58
Number of Observations Used	58

Analysis of Variance

Source	DF	Sum of Squares	Mean Square	F Value	Pr > F
Model	5	7.408043E14	1.481609E14	55.16	<.0001
Error	52	1.396636E14	2.685839E12		
Corrected Total	57	8.80468E14			

Root MSE	1638853	R-Square	0.8414	
Dependent Mean	9394688	Adj R-Sq	0.8261	
Coeff Var	17.44446			

Parameter Estimates

Variable	DF	Parameter Estimate	Standard Error	t Value	Pr > |t|
Intercept	1	493778	1520007	0.32	0.7466
Ratings	1	651375	90775	7.18	<.0001
Price	1	-1833.67081	127.24677	-14.41	<.0001
Num_new_features	1	547167	85234	6.42	<.0001
Stock_market_ind	1	-105.38867	106.01603	-0.99	0.3248
Market_promo_budget	1	100.92891	8.38744	12.03	<.0001

The R-squared and adjusted R-squared values are almost near but not the same, shown here:

R-square	0.8414
Adj R-Sq	0.8261

There is a small difference between the values if R-square and adjusted R-sq. Why is this? In earlier sections, using T-tests, you found that the stock market has no significant impact on the sales of smartphones. Maybe this is the variable that is triggering that small difference between the R-squared and adjusted R-squared values. You will remove the stock market indicator variable and rebuild the model. You will observe that the R-squared and adjusted R-squared values are getting closer.

```
proc reg data= mobiles;
model sales= Ratings Price Num_new_features Market_promo_budget;
run;
```

Table 10-17 shows the output for this code.

Table 10-17. *Output of PROC REG on Mobiles Data Set*

Number of Observations Read	58
Number of Observations Used	58

Analysis of Variance

Source	DF	Sum of Squares	Mean Square	F Value	Pr > F
Model	4	7.381502E14	1.845375E14	68.72	<.0001
Error	53	1.423178E14	2.685241E12		
Corrected Total	57	8.80468E14			

Root MSE	1638671	R-Square	0.8384
Dependent Mean	9394688	Adj R-Sq	0.8262
Coeff Var	17.44252		

Parameter Estimates

Variable	DF	Parameter Estimate	Standard Error	t Value	Pr > |t|
Intercept	1	-804177	778126	-1.03	0.3061
Ratings	1	648559	90721	7.15	<.0001
Price	1	-1855.97681	125.23876	-14.82	<.0001
Num_new_features	1	546713	85224	6.42	<.0001
Market_promo_budget	1	103.06985	8.10532	12.72	<.0001

The R-squared and adjusted R-squared got slightly closer. But there is still a difference without any insignificant variable in the model. This is because of the fewer records. If there are large records and no junk variables, the R-square and adjusted R-squared values can be the same.

When Can You Be Dependent Upon Only R-squared?

You can use R-squared in these cases:

- It is safe to consider adjusted R-squared all the time. In fact, it's recommended.

- If the sample size is adequately large compared to the number of independent variables and if all the independent variables have significant impact, then you may consider R-squared value as the goodness of fit measure. Otherwise, adjusted R-squared is the right measure.

Multiple Regression: Additional Example

The SAT is a standardized test widely used for college admissions in the United States. Let's build a regression model for predicting the SAT scores based on students' high-school marks. Students may need a high proficiency in subjects such as mathematics, general knowledge (GK), science, and general aptitude at the high-school level in order to get high scores in SAT. After collecting some historical data for close to 100 students, say you try to build a model that will predict the SAT score based on the scores obtained in the high-school exams.

The following is the code to import the data:

```
/* importing SAT exam Data*/
PROC IMPORT OUT= WORK.sat_score
            DATAFILE= "C:\Users\VENKAT\Google Drive\Training\Books\Content\
Multiple and Logistic Regression\SAT_Exam.csv"
            DBMS=CSV REPLACE;
     GETNAMES=YES;
     DATAROW=2;
RUN;
```

The following is the code for printing the snapshot of the data file. The data file contains some historical data on the actual marks obtained in the high-school exams.

```
proc print data= sat_score(obs=10) ;
run;
```

In Table 10-18, you can see that there are four independent variables called general knowledge (GK), aptitude (apt), mathematics (math), and science, along with one dependent variable, SAT. The idea is to fit a model using these four independent variables to predict the SAT score.

```
/* Predicting SAT score using rest of the Variables*/

proc reg data=sat_score;
model SAT=General_knowledge Aptitude Mathematics Science;
run;
```

Table 10-18. *Four independent variables to predict the SAT score*

Obs	General knowledge	Aptitude	Mathematics	Science	SAT
1	73	71	74	73	144
2	93	90	102	97	186
3	89	94	97	98	182
4	96	93	115	110	208
5	73	68	87	83	157
6	53	49	36	38	89
7	69	73	71	67	131
8	47	48	55	55	101
9	87	89	66	66	155
10	79	76	83	78	158

Table 10-19 is the output for this code.

Table 10-19. *Output of PROC REG on sat_score Data Set*

Number of Observations Read	96
Number of Observations Used	96

Analysis of Variance

Source	DF	Sum of Squares	Mean Square	F Value	Pr > F
Model	4	99039	24760	2621.54	<.0001
Error	91	859.47379	9.44477		
Corrected Total	95	99899			

Root MSE	3.07323	R-Square	0.9914
Dependent Mean	155.96875	Adj R-Sq	0.9910
Coeff Var	1.97042		

Parameter Estimates

Variable	DF	Parameter Estimate	Standard Error	t Value	Pr > |t|
Intercept	1	-2.07199	2.13436	-0.97	0.3342
General_ knowledge	1	1.16697	0.10003	11.67	<.0001
Aptitude	1	-0.13479	0.09683	-1.39	0.1673
Mathematics	1	-0.11081	0.09887	-1.12	0.2653
Science	1	1.09532	0.09689	11.30	<.0001

Here are observations from the previous output:

- As expected, the analyst would be curious to look at the R-squared value to see whether the model is a good fit. In other words, how much of the variation in the target variable (the SAT score) is explained by independent variables (marks in four high-school subjects)? However, you should first look at the F-value or the P-value of the F-test, which tells whether the model is a significant one. If the P-value is greater than 5 percent, then there is no need to go further down through the output. In that case, you stop at the F-test and say the model is insignificant. If the P-value of the F-test is less than 5 percent, then there is at least one variable that is significant, which means the model may have some impact. In this model, the P-value of the F-test is less than 5 percent; in fact, it is less than 0.0001, as is evident from the output (Analysis of Variance table). The model looks significant, and you can take a further look at other measures like R-squared and T-test.

- The R-squared and adjusted R-squared values are almost same at around 99 percent, which is a really good sign. So, the overall model is explaining almost 99 percent of variations in Y. In other words, if you know a student's marks in aptitude, GK, science, and mathematics, you can precisely predict her SAT score.

- Now let's take a look at the impact of each variable (Table 10-20).

Table 10-20. *The Parameter Estimate Table from the Output of PROC REG on sat_score Data Set*

Parameter Estimates

Variable	DF	Parameter Estimate	Standard Error	t Value	Pr > |t|
Intercept	1	-2.07199	2.13436	-0.97	0.3342
General_knowledge	1	1.16697	0.10003	11.67	<.0001
Aptitude	1	-0.13479	0.09683	-1.39	0.1673
Mathematics	1	-0.11081	0.09887	-1.12	0.2653
Science	1	1.09532	0.09689	11.30	<.0001

From the Parameters Estimates table (Table 10-20), you observe that mathematics and aptitude do not have any significant effect on the dependent variable (SAT score). You will remove these two variables and rebuild the model with just two variables: science and general knowledge (GK).

```
/* Predicting SAT score using two variables only*/

proc reg data=sat_score;
model SAT=General_knowledge Science ;
run;
```

Table 10-21 shows the output for this code.

Table 10-21. *Output of PROC REG on sat_score Data Set*

Number of Observations Read	96
Number of Observations Used	96

Analysis of Variance

Source	DF	Sum of Squares	Mean Square	F Value	Pr > F
Model	2	99012	49506	5190.20	<.0001
Error	93	887.06581	9.53834		
Corrected Total	95	99899			

(*continued*)

Table 10-21. (*continued*)

Root MSE	3.08842	R-Square	0.9911
Dependent Mean	155.96875	Adj R-Sq	0.9909
Coeff Var	1.98015		

Parameter Estimates

Variable	DF	Parameter Estimate	Standard Error	t Value	Pr > |t|
Intercept	1	-2.58369	2.12087	-1.22	0.2262
General knowledge	1	1.03366	0.03375	30.63	<.0001
Science	1	0.98970	0.01800	54.97	<.0001

From the previous output, you can observe the following:

- The p-value of the F-test is less than 0.0001, which is much less than the required magic number of 5 percent, so the model is significant.

- The R-squared and adjusted R-squared values are close to 100 percent.

- The P-values of the T-tests show that all the variables have significant impact on y. For both GK and science, the P-values are less than 0.0001, which again is much superior to the required condition of a P-value less than 5 percent.

- You can go ahead and use the model for predictions.

Some Surprising Results from the Previous Model

In the same SAT score example, when you consult a domain expert, you come to know that math and aptitude are two important "should have" skills to get good scores on the SAT. The domain expert tells you from her experience that the SAT scores of students are dependent on the marks obtained by them in math and aptitude at the high-school level. But the model is telling the opposite story.

Let's build a model using mathematics and aptitude scores alone. If they are really not impacting the SAT scores, then you should see all negative results in F-tests, R-squared values, and T-tests.

```
proc reg data=sat_score;
model SAT=Mathematics Aptitude ;
run;
```

Table 10-22 shows the output of this code.

Table 10-22. *Output of PROC REG on sat_score Data Set*

Number of Observations Read	96
Number of Observations Used	96

Analysis of Variance

Source	DF	Sum of Squares	Mean Square	F Value	Pr > F
Model	2	96693	48347	1402.62	<.0001
Error	93	3205.59628	34.46878		
Corrected Total	95	99899			

Root MSE	5.87101	R-Square	0.9679
Dependent Mean	155.96875	Adj R-Sq	0.9672
Coeff Var	3.76422		

Parameter Estimates

Variable	DF	Parameter Estimate	Standard Error	t Value	Pr > |t|
Intercept	1	-0.45932	4.01646	-0.11	0.9092
Mathematics	1	1.04297	0.03360	31.04	<.0001
Aptitude	1	0.95657	0.06204	15.42	<.0001

Here are some observations from the previous output:

- *F-test*: The P-values are less than 5 percent, so the model is significant. It's strange and surprising.

- *R-squared and adjusted R-squared*: This is 96 percent, which utterly surprising and contrary to what was expected.

- *T-tests* : The P-values of both the variable are less than 5 percent; this is again a shocking result.

- Earlier mathematics and aptitude had negative coefficients; now they have positive coefficients with the same historical data file. The results this time are very much in agreement with what the domain expert suggested.

A model, on the same historical data with all four subjects (xi), showed that mathematics and aptitude scores have no impact at all. Another model, on the same data with two subjects, is showing completely contrary results. Why are mathematics and aptitude insignificant in the presence of science and GK?

- Why did removing science and GK from the model affect the impact of mathematics and aptitude? In generic terms, why did removing some variables affect the impact of other variables, without affecting overall model predictive power?

- Are these independent variables related in some manner? Is there any interrelation between these independent variables that is causing these changes in T-test results?

Let's forget about the dependent variable for some time and observe the intercorrelation between the independent variables. Here is the code:

```
proc corr data=sat_score;
var General_knowledge Aptitude Mathematics Science ;
run;
```

Table 10-23 shows the output of this code.

Table 10-23. *Output of PROC REG on sat_score Data Set*

Variables:	General knowledge Aptitude Mathematics Science

Simple Statistics

Variable	N	Mean	Std Dev	Sum	Minimum	Maximum
General knowledge	96	79.85417	12.23023	7666	46.00000	97.00000
Aptitude	96	79.91667	12.18944	7672	46.00000	101.00000
Mathematics	96	76.68750	22.50488	7362	6.00000	125.00000
Science	96	76.80208	22.92579	7373	3.00000	124.00000

Pearson Correlation Coefficients, N = 96 Prob > |r| under H0: Rho = 0

	General knowledge	Aptitude	Mathematics	Science
General knowledge	1.00000	0.96323	0.64142	0.64078
		<.0001	<.0001	<.0001
Aptitude	0.96323	1.00000	0.60461	0.60748
	<.0001		<.0001	<.0001
Mathematics	0.64142	0.60461	1.00000	0.98977
	<.0001	<.0001		<.0001
Science	0.64078	0.60748	0.98977	1.00000
	<.0001	<.0001	<.0001	

We've formatted the same correlation table for better readability (Table 10-24).

Table 10-24. *Correlation Table for Sat Score Example*

Pearson Correlation Coefficients, N = 96				
Prob > |r| under H0: Rho = 0				
	General knowledge	**Aptitude**	**Mathematics**	**Science**
General knowledge	96%	64%		64%
Aptitude			60%	61%
Mathematics				99%
Science				

The correlation between GK and aptitude is 96 percent; the correlation between mathematics and science is 99 percent. This should be the reason why the mathematics and aptitude variables were insignificant in the presence of science and GK, and they had significant impact without them. This phenomenon of interdependency is called *multicollinearity*. It is not just the pairwise correlation between two variables, but an independent variable can depend on any number of independent variables. This multicollinearity can lead to many false inferences and absurd results. An analyst needs to take this multicollinearity challenge seriously.

Multicollinearity Defined

Multicollinearity is a phenomenon where you see a high interdependency between the independent variables. When you refer to multicollinearity, you talk about independent variables (xi) only. The dependent variable y is nowhere in the picture. Multicollinearity is not just about the relation between a pair of variables and a given independent variable set. Sometimes multiple independent variables together might be related to another independent variable. So, any significant relation or association in the independent variables set is considered as multicollinearity (Figure 10-8).

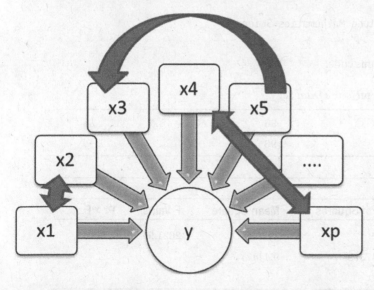

Figure 10-8. *Multicollinearity*

Why Is Multicollinearity a Problem? The Effects of Multicollinearity

In the SAT score example discussed earlier, you saw some indications of what multicollinearity can lead to. Here is a list of challenges that multicollinearity can create:

- Let's discuss the significance of beta coefficients. In a model like $y = 3X1 + 6.5X2 - 0.3X3$, what does each coefficient indicate? A beta coefficient is nothing but the increment in dependent variable (y) for every unit change in independent variable, keeping all other independent variables constant. In this regression line, x1 has a coefficient of 3; it indicates that y increases by 3 units for every unit change in x1, keeping x2 and x3 constant. But what if there is multicollinearity? What if x2 is dependent upon x1? A regression coefficient has no meaning in this case. The increment in y for every unit change in x1 is 3, keeping x2 and x3 constant. But this condition of keeping x2 and x3 constant will not hold good if x2 is dependent upon x1. In other words, x2 also changes because of the change in x1. This is a multicollinearity effect.

- In the presence of multicollinearity, the coefficients that are coming out of the regression model are not stable. Sometimes the coefficients and even their signs might be misleading. In the case of multicollinearity, the regression coefficients will have a high standard deviation, which means that even for small changes in the data (observations), the changes in the regression coefficients may be abnormally high. Sometimes with a small change in the data, the coefficient signs might change. In other words, if a variable shows a positive impact with one set of data, with a small change in the data, it might show a negative impact.

- With multicollinearity in place, the T-test results are not trustworthy. You can't really look at T-test's P-value and make a decision about the impact of any independent variable.

Let's again take a look at the SAT exam's regression model.

```
/* Predicting SAT score using rest of the four variables. General_knowledge, Aptitude,
Mathematics, and Science */

proc reg data=sat_score;
model SAT=General_knowledge Aptitude Mathematics Science;
run;
```

Table 10-25 shows the output for this code.

Table 10-25. *Output of PROC REG on sat_score Data Set*

Number of Observations Read	96
Number of Observations Used	96

Analysis of Variance

Source	DF	Sum of Squares	Mean Square	F Value	Pr > F
Model	4	99039	24760	2621.54	<.0001
Error	91	859.47379	9.44477		
Corrected Total	95	99899			

(continued)

Table 10-25. (*continued*)

Root MSE	3.07323	R-Square	0.9914
Dependent Mean	155.96875	Adj R-Sq	0.9910
Coeff Var	1.97042		

Parameter Estimates

Variable	DF	Parameter Estimate	Standard Error	t Value	Pr > |t|
Intercept	1	-2.07199	2.13436	-0.97	0.3342
General knowledge	1	1.16697	0.10003	11.67	<.0001
Aptitude	1	-0.13479	0.09683	-1.39	0.1673
Mathematics	1	-0.11081	0.09887	-1.12	0.2653
Science	1	1.09532	0.09689	11.30	<.0001

In this example, here are the false implications because of multicollinearity:

- You can see that mathematics and aptitude are negatively impacting the SAT score. In other words, if the mathematics score increases, then the SAT score decreases. If the aptitude score increases, then the SAT score decreases.

- Mathematics and aptitude have no impact on SAT score.

- General knowledge has a higher impact than aptitude and mathematics.

Let's make a small change in the data file and observe the corresponding changes in the beta coefficients. Ideally, if the model is stable, there should be minimal changes in all the coefficient estimates. As shown in Table 10-26, you will change the mathematics score from 102 to 60 in the second row.

Table 10-26. *Highlighting the Changes in Mathematics Score in sat_score Data Set*

Variable	General knowledge	Aptitude	Mathematics	Science	SAT
Old Data Record	93	90	102	97	186
Updated record	93	90	**60**	97	186

Table 10-27 shows the results of the new model built with this update in the data file.

Table 10-27. *Output of PROC REG on sat_score data set with Math Score Changed*

Number of Observations Read	96
Number of Observations Used	96

Analysis of Variance

Source	DF	Sum of Squares	Mean Square	F Value	Pr > F
Model	4	99029	24757	2589.59	<.0001
Error	91	869.98577	9.56028		
Corrected Total	95	99899			

Root MSE	3.09197	R-Square	0.9913
Dependent Mean	155.96875	Adj R-Sq	0.9909
Coeff Var	1.98243		

Parameter Estimates

Variable	DF	Parameter Estimate	Standard Error	t Value	Pr > |t|
Intercept	1	-2.28287	2.15568	-1.06	0.2924
General knowledge	1	1.15481	0.10001	11.55	<.0001
Aptitude	1	-0.12461	0.09698	-1.28	0.2021
Mathematics	1	0.02462	0.06545	0.38	0.7077
Science	1	0.96503	0.06522	14.80	<.0001

The output now has one important and big change. The beta coefficient of mathematics is positive now. With just one value changed from 102 to 60, the coefficient of mathematics turned upside down. This is what we are trying to emphasize as the adverse effect of multicollinearity.

The following is another way of looking at the high standard deviation or coefficient changes in the presence of multicollinearity:

- Let's build a new model: Y vs. X1, X2, and X3. Here you have a significant multicollinearity relationship between X2 and X3.

- Assume that X3 has a near to straight line relationship with X2. And X3 is nearly equal to two times X2, which can be denoted by X3 ~ 2X2.

- Let's assume that the final model equation is Y=X1+20X2-2X2. Please note, there is a negative coefficient for X2 (-2).

Now you will try to establish that in the presence of multicollinearity, the coefficients are so unstable that they might even change their signs without affecting predictive power of the overall model (overall R-square will remain the same).

The model Y=X1+20X2-2X3

The multicollinearity X3 ~ 2X2

The model rewritten Y=X1+ 14X2+6X2-2X3

The model rewritten Y ~ X1+14X2+ 3X3-2X3 (put X2=X3/2)

The model rewritten Y ~ X1+14X2+ X3

So, with multicollinearity, the original model Y=X1+20X2-2X3 finally ends to Y ~ X1+14 X2+ X3. The coefficient of X3 has turned from negative to positive. Similarly, you can play around this model and make the coefficient almost anything. This is exactly what we are talking about—the instability in regression coefficients and high standard deviation in beta coefficient estimates.

In simple terms, if there is an existence of multicollinearity in a model, the regression coefficients can't be trusted for any meaningful analysis.

The multicollinearity challenge raises three questions:

- What are the causes of multicollinearity?

- How do you identify the existence of multicollinearity in my model?

- Once multicollinearity is identified, what is the way out? How can its effect be minimized?

What Are the Causes of Multicollinearity?

Multicollinearity as such is not a result of any mistake in your analysis. If there is some interdependency in the independent variables, there is nothing wrong from the analyst side. An analyst needs to be aware of this, and it should be taken care of when building an accurate regression model. The following are some causes of multicollinearity:

- The way data was collected might result in multicollinearity. Are you choosing all independent, nonassociated variables while collecting the data?

- Too many variables explaining the same piece of information might be one of the causes of multicollinearity. For example, the variables such as average yearly income, average tax paid, and net yearly savings might be related to each other in most cases. All these variables are explaining a person's financial position.

- Specifying the model variables inaccurately might be one more cause. For example, a variable X and its multiple are present in a model. Another example is when X and a polynomial term related to X are present in a model.

- Having too many independent variables can also result into multicollinearity. Having too many independent variables and fewer records has never been a good idea in regression modeling. Sometimes options are not available, and you need to proceed. In such cases, an analyst needs to deal with the multicollinearity challenge.

Identification of Multicollinearity

As you have seen, multicollinearity is a serious challenge for an analyst. It is caused by some known or unknown reasons (or an unknown relation between variables sometimes). Now every time you will be in a position to explain why two variables are related. Analysts need to be alert and identify multicollinearity in the model building phase; otherwise, it may lead to irrational inferences.

Multicollinearity can be identified using correlation techniques. By finding the correlation, you can see the strength of association between two variables. If there is a high correlation, it means that the model has multicollinearity. The existence of correlation is sufficient to say that there is multicollinearity. But the correlation may not be always high even if there is multicollinearity in the model. A simple correlation is not enough to identify all types of multicollinearity. Correlation is just a pairwise association measure, so you will not be able to quantify the association if one independent variable is associated loosely to two or more independent variables. (If the pairwise association is loose, multicollinearity will not be evident in correlation, and you need a different measure.) All the associated variables together can strongly explain the variation in that particular independent variable. Let's take an example where x2, x3, and x5 are related to x4 (Figure 10-9).

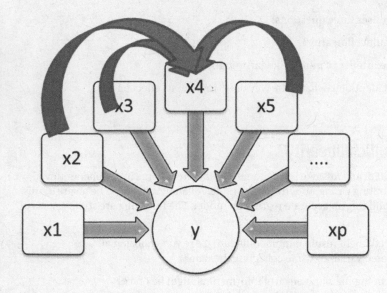

Figure 10-9. *Multicollinearity: x2, x3, and x5 are related to x4*

Forget about Y for some time and see whether the three variables x2, x3, and x5 are really impacting variations in x4. You need a measure that will quantify the relation between multiple variables. How do you get an idea of the combined impact of several independent variables on a dependent variable? In this context, x2, X3, X5 are independent variables, and x4 is the dependent variable. Refer to Figure 10-10.

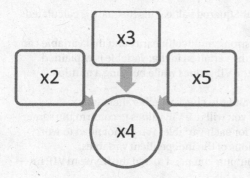

Figure 10-10. *X4 is taken as a dependent variable, and x2, x3, and x5 are related to x4*

We build a regression model using these independent variables and observe the value of R-squared. If the R-square value is high for the model x4 versus x2, x3, and x5, then the variable x4 can be explained by the other three. This will indicate interdependency or multicollinearity.

To detect the multicollinearity within a set of independent variables, you first need to choose the different subsets in the group. Regression lines are built for these subsets, and the R-square value is checked for each line. Here is the overall model:

y vs. y1 , y2, y3.............yp

Here are models for detecting multicollinearity:

x1 vs. x2, x3.......xp → R-squared value (R^21)

x2 vs. x1, x3.......xp → R-squared value (R^22)

....

xp vs. x1, x2, x3.......xp-1 → R-squared value (R^2p)

Finally, note the R^21, R^22 ....R^2p values to detect the multicollinearity.

If the R-square value is high, then it is an indication of multicollinearity. An R-squared value of more than 80 percent is considered as a good indicator for the existence of multicollinearity.

Variance Inflation Factor (VIF)

VIF is a measure that is specifically defined to measure the multicollinearity.

$$VIF = \frac{1}{1-R^2}$$

$$VIF(x_k) = \frac{1}{1-R_k^2}$$

So, the higher the R-square value, the higher the VIF value will be. In fact, VIF will magnify the R-squared value. If the R-squared value is 80 percent, then the VIF value will be 5. Refer to Table 10-28.

Table 10-28. *R-squared and VIF Values*

R-Squared	40%	50%	60%	70%	80%	90%	95%	98%	99%
VIF	1.7	2.0	2.5	3.3	5.0	10.0	20.0	50.0	100.0

Note: Do not confuse this R-Squared with overall model's R-Squared value. This R-square is calculated by building the models between the independent variables.

If the VIF value for a variable is greater than 5, it indicates strong multicollinearity and that variable can be termed as *redundant*. This is because 80 percent or more of the variation in that variable is explained by rest of the independent variables. So, it is mandatory to see the VIF values while building a multiple regression line.

Let's take the SAT score example again and calculate VIF values to check whether there is any multicollinearity. In this model, you already know about it, but you will use VIF values to confirm the same. PROC REG in SAS has a VIF option that will calculate VIF values for each variable. You do not need to worry about finding each VIF value separately for different combinations of the independent variables.

The following is the code for displaying VIF values in the output; you need to add the keyword VIF in the model statement:

```
proc reg data=sat_score;
model SAT=General_knowledge Aptitude Mathematics Science/VIF;
run;
```

Table 10-29 shows the output of this code.

Table 10-29. *Output of PROC REG on sat_score*

Number of Observations Read	96
Number of Observations Used	96

Analysis of Variance

Source	DF	Sum of Squares	Mean Square	F Value	Pr > F
Model	4	99039	24760	2621.54	<.0001
Error	91	859.47379	9.44477		
Corrected Total	95	99899			

Root MSE	3.07323	R-Square	0.9914
Dependent Mean	155.96875	Adj R-Sq	0.9910
Coeff Var	1.97042		

Parameter Estimates

Variable	DF	Parameter Estimate	Standard Error	t Value	Pr > |t|	Variance Inflation
Intercept	1	-2.07199	2.13436	-0.97	0.3342	0
General_knowledge	1	1.16697	0.10003	11.67	<.0001	15.05316
Aptitude	1	-0.13479	0.09683	-1.39	0.1673	14.01316
Mathematics	1	-0.11081	0.09887	-1.12	0.2653	49.79461
Science	1	1.09532	0.09689	11.30	<.0001	49.63107

Though the output looks the same as any other multiple regression output, there is a new column added in the Parameter Estimates table: Variance Inflation. This is just the VIF value. So, you see that column and check whether there are any variables with VIF more than 5.

In this output, all the variables have VIF values greater than 5, but it doesn't imply that all four variables are interrelated. Generally, VIF values appear in pairs. If you remove one variable from the pair, the other one is automatically corrected. For example, let's remove science from the model and check the multicollinearity again.

```
proc reg data=sat_score;
model SAT=General_knowledge Aptitude Mathematics /VIF ;
run;
```

Please refer to Table 10-30 for the output.

Table 10-30. *Output of PROC REG on sat_score*

Number of Observations Read	96
Number of Observations Used	96

Analysis of Variance

Source	DF	Sum of Squares	Mean Square	F Value	Pr > F
Model	3	97832	32611	1451.85	<.0001
Error	92	2066.46278	22.46155		
Corrected Total	95	99899			

Root MSE	4.73936	R-Square	0.9793
Dependent Mean	155.96875	Adj R-Sq	0.9786
Coeff Var	3.03866		

Parameter Estimates

Variable	DF	Parameter Estimate	Standard Error	t Value	Pr > |t|	Variance Inflation
Intercept	1	-4.02414	3.28069	-1.23	0.2231	0
General_ knowledge	1	1.09637	0.15395	7.12	<.0001	14.99448
Aptitude	1	-0.04114	0.14878	-0.28	0.7828	13.91059
Mathematics	1	0.98752	0.02822	34.99	<.0001	1.70601

You can see a massive change in the VIF value for mathematics. In the same way, if you remove mathematics, you will see the science variable VIF changed to 1.7.

The following facts are worth noting in case of multicollinearity:

- .VIF is a measure that helps in detecting multicollinearity. The correlation matrix can also help sometimes, but VIF takes care of all the variables.

- The F-test and T-test behaving in a contrasting manner is also an indication of multicollinearity. A high F-statistic and R-squared value will make you believe that the overall model is a good fit. The T-tests, on other hand, may show that most of the variables are not having any impact on y. This situation is caused by multicollinearity.

- The wrong signs for the coefficients or counterintuitive estimates for the known variables is another sign of multicollinearity.

- You can make a small change in the sample or the input data and observe the changes in the regression coefficients. If these changes are abnormally high, then it is an indication of multicollinearity.

- The condition number is another way of identifying high standard deviations in the beta coefficients. Instead of directly finding the associations between the independent variables, the condition number looks at the expected variance in the beta coefficient. If the condition number is high, it is an indicator of multicollinearity. As a rule of thumb, if the condition number is more than 30, it is a sign of multicollinearity. A more detailed discussion on this topic is beyond the scope of this book.

- Tolerance is another measure, which is used as an indicator of the multicollinearity. Tolerance is nothing but the inverse of VIF. So, nothing really is new in it. $Tolerance = 1/VIF$.

Up to now, we have discussed the challenges that occur because of multicollinearity. We have also discussed ways to detect it. In the following section, we will discuss how to treat multicollinearity while building a regression model.

Redemption of Multicollinearity (Treating Multicollinearity)

Multicollinearity is seen most of the times as a redundancy. This means that all of the variables involving multicollinearity are not required in the model. Other truly independent variables in the model are sufficient to explain the variations in the final dependent variable. Consider building a model for Y using X1, X2, X3, X4, and X5. If X2, X3, and X5 are explaining more than 80 percent of variations in X4, then there is no need to keep X4 in the model. You can very well drop X4 from the model specification and rebuild an accurate enough model for Y using X1, X2, X3, and X5 alone.

Dropping Troublesome Variables

In most of the cases you go ahead and drop the troublesome variables from the independent variable list. But you need to be careful. You can't simply drop all the variables having a VIF value greater than 5. As discussed earlier, the VIF values come in pairs. If you drop a variable, the other one is adjusted automatically.

Let's look at the VIF values for the SAT exam data (Table 10-31).

Table 10-31. *VIF Values for SAT Exam Data*

Parameter Estimates

Variable	DF	Parameter Estimate	Standard Error	t Value	Pr > |t|	Variance Inflation
Intercept	1	-2.07199	2.13436	-0.97	0.3342	0
General_ knowledge	1	1.16697	0.10003	11.67	<.0001	15.05316
Aptitude	1	-0.13479	0.09683	-1.39	0.1673	14.01316
Mathematics	1	-0.11081	0.09887	-1.12	0.2653	49.79461
Science	1	1.09532	0.09689	11.30	<.0001	49.63107

VIF values for all the variables are greater than 5. But as expected, they all are in pairs (with two values close to each other). You take the highest pair and drop a variable from there. Here mathematics has a slightly higher VIF, and you can drop it. Here both math and science have almost the same VIF values, so you keep the most important variable (in the context of the business problem). Otherwise, you can go ahead and drop the variable with the highest VIF.

Here is the model-building code after dropping mathematics:

```
proc reg data=sat_score;
model SAT=General_knowledge Aptitude Science/VIF;
run;
```

Please refer to Table 10-32 for the output.

Table 10-32. *VIF Values for SAT Exam Data After Dropping Mathematics*

Parameter Estimates

Variable	DF	Parameter Estimate	Standard Error	t Value	Pr > |t|	Variance Inflation
Intercept	1	-2.19969	2.13428	-1.03	0.3054	0
General knowledge	1	1.15437	0.09953	11.60	<.0001	14.86326
Aptitude	1	-0.12438	0.09652	-1.29	0.2008	13.88427
Science	1	0.98860	0.01796	55.05	<.0001	1.70041

The variable science looks fine now. VIF is still high for GK and aptitude. Let's drop GK and rebuild the model. Here is the code:

```
proc reg data=sat_score;
model SAT= Aptitude  Science/VIF;
run;
```

Please refer to Table 10-33 for the output.

Table 10-33. *Output of PROC REG on sat_score with Only Aptitude and Science*

Number of Observations Read	96
Number of Observations Used	96

Analysis of Variance

Source	DF	Sum of Squares	Mean Square	F Value	Pr > F
Model	2	97754	48877	2118.79	<.0001
Error	93	2145.34522	23.06823		
Corrected Total	95	99899			

Root MSE	4.80294	R-Square	0.9785	
Dependent Mean	155.96875	Adj R-Sq	0.9781	
Coeff Var	3.07942			

Parameter Estimates

Variable	DF	Parameter Estimate	Standard Error	t Value	Pr > |t|	Variance Inflation
Intercept	1	1.61044	3.29119	0.49	0.6258	0
Aptitude	1	0.92924	0.05089	18.26	<.0001	1.58487
Science	1	1.04290	0.02706	38.54	<.0001	1.58487

Everything seem to be perfect with this model. There is no multicollinearity, so you can trust these coefficients. The coefficient signs are also intuitively correct (from the business knowledge angle) for both the variables.

Other Ways of Treating Multicollinearity

Though dropping troublesome variables is the most widely used method, there are a few other ways of dealing the multicollinearity.

- You can use principal components instead of variables. Principal components are linear combinations of variables, which will be explaining maximum variance in the data. If some variables are intercorrelated, you can use noncorrelated linear combination of variables instead of directly using the variables.

- On having a good understanding of the causes of multicollinearity, an analyst can reduce it by collecting more data and a better unbiased sample.

- If prediction is the only motto and the relationship with Y and xi is not of interest, the same model with interdependent independent variables may be used. If the model with correlated Xi still has a high R-squared value, it may be good enough for prediction purposes. The challenges comes only when the interest is in analyzing a one-to-one relationship between independent and dependent variables.

- Ridge regression is another way to treat the multicollinearity. The main philosophy behind the ridge regression is that it's better to get a biased estimated of betas with less standard deviation instead of unbiased beta estimates with a high standard deviation. So, ridge regression does some tweaking to the optimization matrix while finding the least square estimates of regression coefficients. The details are beyond the scope of this book.

- Data transformation may yield good results sometimes.

How to Analyze the Output: Linear Regression Final Check List

In this chapter you learned several measures and several challenges that need careful treatment. This check list will help you; you can remember it with the acronym FRAVT, which stands for F-test, R-squared, adjusted R-squared, VIF, and T-tests.

Double-Check for the Assumptions of Linear Regression

You have to make sure that all the regression assumptions are religiously followed by the data (observations) before attempting to build a model. Generally this process takes a lot of time. Many analysts tend to ignore this step and take it for granted that all the regression assumptions are followed. Generally, a scatterplot is drawn between the Xi and Y variables to verify the linearity assumption. Here you get almost 90 percent of an idea about the existence of outliers, nonlinearity, heteroscedasicity, and so on. If all regression assumptions are followed, only then can you move on to the next check point, the F-test. Most of the time analysts come to know about possible assumption violations when they are right in the middle of analysis and something goes wrong.

F-test

The first measure to look at in this order is the F-test. It gives you an idea about the overall significance of a model. If an F-test shows that the model is not significant, there is no need to go any further in the model-building process. You can simply stop the model building and look for other impacting variables to predict y. You can search for more data or do some more research to check whether there are any vital errors at any stage in the overall model-building process. If F-test is passed, in other words, the model is established as significant, you can move on to the next check point of the R-squared value.

R-squared

The R-squared value comes after conforming the fact that the model is significant. R-squared will tell you how significant the model is. A higher R-squared value (greater than 80 percent) indicates that model is explaining the maximum variation in the dependent variable.

Adjusted R-Squared

Because R-squared has some downsides while using multiple regression methodology, you also have to look at the adjusted R-squared value and make sure that there are no junk variables or the model is over specified with too many independent variables.

VIF

The next step is to make sure that there is no multicollinearity within the independent variables. You can check it using the VIF values of each variable. If multicollinearity is detected, then proper treatment needs to be given to the independent variables in order to prepare an independent set of predictor variables.

T-test for Each Variable

The final item in the list is the T-test. The T-test results tell you about the most impacting variables. You can safely drop all the nonimportant variables and keep only the most impacting ones. Sometimes a number of model iterations are required to identify and keep few most-impacting variables. This may involve compromising a little bit on the R-squared value to reduce number of variables.

Analyzing the Regression Output: Final Check List Example

Let's, once again, observe the output of smartphone sales data. You can assume that the analyst has already validated the data against all the assumptions of linear regression. The following is the code, SAS output (Table 10-34), and final checklist steps:

```
proc reg data= mobiles;
model sales= Ratings Price Num_new_features Stock_market_ind Market_promo_budget/vif;
run;
```

Please refer to Table 10-34 for the output.

Table 10-34. Output of PROC REG on Mobiles Data Set

Number of Observations Read	58
Number of Observations Used	58

Analysis of Variance

Source	DF	Sum of Squares	Mean Square	F Value	Pr > F
Model	5	7.408043E14	1.481609E14	55.16	<.0001
Error	52	1.396636E14	2.685839E12		
Corrected Total	57	8.80468E14			

Root MSE	1638853	R-Square	0.8414
Dependent Mean	9394688	Adj R-Sq	0.8261
Coeff Var	17.44446		

(*continued*)

Table 10-34. (*continued*)

Parameter Estimates

Variable	DF	Parameter Estimate	Standard Error	t Value	Pr > |t|	Variance Inflation
Intercept	1	493778	1520007	0.32	0.7466	0
Ratings	1	651375	90775	7.18	<.0001	1.25364
Price	1	-1833.67081	127.24677	-14.41	<.0001	2.70525
Num_new_features	1	547167	85234	6.42	<.0001	1.49478
Stock_market_ind	1	-105.38867	106.01603	-0.99	0.3248	1.07486
Market_promo_budget	1	100.92891	8.38744	12.03	<.0001	2.02419

Here is the check list:

1. *Assumptions of regression*: You already tested that the data doesn't violate any of the linear regression assumptions.

2. *F-test*: This looks good; the model is significant.

3. *R-squared*: More than 80 percent of variance in y is explained by xi. Hence, the model is a good fit.

4. *Adjusted R-squared*: This is slightly less than R-squared, indicating some junk variable is in the data. Are there any insignificant variables? Yes, the stock market indicator is insignificant; you can drop it and rebuild the model.

```
proc reg data= mobiles;
model sales= Ratings Price Num_new_features Market_promo_budget/vif;
run;
```

Please refer to Table 10-35 for the output.

Table 10-35. *Output of Regression Model on Mobiles Data Set After Dropping Stock_market_ind*

Number of Observations Read	58
Number of Observations Used	58

Analysis of Variance

Source	DF	Sum of Squares	Mean Square	F Value	Pr > F
Model	4	7.381502E14	1.845375E14	68.72	<.0001
Error	53	1.423178E14	2.685241E12		
Corrected Total	57	8.80468E14			

(*continued*)

Table 10-35. (*continued*)

Root MSE	1638671	R-Square	0.8384
Dependent Mean	9394688	Adj R-Sq	0.8262
Coeff Var	17.44252		

Parameter Estimates

Variable	DF	Parameter Estimate	StandardError	t Value	Pr > |t|	Variance Inflation
Intercept	1	-804177	778126	-1.03	0.3061	0
Ratings	1	648559	90721	7.15	<.0001	1.25242
Price	1	-1855.97681	125.23876	-14.82	<.0001	2.62113
Num_new_features	1	546713	85224	6.42	<.0001	1.49474
Market_promo_ budget	1	103.06985	8.10532	12.72	<.0001	1.89073

Here is the check list:

1. *Assumptions of regression*: It is already given that the data doesn't violate regression assumptions.

2. *F-test*: This looks good; the model is significant.

3. *R-squared*: More than 80 percent of variance of explanation is a good fit.

4. *Adjusted R-squared*: This is almost close to R-squared, so no junk values are in the model.

5. The VIF values are all within the limits; there are no multicollinearity threats.

6. All variables pass the T-test and show that all of them have significant impact on sales.

The model is ready to be used for smartphone sales perditions. Given the values of Ratings, Price, Num_new_features, and Market_promo_budget, the accuracy will be more than 80 percent. The following is the final model equation for predictions:

$$Sales = -804177 + 648559*Ratings - 1855.97*Price + 546713*Num\,new\,features$$
$$+ 103.06*Market\_promo\_budget$$

Conclusion

In this chapter, you started with multiple linear regressions to tackle the predictions where more than one independent variable is used. You also learned the goodness of fit measures for multiple regression. The multiple regression has several independent variables, and their interdependency may lead to absurd results. You learned how to handle the multicollinearity issue. Finally, you saw a checklist to be used while analyzing the multiple regression output. Several concepts were simplified and dealt with at a basic level. You may need to refer to dedicated text books on regression to get some in-depth theory behind these concepts.

We talked about linear regression. Obviously you can't expect all the relations in this world to be linear. What if you come across a nonlinear relationship between dependent (y) and independent (xi) variables? How do you build a nonlinear regression line? What are the changes in the assumptions? What are the goodness of fit measures? What are the other challenges involved in the process? Nonlinear regression is the topic of the next chapter. Stay tuned!

CHAPTER 11

■ ■ ■

Logistic Regression

In previous chapters, we covered correlation and linear regression modeling in detail. If you look to quantify the relationship between two variables, you use the correlation coefficient. For example, you can quantify the relation between salary and expenses using correlation. If you needed to predict a response variable based upon some other item, you could use linear regression modeling, provided the relationship is linear. For example, if you want to predict exactly what a person's expenses will be when his salary is $10,000, you can use linear regression modeling, provided the expense and salary fit on a straight-line graph. In some cases, this relationship is not actually linear, but you can make it linear by applying some simple mathematical transformations; still, you can use linear regression modeling.

Is linear regression modeling the solution for all real-life prediction problems? Or do you need to apply different techniques in some cases? Let's explore these questions using a simple example.

Predicting Ice-Cream Sales: Example

Table 11-1 shows a snapshot of a table that contains the ice-cream sales data for an ice-cream shop. For now, you are recording just two variables: the age of the customer and whether she buys ice cream. The data set has an indicator variable called buy_ind, which takes a value of 0 if a customer buys ice cream and a value of 1 if not. You are asked to predict ice-cream sales (in other words, the buy_ind variable) using age as an independent variable.

Table 11-1. Ice-Cream Sales Data

Age	buy_ind
6	0
25	0
32	1
44	1
34	1
43	1
72	1
67	0
58	1
.	.
.	.
58	1

The data set has just two variables: buy_ind and Age. The following is the SAS code for importing the data and building a simple linear regression line for the response variable:

```
/* Importing ice cream sales data*/
PROC IMPORT OUT= WORK.ice_cream_sales
            DATAFILE= "C:\Users\VENKAT\Google Drive\Training\Books\Conte nt\8.
Logistic Regression\Data Logistic Regression\IceCream_sales.csv"
            DBMS=CSV REPLACE;
    GETNAMES=YES;
    DATAROW=2;
RUN;

/* Fitting a simple regression line to predict buy_ind */
proc reg data=ice_cream_sales;
model buy_ind=age;
run;
```

Table 11-2 shows the output of the previous code.

Table 11-2. *Output of PROC REG on Ice-Cream Sales Data*

Number of Observations Read	50
Number of Observations Used	50

Analysis of Variance

Source	DF	Sum of Squares	Mean Square	F Value	Pr > F
Model	1	4.69353	4.69353	38.80	<.0001
Error	48	5.80647	0.12097		
Corrected Total	49	10.50000			

Root MSE	0.34780	R-Square	0.4470
Dependent Mean	0.70000	Adj R-Sq	0.4355
Coeff Var	49.68641		

Parameter Estimates

0Variable	DF	Parameter Estimate	Standard Error	t Value	Pr > |t|
Intercept	1	0.05851	0.11413	0.51	0.6105
Age	1	0.01489	0.00239	6.23	<.0001

The following are the observations and inferences for the output in Table 11-2:

- There are a total of 50 observations used for building this model.

- The P-value of the F-testis less than 5 percent, which suggests that the overall model is significant.

- The R-squared value is on the lower side, which shows that the model is not a good fit. In other words, you can't expect it to deliver respectable predictions.

- The Age variable seems to significantly impact buying. This has been indicated by the P-value of T-test for Age.

- To summarize, you can't use this model for your predictions. It is not a good fit.

By this time, having learned so much about regression, the model fitting for the ice-cream sales data may appear to be a routine exercise. You can easily conclude that the fit is not good for the data. The variable Age can't explain much of the variation in buying, which means whether a customer buys the ice-cream or not can't be estimated simply by the variable Age. In other words, simply by knowing a customer's specific age, you can't tell anything about ice-cream sales.

On second thought, when you take a close look at the data, you can easily observe a pattern in it. You can see that when customer is younger, the buy_ind variable is 0 most of the time, and when customer is older, buy_ind is 1 most of the time. This shows that older people are not buying ice cream. A scatterplot between the two variables reiterates the same (Figure 11-1).

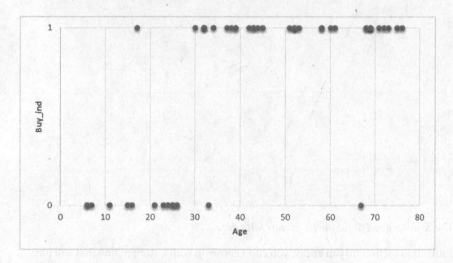

Figure 11-1. *Scatterplot on ice-cream sales data*

Figure 11-1 shows that almost all customers younger than 30 have bought ice cream, and almost all the customers older than 30 haven't. The data has some apparent pattern, but you are not able to capture it using the linear regression models. Maybe a linear regression line is not a good fit for this data. See Figures 11-2 and 11-3.

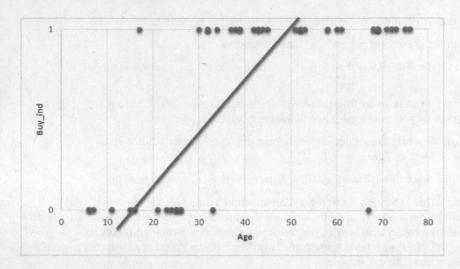

Figure 11-2. *A straight line is not a good fit for the ice-cream sales data*

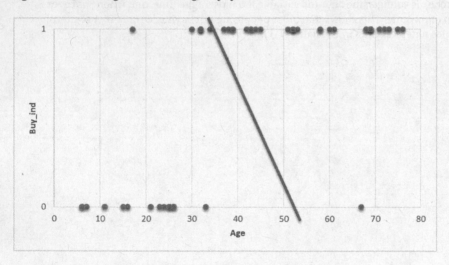

Figure 11-3. *A straight line is not a good fit for the ice-cream sales data*

No matter what optimization technique you apply, you can't come up with a straight line that will pass through the core of the data. If you recall the regression assumptions from regression in Chapter 9, you will realize that when the relation between the dependent (buy_ind) and independent (Age) variables is not linear, you can't apply a linear regression model. This is the case here.

Nonlinear Regression

If the relation between x and x is nonlinear and if it cannot be converted to a linear relationship by applying mathematical transformations, then you need to fit the best applicable nonlinear curve. Anything other than a straight-line relationship is termed *nonlinear*. Since you could not fit a linear regression line to the ice-cream sales data, let's see if you can fit a nonlinear curve. Figures 11-4 to 11-8 are some examples of nonlinear expressions.

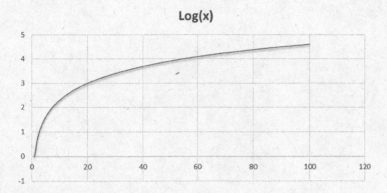

Figure 11-4. *Log(x)*

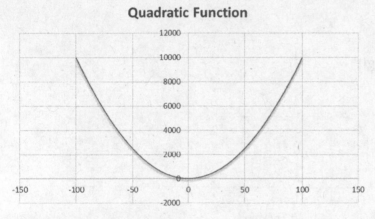

Figure 11-5. *Quadratic function*

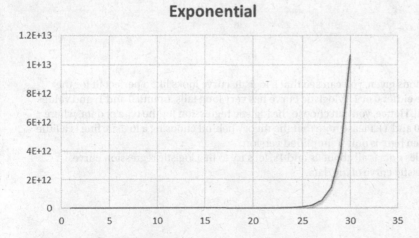

Figure 11-6. *Exponential function*

405

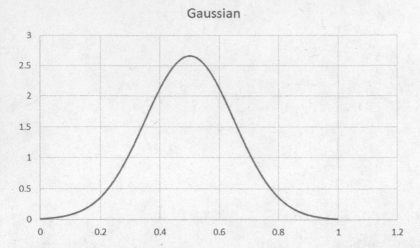

Figure 11-7. *Gaussian function*

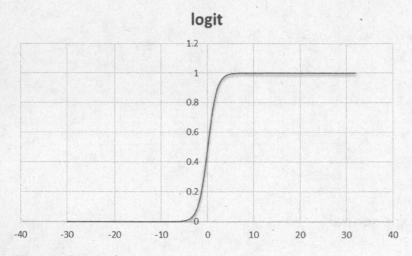

Figure 11-8. *Logit function*

Out of all nonlinear functions given, you can see that a logistic curve looks like a perfect fit for this data, where all the Y values are either 0 or 1. A logistic curve has very long tails around 0 and 1, and values other than 0 and 1 are minimal. Hence, you can choose the logistic regression for the type of data, where you have only two outcomes (0 and 1). Please note that the theory behind choosing a logistic line is a little complicated, and the logic given here is only a simplified version.

Because the ice-cream sales data is all about 0s and 1s, let's try to fit a logistic regression curve. Figure 11-9 is an imaginary logistic curve of the data.

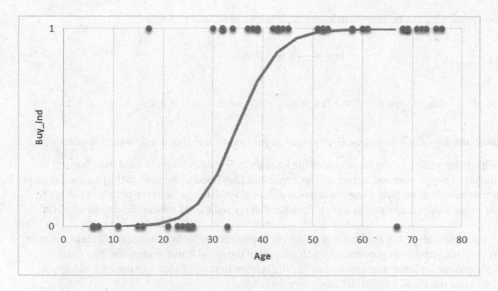

Figure 11-9. *An imaginary logistic curve through the ice-cream sales data*

You have the straight equation $y = \beta_0 + \beta_1 x_1 + \beta_2 x_2$ for linear regression. In a similar fashion, you have an equation for the logistic line as well.

$$p(y/x) = \frac{e^{\beta_0 + \beta_1 x_1}}{1 + e^{\beta_0 + \beta_1 x_1}}$$

The logistic function, explained by the earlier equation, is never less than 0 and never greater than 1. The maximum value is 1; $e^{\beta_0 + \beta_1 x_1}$ tends to give very high values, or it tends to reach infinity. For example, if the numerator is 1,000,000, then the denominator will take a value of 1,000,001. In such a case, the probability is almost 1. The lower limit is 0; imagine a numerator of 0.0000000001. The denominator is 1.0000000001, and then the probability value is almost 0.

Please note that you are not trying to predict Y, but you are trying to predict the probability of Y=0 or Y=1. Since Y takes just two values, there is hardly any variation in Y. So, it makes more sense to predict the probabilities rather than direct values.

Logistic Regression

Logistic regression may be a perfect choice when you have just two outcomes in your response variable. You encounter only two outcomes situations in many cases like these: yes versus no in response to a question, win versus loss in a game, buy versus no buy in customer sales, response versus no response in cold calling, click versus no click in web analytics, default versus no default in credit risk analysis, satisfied versus not satisfied in survey analytics, and so on. It is not possible to predict these categorical outcomes using linear regression because it expects numeric dependent variables.

Here is the logistic line equation discussed in the previous section. It's for a single independent variable.

$$p(y/x) = \frac{e^{\beta_0 + \beta_1 x_1}}{1 + e^{\beta_0 + \beta_1 x_1}}$$

Here is the logistic line equation for multiple independent variables:

$$p(y/x) = \frac{e^{\beta_0 + \beta_1 x_1 + \beta_2 x_2 + \beta_3 x_3 + \beta_4 x_4 + \ldots\ldots\ldots + \beta_p x_p}}{1 + e^{\beta_0 + \beta_1 x_1 + \beta_2 x_2 + \beta_3 x_3 + \beta_4 x_4 + \ldots\ldots\ldots + \beta_p x_p}}$$

The previous equations can be written in logarithmic teams as follows.

$$\log(\frac{p}{1-p}) = \beta_0 + \beta_1 x_1$$

$$\log(\frac{p}{1-p}) = \beta_0 + \beta_1 x_1 + \beta_2 x_2 + \beta_3 x_3 + \beta_4 x_4 + \ldots\ldots\ldots + \beta_p x_p$$

$\log(\frac{p}{1-p})$ is called the *log odds probability*. Here you are trying to fit a line that predicts the log odds of Y rather than its direct values. In the same way as the least squares methodology is used to fit linear regression coefficients (please refer to Chapter 9), the maximum likelihood estimation (MLE) method is used to find the beta coefficients if you have a logistic regression line. We will not get into the details of the MLE method, but you can consider it an optimization technique that is used to find the coefficients in a logistic regression line.

An analyst has to first sense the applicability of logistic regression by looking at the dependent variable and applying the logistic regression procedure in SAS instead of the usual linear regression. You used the PROC REG procedure for linear regression. For finding out the beta coefficients in the case of logistic regression, you will use the PROC LOGISTIC procedure from SAS.

Logistic Regression Using SAS

You have discovered that linear regression is not the right regression line to fit the ice-cream sales case. You also discovered that a logistic regression line would be a better fit. Now you will create a logistic regression model using SAS.

The following is the SAS code to fit the desired logistic regression line:

```
Proc logistic data=ice_cream_sales;
model buy_ind=age;
run;
```

Here is an explanation of the code:

- *PROC LOGISTIC*: This calls the logistic regression line procedure in SAS.

- *Data*: This indicates the ice_cream_sales data set.

- *Model*: This is similar to a linear regression line. The dependent variable is on the left side of the equal sign, and a list of independent variables is on the right side. Multiple independent variables (if available) are separated by a single space between them.

The previous code tries to fit a logistic regression line to the ice-cream sales data, which means that you are trying to estimate the beta coefficients in the logistic regression equation, given here:

$$p(y/x) = \frac{e^{\beta_0 + \beta_1 x_1}}{1 + e^{\beta_0 + \beta_1 x_1}}$$

This code generates the output shown in Table 11-3.

Table 11-3. *Output of PROC LOGISTIC on Ice-Cream Sales Data*

The SAS System

The LOGISTIC Procedure

Model Information

Data Set	WORK.ICE_CREAM_SALES
Response Variable	buy_ind
Number of Response Levels	2
Model	binary logit
Optimization Technique	Fisher's scoring

Number of Observations Read	50
Number of Observations Used	50

Response Profile

Ordered Value	buy_ind	Total Frequency
1	0	15
2	1	35

Probability modeled is buy_ind='0'.

Model Convergence Status

Convergence criterion (GCONV=1E-8) satisfied.

Model Fit Statistics

Criterion	Intercept Only	Intercept and Covariates
AIC	63.086	36.114
SC	64.998	39.938
-2 Log L	61.086	32.114

Testing Global Null Hypothesis: BETA=0

Test	Chi-Square	DF	Pr > ChiSq
Likelihood Ratio	28.9720	1	<.0001
Score	22.3501	1	<.0001
Wald	10.8989	1	0.0010

Analysis of Maximum Likelihood Estimates

Parameter	DF	Estimate	Standard Error	Wald Chi-Square	Pr > ChiSq
Intercept	1	3.8982	1.3446	8.4044	0.0037
Age	1	-0.1353	0.0410	10.8989	0.0010

Odds Ratio Estimates

Effect	Point Estimate	95% Wald Confidence Limits	
Age	0.873	0.806	0.947

Association of Predicted Probabilities and Observed Responses

Percent Concordant	92.0	Somers' D	0.851
Percent Discordant	6.9	Gamma	0.861
Percent Tied	1.1	Tau-a	0.365
Pairs	525	C	0.926

SAS Logistic Regression Output Explanation

The output has several measures and tables. The following sections explain.

Output Part 1: Response Variable Summary

Table 11-4 shows the LOGISTIC procedure.

Table 11-4. *Model Information, Output of PROC LOGISTIC*

Model Information	
Data Set	WORK.ICE_CREAM_SALES
Response Variable	buy_ind
Number of Response Levels	2
Model	binary logit
Optimization Technique	Fisher's scoring

Here is an explanation of Table 11-4:

- *Data set*: This is the data source.

- *Response variable*: This is the dependent variable or the Y variable, the buying indicator in this example.

- *Number of response levels*: This is the number of levels in the dependent variable (mostly Yes/No); it's 1 or 0 in this example.

- *Model*: This is the binary logistic regression; it's the same as binary logit.

- *Optimization technique*: Which optimization technique is used to find the regression coefficients? SAS chooses the most appropriate technique.

Now consider more tables from the output in Table 11-5.

Table 11-5. *Output Tables of PROC LOGISTIC*

Number of Observations Read	50
Number of Observations Used	50

Response Profile

Ordered Value	buy_ind	Total Frequency
1	0	15
2	1	35

Probability modeled is buy_ind='0'.

Here is an explanation of Table 11-5:

- *Number of observations read*: This is the count of observations read from the data set.

- *Number of observations used*: This is the number of observations used for creating the model. There are no missing values or default values in the data. If there are missing values, SAS will give a count of those missing values separately.

- *Response profile ordered values*: This is how SAS ordered the values that you supplied. It considered the 0 category the first level and the 1 category the second level.

- *Probability modeled is buy_ind='0'*: The logistic regression will finally give the probability of Y, which can take a value of 0 or 1. In this output, SAS is informing you that the model is built for 0; in other words, the output probability will be given for the occurrence of Y being 0. By default SAS builds the model for smaller values. It doesn't really matter when you have only two levels. If the probability of 0 is 65 percent, then the probability of 1 will obviously be 35 percent. You can use the descending option (refer to the case study later in this chapter, where SAS code with this option has been used) to directly model for the higher-order value. As per the output, in the ice-cream sales example, the model is built for the probability of Y being 0, which means a buy decision.

- *Total frequency*: This is the frequency of each category in a dependent variable.

Output Part 2: Model Fit Summary

In output part 2, we discuss the model convergence status and the model fit statistics. Refer to Table 11-6.

Table 11-6. *Model Fit Summary, the Output Tables of PROC LOGISTIC*

Model Convergence Status

Convergence criterion (GCONV=1E-8) satisfied.

Model Fit Statistics

Criterion	Intercept Only	Intercept and Covariates
AIC	63.086	36.114
SC	64.998	39.938
-2 Log L	61.086	32.114

Here is an explanation of the Table 11-6:

- *Model convergence status*: This is related to your model's optimization convergence and precision. A detailed explanation on this is beyond the scope for this book.

- *Model fit statistics*:

  - *AIC*: Akaike Information Criterion is a measure that is used to compare two models so as to pick the best one. Generally, a model with less AIC is desired. A stand-alone AIC value is not of much significance, so you can ignore it for now.

  - *SC*: Schwarz Criterion is similar to AIC, used for comparing two models. AIC and SC values will be less for models that have a fewer number of predictor variables and high accuracy in line fitting. For example, if you are comparing two models to find which one is better, you should go for the model that has minimum AIC and SBC. Generally, AIC and SC are less for the models, which have high accuracy with fewer predictor variables.

Output Part 3: Test for Regression Coefficients

In part 3, you are testing the global null hypothesis for beta = 0. Table 11-7 shows the test results for the null hypothesis (that the coefficients of all the independent variables are equal to zero) versus the alternative hypothesis (that at least one of the confidents is nonzero). In other words, you are testing whether all the independent (X) variables are insignificant versus at least one of them is significant.

H_0: $\beta_1 = \beta_2 = \beta_3 = \beta_4 = \beta_5 = \beta_6 \ldots\ldots = \beta_p = 0$

vs

H_1: $\beta_1 \neq 0 \, or \, \beta_2 \neq 0 \, or \, \beta_3 \neq 0 \, or \, \beta_4 \neq 0 \, or \, \beta_5 \neq 0 \, or \, \beta_6 \neq 0 \, or \ldots \beta_p \neq 0$

Table 11-7. *Testing Global Null Hypothesis, the Output Tables of PROC LOGISTIC*

Testing Global Null Hypothesis: BETA = 0			
Test	Chi-Square	DF	Pr > ChiSq
Likelihood Ratio	28.9720	1	<.0001
Score	22.3501	1	<.0001
Wald	10.8989	1	0.0010

All these three tests (Likelihood, Score, and Wald) are testing the previous hypothesis only. If all the tests show that the P-value (Pr > ChiSq) is greater than 5 percent, then there is not much evidence to reject the null hypothesis. This means there is not even a single variable that has a significant impact (on Y). If in any of the tests the P-value is less than 5 percent, then there is at least one variable that has a significant impact on the dependent variable. Mostly all three tests show the same result. Here in this example, the Chi-square tests show that the null hypothesis is rejected. This means there is at least one independent variable whose coefficient is not equal to zero.

Output Part 4: The Beta Coefficients and Odds Ratio

The coefficients, the regression parameters, or the beta values of the independent variable are given in Table 11-8.

Table 11-8. *Analysis of Maximum Likelihood Estimates, the Output Tables of PROC LOGISTIC*

Analysis of Maximum Likelihood Estimates					
Parameter	DF	Estimate	Standard Error	Wald Chi-Square	Pr > ChiSq
Intercept	1	3.8982	1.3446	8.4044	0.0037
Age	1	-0.1353	0.0410	10.8989	0.0010

The table (Analysis of Maximum Likelihood Estimates) shows the MLE estimates for the independent variable coefficients, as given by the following equation:

$$p(y/x) = \frac{e^{\beta_0 + \beta_1 x_1}}{1 + e^{\beta_0 + \beta_1 x_1}}.$$

The value of β_0 is 3.8982, and the value of β_1 is -0.1353.

The following is the logistic regression line equation based on the estimates given by SAS:

$$p(y/x) = \frac{e^{\beta_0 + \beta_1 x_1}}{1 + e^{\beta_0 + \beta_1 x_1}}$$

$$p(y/x) = \frac{e^{3.8982 - 0.1353 * \text{Age}}}{1 + e^{3.8982 - 0.1353 * \text{Age}}}$$

You can substitute the values of age in this equation to find the probability of buying for each customer.

The odds ratio estimates (Table 11-9) are a little different from the previous estimates. These estimates are used to see the exact impact of each individual variable on the odds of the positive outcome of the model. For instance, you are predicting the probability of buying ice cream in this example. The odds ratio estimates tell you what the impact of age is on the odds of buying (in other words, the chances of buying). What is the change in the odds when there is a unit change in the independent variable? In this model, the change in odds is 0.873 whenever there is a unit change in the age.

Table 11-9. *Odds Ratio Estimates, the Output Tables of PROC LOGISTIC*

Odds Ratio Estimates			
Effect	Point Estimate	95% Wald Confidence Limits	
Age	0.873	0.806	0.947

The following is the regression line for ice-cream sales data; you substitute the age to find the predicted probability of buying:

$$P(y/x) = \frac{e^{3.8982-0.1353*Age}}{1+e^{3.8982-0.1353*Age}}$$

For example, when the age is 6, the probability of buying is 95.6 percent, and the probability of not buying is 4.3 percent. The odds of buying over not buying are 95 percent/5 percent, that is, 21.8981.

When the age is increased by one unit, such as increased to 7 from 6, the probability of buying is 95.03 percent and not buying is 4.968 percent. The odds of buying over not buying are 95.03 percent/4.968 percent, that is, 19.12698.

When you changed the age by one unit, from 6 to 7, the odds also changed, from 21.8981 to 19.12698. The odds ratio is 0.8734 (odds of buying over not buying [age=7]/the odds of buying over not buying [age=6]), which is the same as given in the odds ratio estimate table in SAS. The odds ratio estimate gives you the estimate of exact change in the odds when there is a unit change in the independent variable. Table 11-10 illustrates the same.

Table 11-10. *Change in odds with unit change age*

Age	Predicted Probability of Buying(A)	Predicted Probability of Not Buying(B)	Odds of Buying over Not buying (A/B)	Change in Odds with Unit Change Age
6	96%	4%	21.89810	NA
7	95%	5%	19.12698	0.873454
8	94%	6%	16.70654	0.873454
9	94%	6%	14.59239	0.873454
10	93%	7%	12.74578	0.873454
11	92%	8%	11.13285	0.873454
12	91%	9%	9.72403	0.873454
13	89%	11%	8.49349	0.873454
14	88%	12%	7.41867	0.873454
15	87%	13%	6.47987	0.873454
16	85%	15%	5.65986	0.873454
17	83%	17%	4.94363	0.873454
18	81%	19%	4.31803	0.873454
19	79%	21%	3.77160	0.873454
20	77%	23%	3.29432	0.873454
21	74%	26%	2.87744	0.873454

(continued)

Table 11-10. (*continued*)

Age	Predicted Probability of Buying(A)	Predicted Probability of Not Buying(B)	Odds of Buying over Not buying (A/B)	Change in Odds with Unit Change Age
22	72%	28%	2.51331	0.873454
23	69%	31%	2.19526	0.873454
24	66%	34%	1.91746	0.873454
25	63%	37%	1.67481	0.873454
26	59%	41%	1.46287	0.873454
27	56%	44%	1.27775	0.873454
28	53%	47%	1.11605	0.873454

From Table 11-10, you can see that the odds ratio is exactly 0.8734 whenever there is a change of 1 unit in the independent variable, such as the age.

Output Part 5: Validation Statistics

Table 11-11 is an important table because you can use it for validating the model. This gives you an idea about misclassification errors and the effectiveness of the model. A higher percent of concordance is always desired. Somers' D, Gamma, Tau-a, and C are derived from misclassification versus classification only.

Table 11-11. *Association of Predicted Probabilities and Observed Responses, the Output Tables of PROC LOGISTIC*

Association of Predicted Probabilities and Observed Responses			
Percent Concordant	92.0	Somers' D	0.851
Percent Discordant	6.9	Gamma	0.861
Percent Tied	1.1	Tau-a	0.365
Pairs	525	C	0.926

Classifying 0 as 0 or giving a higher probability to 0 when the actual dependent variable value is 0 is *classification without error*. If the model is classifying 0 as 1 and 1 as 0 or giving a higher probability to 0 when the actual dependent value is 1, or vice versa, it is *misclassification*. We will explain classification and misclassification in detail in the "Goodness of Fit for Logistic Regression" section.

Individual Impact of Independent Variables

In linear regression models, you used the P-value to check whether an independent variable has a significant impact on a dependent variable. The beta coefficients in linear regression follow T-distribution, so you did a T-test to see the impact of each variable. Here in logistic regression, the beta coefficients follow the Chi-square distribution. So, the probability value (P-value) of the Chi-square test tells you about the impact of the independent variables in logistic regression models.

A Chi-square test in logistic regression tests the hypotheses here:

H_0: The independent variable has no significant impact on the dependent variable.

H_1: The independent variable has significant impact on the dependent variable.

You will look at the P-value of the Chi-square test to make a decision on acceptance or rejection of the hypothesis.

- If the P-value of the Chi-square test is less than 5 percent, you reject the null hypotheses; you reject the null hypothesis that the variable has no significant impact on the dependent variable. This means that the variable has some significant impact; hence, you keep it in the model.

- If the P-value of the Chi-square test is greater than 5 percent, then there is not enough evidence to reject the null hypotheses. So, you accept the null hypothesis that the variable has no significant impact on the dependent variable; hence, you drop it from the model. Dropping such insignificant variables from the model will have no influence on model accuracy.

You can take a look at the Wald Chi-square value when you are comparing two independent variables to decide which variable has a greater impact. If Wald Chi-square value is high, then the P-value is low. For example, if you are comparing the impact of two independent variables such as income and number of dependents on the response variable, then the variable with the higher Wald Chi-square value will be chosen because it has a higher impact on the dependent variable (or response variable).

Table 11-12 is the Chi-square test table from the ice-cream sales example.

Table 11-12. *Chi-square Test, the Output Tables of PROC LOGISTIC*

Analysis of Maximum Likelihood Estimates

Parameter	DF	Estimate	Standard Error	Wald Chi-Square	Pr > ChiSq
Intercept	1	3.8982	1.3446	8.4044	0.0037
Age	1	-0.1353	0.0410	10.8989	0.0010

- H_0: Age has no significant impact on ice-cream purchase.
- H_1: Age has a significant impact on ice-cream purchase

The P-value for the Age variable is 0.0010, which is less than 0.05 (5 percent). So, you can reject the null hypothesis, which would mean that age has a significant impact on ice-cream purchase. Logistic regression:independent variables

Goodness of Fit for Logistic Regression

In this section, we will discuss the Chi-square test in detail, and concordance.

Chi-square Test

The Chi-square value and the associated Chi-square test is a basic measure of goodness of fit in logistic regression. The Chi-square test is used for testing the null hypothesis that all the independent variables' regression coefficients are zero versus at least one of the independent variables is significant. The Chi-square test is similar to F-test in linear regression.

H_0:

- $\beta_1 = \beta_2 = \beta_3 = \beta_4 = \beta_5 = \beta_6 \ldots\ldots = \beta_p = 0$.

- This is same as $\beta_1 = 0 \, and \, \beta_2 = 0 \, and \, \beta_3 = 0 \, and \, \beta_4 = 0 \ldots\ldots and \, \beta_p = 0$.

- This means that the overall model is insignificant or it has no explanatory power.

H_1:

- $\beta_1 \neq 0 \, or \, \beta_2 \neq 0 \, or \, \beta_3 \neq 0 \, or \, \beta_4 \neq 0 \, or \, \beta_5 \neq 0 \, or \, \beta_6 \neq 0 \, or \ldots \beta_p \neq 0$.

- This is same as at least one $\beta \neq 0$.

- This means that at least one variable is significant, or the model has some explanatory power.

If the P-value of the Chi-square test is greater than 5 percent, then the model is at risk. If the P-value is more than 5 percent, then you have to accept the null hypothesis, which means the overall model is insignificant. There is no point in proceeding further, and none of the independent variables has significant impact on the dependent variable. It forces you to collect more data and look for more impactful independent variables. In the SAS output, you see three Chi-square tests—Likelihood Ratio, Score, and Wald—all of them for the same purpose. Generally, all these tests show the same result.

Concordance

As discussed earlier, the concordance measure gives you an estimate of accuracy or the goodness of fit of your logistic regression model. You will consider an example in this section to understand concordance and discordance.

Imagine a simple data set where there are just 10 observations. Five records of the dependent variable take a value of 1; the other five are at 0. After building the model, you will determine the predicted probability of the dependent variable for each record. It looks like Table 11-13.

Table 11-13. *Predicted Probability of Dependent Variable*

DEPENDENT VARIABLE	PREDICTED PROBABILITY
0	P1
0	P2
0	P3
0	P4
0	P5
1	P6
1	P7
1	P8
1	P9
1	P10

You can make as many as 25 pairs using five 0s and five 1s along with their corresponding probabilities.

(0 P1 – 1 P6), (0 P1 – 1 P7), (0 P1 – 1 P8), (0 P1 – 1 P9), (0 P1 – 1 P10)
(0 P2 – 1 P6), (0 P2 – 1 P7), (0 P2 – 1 P8), (0 P2 – 1 P9), (0 P2 – 1 P10)
(0 P3 – 1 P6), (0 P3 – 1 P7), (0 P3 – 1 P8), (0 P3 – 1 P9), (0 P3 – 1 P10)
(0 P4 – 1 P6), (0 P4 – 1 P7), (0 P4 – 1 P8), (0 P4 – 1 P9), (0 P4 – 1 P10)
(0 P5 – 1 P6), (0 P5 – 1 P7), (0 P5 – 1 P8), (0 P5– 1 P9), (0 P5 – 1 P10)

Take the first pair (0 P1 – 1 P6). If you built a model to predict the value 0, then P1 should be higher than P6. A good model should give higher probabilities to zero and lower probabilities to 1. Hence, P1 should be greater than P6 in the first pair, P1 should be greater than P7 in the second pair, and P5 should be greater than P10 in the last pair. This accurate classification is called *concordance*. *Percent concordance* is the percentage of concordance cases in all possible pairs. The higher the percent concordance, the higher the accuracy of the model.

If in the first pair (0 P1 – 1 P6) if P1 is less than P6, then it is not a good sign. You built a model to predict zero, and the model is finally predicting higher probabilities for 1 and lower probability for 0. This misclassification is called *discordance*. The percentage of discordant pairs in overall (all possible pairs) is called *percent discordance*.

For example, take the data pair from the ice-cream sales data.

Age	Buy
6	0
45	1

The pair is (0 P1 – 1 P2). If you substitute the values of age in the model, you get p1 = 95 percent, p2 = 0.6 percent. So, you are correct in predicting that if a person is 6 years old, he has a higher probability of buying the ice cream (p1 = 95 percent), whereas a 45-year-old person has almost zero probability of buying (p2 = 0.6 percent). The predictions are actually matching with the original observations from the table.

This is what you expect from a good model; when you observe 0 (in actual values) and still try to predict it, the model probability should suggest the same. Similarly, when you take 1 and try to predict it using the model, the models should suggest it. When you take a pair (0, 1) and the probabilities are (0 P1 – 1 P2), then at least you expect the model to give you predictions P1 and P2 as P1 > P2. This is concordance. The opposite of this is discordance.

Tied cases are those for which our model gives the same probability of 0.5 for 0 and 0.5 for 1.

Table 11-14 is the output for ice-cream sales example.

Table 11-14. *Association of Predicted Probabilities and Observed Responses, the Output Tables of PROC LOGISTIC on Ice-Cream Sales Data*

Association of Predicted Probabilities and Observed Responses			
Percent Concordant	92.0	Somers' D	0.851
Percent Discordant	6.9	Gamma	0.861
Percent Tied	1.1	Tau-a	0.365
Pairs	525	c	0.926

Total pairs = 15*35 =525 (15 zeros, 35 ones)

- Percent concordant = Percent of right classification = 92.0

- Percent discordant = Percent of wrong classification = 6.9

- Percent tied = 1.1 (100–(92.0+6.9))

The ice-cream sales model that you built is a good one for prediction since the percent concordance is 92 percent. Any concordance greater than 70 percent is good; greater than 80 percent is ideal.

Prediction Using Logistic Regression

The logistic regression line can be stated by substituting the beta coefficients in the following equation:

$$p(y/x) = \frac{e^{\beta_0+\beta_1 x_1+\beta_2 x_2+\beta_3 x_3+\beta_4 x_4+\ldots\ldots+\beta_p x_p}}{1+e^{\beta_0+\beta_1 x_1+\beta_2 x_2+\beta_3 x_3+\beta_4 x_4+\ldots\ldots+\beta_p x_p}}$$

In the ice-cream sales example, you can manually substitute the values of age and get the probability of buying, that is, the probability of y=0. There is a note in the SAS output that says "Probability modeled is buy_ind='0.'" This means that the probability values are for buy_ind=0, or you are getting the probability of 0 when you substitute the value of Age.

$$p(y/x) = \frac{e^{3.8982-0.1353*\text{Age}}}{1+e^{3.8982-0.1353*\text{Age}}}$$

For example, say you want to find the probability that a person who is 12 will buy the ice cream. The following is the substitution:

$$p(y=0/x=12) = \frac{e^{3.8982-0.1353*12}}{1+e^{3.8982-0.1353*12}}$$

$P(y=0/x=12) = 0.907$ (almost 91%)

Multicollinearity in Logistic Regression

If you have multiple independent variables, then, as expected, you will have multiple beta coefficients.

$$p(y/x) = \frac{e^{\beta_0+\beta_1 x_1+\beta_2 x_2+\beta_3 x_3+\beta_4 x_4+\ldots\ldots+\beta_p x_p}}{1+e^{\beta_0+\beta_1 x_1+\beta_2 x_2+\beta_3 x_3+\beta_4 x_4+\ldots\ldots+\beta_p x_p}}$$

We discussed multicollinearity in regression in Chapter 10 . It is an interdependency of the independent variables that upsets the model accuracy by increasing the standard deviation of the beta coefficients. We discussed the effects of multicollinearity and how to detect and remove it in linear regression. Is multicollinearity a problem in multiple logistic regression too? How different is multicollinearity in logistic regression?

Figure 11-10 shows multicollinearity in linear regression. X1, X2, X3, X4, X5...Xp are independent variables, and Y is a dependent variable that is linearly dependent on all of these.

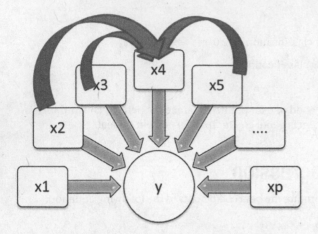

Figure 11-10. *Multicollinearity in linear regression*

You need to take the note of one fact here; multicollinearity is connected with independent variables, and their relation with the dependent variable has nothing to do with it. Whether it is a linear regression or a nonlinear regression, you can have scenarios where the independent variables are interdependent (Figure 11-11).

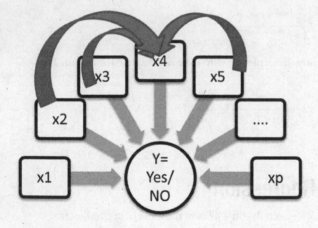

Figure 11-11. *Multicollinearity in nonlinear regression*

In simple terms, multicollinearity is an issue with logistic regression as well. The dependent variable is nowhere on the scene while you are dealing with multicollinearity. So, it doesn't really matter whether you are building a linear or nonlinear regression model. Multicollinearity dents the accuracy of a logistic regression model as well. You have to be careful of multicollinearity while fitting a multiple logistic regression line. The steps are same as discussed in linear regression multicollinearity, shown here:

1. Detect or conform multicollinearity using VIF values. If VIF is more than 5, then there is some multicollinearity within the independent variables.

2. Once you find the existence of multicollinearity, you can drop one of the troublesome variable by looking at the individual variable's Wald Chi-square value. Keep the variables that have a higher Wald Chi-square value.

No VIF Option in PROC LOGISTIC

Multicollinearity has nothing to do with logistic or linear regression. This phenomenon is entirely centered on independent variables. Accordingly, SAS doesn't have a separate option of VIF in logistic regression. You have to follow the same steps as you followed with linear regression as far as the detection of multicollinearity is concerned. For variable selection or dropping, you can use logistic regression. In simple words, PROC LOGISTIC doesn't have a VIF option; only PROC REG has that option.

If you write the VIF option in logistic regression, then SAS gives you an error. The following is one such example:

```
NOTE: The SAS System stopped processing this step because of errors.
704  proc logistic data=loans_data;
705  model SeriousDlqin2yrs = utilization Age Num_loans Num_dependents MonthlyIncome
705! Num_Savings_Acccts DebtRatio/vif;
                          ---
                          22
                          202
ERROR 22-322: Syntax error, expecting one of the following: ;, ABSFCONV, AGGREGATE, ALPHA,
              BACKWARD, BEST, BINWIDTH, BUILDRULE, CL, CLODDS, CLPARM, CODING, CONVERGE,
              CORRB, COVB, CT, CTABLE, DETAILS, DSCALE, EXPB, FAST, FCONV, FIRTH, GCONV,
              HIERARCHY, INCLUDE, INFLUENCE, IPLOTS, ITPRINT, L, LACKFIT, LINK, MAXITER,
              MAXSTEP, NOCHECK, NODUMMYPRINT, NOFIT, NOINT, NOLOGSCALE, OFFSET, OUTROC,
              PARMLABEL, PEVENT, PL, PLCL, PLCONV, PLRL, PPROB, PSCALE, RIDGING, RISKLIMITS,
              ROCEPS, RSQUARE, SCALE, SELECTION, SEQUENTIAL, SINGULAR, SLE, SLENTRY, SLS,
              SLSTAY, START, STB, STEPWISE, STOP, STOPRES, TECHNIQUE, WALDCL, WALDRL, XCONV.
ERROR 202-322: The option or parameter is not recognized and will be ignored.
706  run;
```

So, in the case of logistic regression, you have to use PROC REG and follow the procedure explained in Chapter 10.

Logistic Regression Final Check List

Up to this point, we have discussed all the important concepts related to regression analysis. Logistic regression is just another type of regression. It is one of the most widely used nonlinear regression models across numerous industry verticals. The following is a check list to be used while building a logistic regression model:

1. *Applicability*: Look at the data and the dependent variables. Is it categorical? Yes/No, 0/1, Win/Loss, and so on, are the types of response variable outcomes where you can apply logistic regression.

2. *Chi-square value*: Look at the overall Chi-square value to decide whether a model is significant. If the Chi-square test fails, then stop it right there; the overall model itself is not significant. Chi-square is not going to tell you anything about the precise accuracy of a model.

3. *VIF*: The multicollinearity issue needs to be solved in same fashion as in linear regression models (using PROC REG with the VIF option). Identify and resolve it by dropping the troublesome variables or by using PCA or FA, as explained in regression analysis in Chapter 10.

4.   *Overall accuracy/concordance*: Determine the accuracy of a mode by looking at its concordance and C values. The higher the value, the better the model. If the values of concordance and C are not satisfactory, then you may think of collecting some more data or adding better, more impactful independent variables, which will improve the overall model performance.

5.   *Individual impact/Wald Chi-square value*: Look at the individual impact of each variable by looking at the Wald Chi-square value. Drop the insignificant variables

Loan Default Prediction Case Study

As we have done throughout this book, we will revise the concepts learned in this chapter using a real-life case study.

Background and Problem Statement

ABC Bank has a personal loan product. It needs to be very sure about the payback capability of customers before it can approve a loan. The bank gets thousands of loan applications from its prospective customers, and it needs to have a flawless scientific method to approve or reject the loan applications. The bank decides to use data analytics to judge the risk associated with each customer.

ABC Bank has tons of historical data on customer profiles, their previous loans, their credit card spending, their bank accounts, and so on. The bank has historic data of both creditworthy and non-credit-worthy customers. Identifying the non-credit-worthy applicants before approving a loan saves millions of dollars for the bank. So, it is important to quantify the risk associated with each customer at the time of loan approval.

Objective

Analyze the historical loans data and build a model that will help the bank segregate the customers, based on the associated risk, at the time of loan approval.

Data Set

The bank's historical data contains the records of both defaults and nondefaulters. It is collected over a period of two years. The data set has a record of 150,000 customers. It also has a response/dependent variable, which indicates the creditworthiness of each customer. Some predictor/independent variables are also there in each record, which indicate the customer's financial health. The data set given here is reasonably clean for the model-building exercise. The data validation and sanitization have already been done. Also, there are no outliers or missing values in the variables.

Table 11-15 contains the details of the data set (the data dictionary).

Table 11-15. *The Details of Data Set Used in the ABC Bank Example*

Variable Name	Description	Type
Cust_id	Customer ID; unique for each customer.	Number
SeriousDlqin2yrs	Person experienced BK/CO.	1/0
	BK: Bankruptcy. This is when a debtor realizes that she is unable to pay her bills. When any individual or business is unable to pay their *creditors*, bankruptcy is a legal option that can help them to get relief from their due payments and debt.	
	CO: A customer fails to make the minimum payment due for a period of six months or more.	
	1 – Yes BK or CO: Bad customer.	
	0 – No BK or CO: Good customer.	
Utilization	Average monthly utilization of credit limit.	Percentage
Age	Age of the customer.	Number
Num_loans	Number of loans including mortgage, vehicle, and personal loans.	Number
Num_dependents	Number of dependents.	
MonthlyIncome	Average monthly income of the customer.	Number
Num_Savings_Acccts	Number of savings accounts a customer has.	Number
DebtRatio	Monthly debt payments, alimony, living costs divided by monthly gross income.	Percentage

Data Import

As a first step, you will import the data and do a quick check on the number of records, variables, formats, and so on.

The following is the code for importing the data set Customer_Loans.csv:

```
PROC IMPORT OUT= WORK.loans_data
            DATAFILE= "C:\Users\VENKAT\Google Drive\Training\Books\Content\11.
Logistic Regression\Data Logistic Regression\Customer_Loans.csv"
            DBMS=CSV REPLACE;
     GETNAMES=YES;
     DATAROW=2;
RUN;
```

Here are the notes from the log file for the previous code:

```
NOTE: The infile 'C:\Users\VENKAT\Google Drive\Training\Books\Content\11. Logistic
    Regression\Data Logistic Regression\Customer_Loans.csv' is:

    Filename=C:\Users\VENKAT\Google Drive\Training\Books\Content\11. Logistic
    Regression\Data Logistic Regression\Customer_Loans.csv,
    RECFM=V,LRECL=32767,File Size (bytes)=6918139,
    Last Modified=28Aug2014:11:52:33,
    Create Time=28Aug2014:11:12:22
```

```
NOTE: 150000 records were read from the infile 'C:\Users\VENKAT\Google
      Drive\Training\Books\Content\11. Logistic Regression\Data Logistic
      Regression\Customer_Loans.csv'.
      The minimum record length was 26.
      The maximum record length was 48.
NOTE: The data set WORK.LOANS_DATA has 150000 observations and 9 variables.
NOTE: DATA statement used (Total process time):
      real time            1.23 seconds
      cpu time             1.24 seconds

150000 rows created in WORK.LOANS_DATA from C:\Users\VENKAT\Google
Drive\Training\Books\Content\11. Logistic Regression\Data Logistic
Regression\Customer_Loans.csv.

NOTE: WORK.LOANS_DATA data set was successfully created.
NOTE: PROCEDURE IMPORT used (Total process time):
      real time            1.49 seconds
      cpu time             1.51 seconds
```

The following is the code for PROC CONTENTS and print data to see the variables and the snapshot of the data:

```
proc contents data=loans_data varnum;
run;

proc print data=loans_data(obs=20);
run;
```

Tables 11-16 and 11-17 list the output of these code snippets.

Table 11-16. The Output of proc contents Procedure on loans_data Data Set

The SAS System

The CONTENTS Procedure

Data Set Name	WORK.LOANS_DATA	Observations	150000
Member Type	DATA	Variables	9
Engine	V9	Indexes	0
Created	Tuesday, July 01, 2008 12:00:00 AM	Observation Length	72
Last Modified	Tuesday, July 01, 2008 12:00:00 AM	Deleted Observations	0
Protection		Compressed	NO
Data Set Type		Sorted	NO
Label			
Data Representation	WINDOWS_32		
Encoding	wlatin1 Western (Windows)		

Engine/Host Dependent Information

Data Set Page Size	8192
Number of Data Set Pages	1328
First Data Page	1
Max Obs per Page	113
Obs in First Data Page	85
Number of Data Set Repairs	0
Filename	C:\Users\VENKAT\AppData\Local\Temp\SAS Temporary Files_TD4892\loans_data.sas7bdat
Release Created	9.0201M0
Host Created	W32_VSPRO

Variables in Creation Order

#	Variable	Type	Len	Format	Informat
1	Cust_id	Num	8	BEST12.	BEST32.
2	SeriousDlqin2yrs	Num	8	BEST12.	BEST32.
3	Utilization	Num	8	BEST12.	BEST32.
4	Age	Num	8	BEST12.	BEST32.
5	Num_loans	Num	8	BEST12.	BEST32.
6	Num_dependents	Num	8	BEST12.	BEST32.
7	MonthlyIncome	Num	8	BEST12.	BEST32.
8	Num_Savings_Acccts	Num	8	BEST12.	BEST32.
9	DebtRatio	Num	8	BEST12.	BEST32.

Table 11-17. The Output of proc print Procedure on loans_data Data Set (obs=20)

Obs	Cust_id	Serious Dlqin2yrs	utilization	Age	Num_ loans	Num_ dependents	Monthly Income	Num_ Savings_ Acccts	Debt Ratio
1	100001	1	0.766126609	45	6	2	9120	2	0.852982129
2	100002	0	0.957151019	40	0	1	4600	2	0.121876201
3	100003	0	0.65818014	38	0	0	5042	2	0.085113375
4	100004	0	0.233809776	30	0	0	5300	2	0.036049682
5	100005	0	0.9072394	49	1	0	6357	2	0.024925695
6	100006	0	0.213178682	74	1	1	5500	2	0.375606969
7	100007	0	0.305682465	57	3	0	6357	2	0.274502

(*continued*)

Table 11-17. (*continued*)

Obs	Cust_id	Serious Dlqin2yrs	utilization	Age	Num_ loans	Num_ dependents	Monthly Income	Num_ Savings_ Acccts	Debt Ratio
8	100008	0	0.754463648	39	0	0	5500	2	0.209940017
9	100009	0	0.116950644	27	0	0	6357	2	0.274502
10	100010	0	0.189169052	57	4	2	6357	2	0.606290901
11	100011	0	0.644225962	30	0	0	4500	2	0.30947621
12	100012	0	0.01879812	51	2	2	8501	2	0.53152876
13	100013	0	0.010351857	46	2	2	14454	2	0.298354075
14	100014	1	0.964672555	40	1	2	13700	2	0.432964747
15	100015	0	0.019656581	76	1	0	6357	2	0.274502
16	100016	0	0.548458062	64	1	2	13362	2	0.209891754
17	100017	0	0.061086118	78	2	0	6357	2	0.274502
18	100018	0	0.166284079	53	0	0	10800	1	0.18827406
19	100019	0	0.221812771	43	1	2	5280	2	0.527887839
20	100020	0	0.602794411	25	0	0	2333	2	0.065868263

You have all the variables in the data, and it looks like all the variables are populated. The data looks clean. If not, you need to first explore the data, validate it for the accuracy, and finally clean it by treating missing values and outliers in the data. You need to use SAS procedures such as PROC CONTENTS, PROC FREQ, and PROC UNIVARIATE, which have been discussed at length in data exploration in Chapter 7.

Model Building

The response variable in the data set is binary. SeriousDlqin2yrs is the dependent variable that you are planning to predict. It takes two values:

- 1, Yes BK or CO: Bad customer

- 0, No BK or CO: Good customer

Logistic regression is a good choice for predicting the default probability of a customer based on characteristics such as number of loans, debt ratio, monthly income, and so on.

The following is the code for building a logistic regression line:

```
proc logistic data=loans_data;
model SeriousDlqin2yrs = utilization Age Num_loans Num_dependents MonthlyIncome
Num_Savings_Acccts DebtRatio;
run;
```

The following are the messages from the log file when the previous code was executed:

```
701  proc logistic data=loans_data;
702  model SeriousDlqin2yrs = utilization Age Num_loans Num_dependents MonthlyIncome
702! Num_Savings_Acccts DebtRatio;
703  run;

NOTE: Writing HTML Body file: sashtml44.htm
NOTE: PROC LOGISTIC is modeling the probability that SeriousDlqin2yrs='0'. One way to change
      this to model the probability that SeriousDlqin2yrs='1' is to specify the response
      variable option EVENT='1'.
NOTE: Convergence criterion (GCONV=1E-8) satisfied.
NOTE: There were 150000 observations read from the data set WORK.LOANS_DATA.
NOTE: PROCEDURE LOGISTIC used (Total process time):
      real time            4.92 seconds
      cpu time             3.58 seconds
```

Table 11-18 shows the SAS output for this code.

Table 11-18. *The Output of proc logistic on loans_data*

The SAS System

The LOGISTIC Procedure

Model Information

Data Set	WORK.LOANS_DATA
Response Variable	SeriousDlqin2yrs
Number of Response Levels	2
Model	binary logit
Optimization Technique	Fisher's scoring

Number of Observations Read	150000
Number of Observations Used	150000

Response Profile

Ordered Value	SeriousDlqin2yrs	Total Frequency
1	0	139974
2	1	10026

Probability modeled is SeriousDlqin2yrs='0'.

Model Convergence Status

Convergence criterion (GCONV=1E-8) satisfied.

Model Fit Statistics

Criterion	Intercept Only	Intercept and Covariates
AIC	73618.167	57641.254
SC	73628.085	57720.601
-2 Log L	73616.167	57625.254

Testing Global Null Hypothesis: BETA=0

Test	Chi-Square	DF	Pr > ChiSq
Likelihood Ratio	15990.9128	7	<.0001
Score	14078.3178	7	<.0001
Wald	11514.4825	7	<.0001

Analysis of Maximum Likelihood Estimates

Parameter	DF	Estimate	Standard Error	Wald Chi-Square	Pr > ChiSq
Intercept	1	1.1431	0.0622	337.3241	<.0001
utilization	1	-1.9713	0.0312	3999.3621	<.0001
Age	1	0.0108	0.000832	167.2586	<.0001
Num_loans	1	-0.2464	0.0106	538.4191	<.0001
Num_dependents	1	-0.1941	0.00960	408.6747	<.0001
MonthlyIncome	1	0.000426	5.983E-6	5078.5897	<.0001
Num_Savings_Acccts	1	-0.00945	0.00517	3.3408	0.0676
DebtRatio	1	-1.1702	0.0510	526.4205	<.0001

Odds Ratio Estimates

Effect	Point Estimate	95% Wald Confidence Limits	
Utilization	0.139	0.131	0.148
Age	1.011	1.009	1.012
Num_loans	0.782	0.766	0.798
Num_dependents	0.824	0.808	0.839
MonthlyIncome	1.000	1.000	1.000
Num_Savings_Acccts	0.991	0.981	1.001
DebtRatio	0.310	0.281	0.343

Association of Predicted Probabilities and Observed Responses

Percent Concordant	82.8	Somers' D	0.664
Percent Discordant	16.5	Gamma	0.668
Percent Tied	0.7	Tau-a	0.083
Pairs	1403379324	c	0.832

Now we will discuss the output in detail.

Some 150,000 customer records are read from the data set (Table 11-19) and are being used for this analysis. There are 139,974 good accounts and 10,026 bad accounts (Table 11-20). In the real-life bank loan portfolios, you would expect more than 90 percent of customers are good.

Table 11-19. Number of Observations in loans_data

Number of Observations Read	150000
Number of Observations Used	150000

Table 11-20. Frequency of 0 and 1 in SeriousDlqin2yr Variable

Response Profile

Ordered Value	SeriousDlqin2yrs	Total Frequency
1	0	139974
2	1	10026

Probability modeled is SeriousDlqin2yrs='0

You may be wondering what you are trying to predict if good and bad accounts are already given. Why don't you directly use them? Remember, this is historical data, where you already know who is good and who is bad. You are trying to use this data to fit a model, which will predict the probability of default (by new customers), given independent variables such as income, loans, debits, and utilization. You can use this model for future loan approvals for the new customers, even if they are not in the historic data. For all future customers, you will solicit the data on independent variables at the time of loan application, which will be used to assess the risk associated with each customer.

Of the following tables, Table 11-21 talks about the model convergence and Table 11-22 gives the Chi-square test results.

Table 11-21. *Model Convergence Status*

Model Convergence Status
Convergence criterion (GCONV=1E-8) satisfied.

Table 11-22. *Chi-square Test Results for the Model*

Testing Global Null Hypothesis: BETA=0			
Test	Chi-Square	DF	Pr > ChiSq
Likelihood Ratio	15990.9128	7	<.0001
Score	14078.3178	7	<.0001
Wald	11514.4825	7	<.0001

Looking at the Chi-square test results of the output (Table 11-22), you can confirm that the model is significant. All the P-values for the Chi-square tests are less than 5 percent (0.05). So, it is safe to conclude that the model is significant. In other words, at least one variable in the independent variables list has a significant impact on the response/dependent variable (SeriousDlqin2yrs). Take a look at the goodness of fit discussed earlier in the chapter to get a clear idea of how to interpret Table 11-22.

Table 11-23 is about variable coefficient estimates. The variable signs indicate that some of them have positive impact on dependent variable, while others impact negatively. Before looking at the sign of coefficients and their values or making any attempt to write the model equation, you need to make sure there is no interdependency within the independent variables. You are well aware that multicollinearity is a concern in logistic regression as well. You can't trust these coefficients unless you make sure that there is no multicollinearity in the model.

Table 11-23. *Variable Coefficient Estimates*

Analysis of Maximum Likelihood Estimates					
Parameter	DF	Estimate	Standard Error	Wald Chi-Square	Pr > ChiSq
Intercept	1	1.1431	0.0622	337.3241	<.0001
utilization	1	-1.9713	0.0312	3999.3621	<.0001
Age	1	0.0108	0.000832	167.2586	<.0001
Num_loans	1	-0.2464	0.0106	538.4191	<.0001
Num_dependents	1	-0.1941	0.00960	408.6747	<.0001
MonthlyIncome	1	0.000426	5.983E-6	5078.5897	<.0001
Num_Savings_Acccts	1	-0.00945	0.00517	3.3408	0.0676
DebtRatio	1	-1.1702	0.0510	526.4205	<.0001

Handling Multicollinearity

To reiterate, multicollinearity is an issue affected by only the list of independent variables. It has nothing to do with the relation of the independent variables to the dependent variables. Whether you are building a linear or logistic regression line, the treatment of the multicollinearity issue remains the same.

The following is the code for detecting multicollinearity using the VIF option:

```
proc logistic data=loans_data;
model SeriousDlqin2yrs = utilization Age Num_loans Num_dependents MonthlyIncome
Num_Savings_Acccts DebtRatio/vif;
run;
```

Remember that PROC LOGISTIC has no VIF option. If a VIF option is used, this PROC code throws an error, as shown here:

```
NOTE: The SAS System stopped processing this step because of errors.
704  proc logistic data=loans_data;
705  model SeriousDlqin2yrs = utilization Age Num_loans Num_dependents MonthlyIncome
705! Num_Savings_Acccts DebtRatio/vif;
                       ---
                       22
                       202
ERROR 22-322: Syntax error, expecting one of the following: ;, ABSFCONV, AGGREGATE, ALPHA,
              BACKWARD, BEST, BINWIDTH, BUILDRULE, CL, CLODDS, CLPARM, CODING, CONVERGE,
              CORRB, COVB, CT, CTABLE, DETAILS, DSCALE, EXPB, FAST, FCONV, FIRTH, GCONV,
              HIERARCHY, INCLUDE, INFLUENCE, IPLOTS, ITPRINT, L, LACKFIT, LINK, MAXITER,
              MAXSTEP, NOCHECK, NODUMMYPRINT, NOFIT, NOINT, NOLOGSCALE, OFFSET, OUTROC,
              PARMLABEL, PEVENT, PL, PLCL, PLCONV, PLRL, PPROB, PSCALE, RIDGING, RISKLIMITS,
              ROCEPS, RSQUARE, SCALE, SELECTION, SEQUENTIAL, SINGULAR, SLE, SLENTRY, SLS,
              SLSTAY, START, STB, STEPWISE, STOP, STOPRES, TECHNIQUE, WALDCL, WALDRL, XCONV.
ERROR 202-322: The option or parameter is not recognized and will be ignored.
706  run;
```

So, you need to use PROC REG for the treatment of multicollinearity here as well.

```
proc reg data=loans_data;
model SeriousDlqin2yrs = utilization Age Num_loans Num_dependents MonthlyIncome
Num_Savings_Acccts DebtRatio/vif;
run;
```

Table 11-24 shows the output for the previous code.

Table 11-24. Output of proc reg on loans_data

The SAS System

The REG Procedure

Model: MODEL1

Dependent Variable: SeriousDlqin2yrs

Number of Observations Read	150000
Number of Observations Used	150000

Analysis of Variance

Source	DF	Sum of Squares	Mean Square	F Value	Pr > F
Model	7	87809867	125.44267	2219.38	<.0001
Error	149992	8477.76349	0.05652		
Corrected Total	149999	9355.86216			

Root MSE	0.23774	**R-Square**	0.0939
Dependent Mean	0.06684	**Adj R-Sq**	0.0938
Coeff Var	355.68877		

Parameter Estimates

Variable	DF	Parameter Estimate	Standard Error	t Value	Pr > |t|	Variance Inflation
Intercept	1	0.11651	0.00342	34.11	<.0001	0
utilization	1	0.14700	0.00194	75.88	<.0001	1.11193
Age	1	-0.00057428	0.00004428	-12.97	<.0001	1.13520
Num_loans	1	0.00849	0.00067032	12.67	<.0001	1.45625
Num_dependents	1	0.01100	0.00058660	18.75	<.0001	1.11697
MonthlyIncome	1	-0.00001339	1.931776E-7	-69.33	<.0001	1.22390
Num_Savings_Acccts	1	0.00050943	0.00028330	1.80	0.0722	1.00006
DebtRatio	1	0.06312	0.00356	17.74	<.0001	1.33570

In this case, you don't really care about R-square or the adjusted R-square. The only table that you are going to focus on in this output is the Parameter Estimates table. In particular, you are interested only in VIF values, which give you an indication about the multicollinearity. None of the variables has a VIF value of more than 5, which indicates that there is no interdependency in the independent variables. So, you can keep all the independent variables while building the logistic regression model.

Predicting Delinquency

You can see a note in the output saying "Probability modeled is SeriousDlqin2yrs='0,'" which means the resultant probability from the model will be the probability of good. This would mean the higher the probability, the better the customer. If you want to predict the probability of bad instead of the probability of good, you have to use the descending option in the PROC LOGISTIC code. Most of the time, you try to predict the probability of the default rather than the probability of the nondefault. The following is the code for the final model. Remember, this will change the coefficients as well.

```
proc logistic data=loans_data descending;
model SeriousDlqin2yrs = utilization Age Num_loans Num_dependents MonthlyIncome
Num_Savings_Acccts DebtRatio;
run;
```

Table 11-25 shows the output of this code.

Table 11-25. Output of proc logistic on loans_data

The SAS System	
The LOGISTIC Procedure	
Model Information	
Data Set	WORK.LOANS_DATA
Response Variable	SeriousDlqin2yrs
Number of Response Levels	2
Model	binary logit
Optimization Technique	Fisher's scoring

Number of Observations Read	150000
Number of Observations Used	150000

Response Profile		
Ordered Value	SeriousDlqin2yrs	Total Frequency
1	1	10026
2	0	139974

Probability modeled is SeriousDlqin2yrs='1'.

Model Convergence Status

Convergence criterion (GCONV=1E-8) satisfied.

Model Fit Statistics

Criterion	Intercept Only	Intercept and Covariates
AIC	73618.167	57641.254
SC	73628.085	57720.601
-2 Log L	73616.167	57625.254

Testing Global Null Hypothesis: BETA=0

Test	Chi-Square	DF	Pr > ChiSq
Likelihood Ratio	15990.9128	7	<.0001
Score	14078.3178	7	<.0001
Wald	11514.4825	7	<.0001

Analysis of Maximum Likelihood Estimates

Parameter	DF	Estimate	Standard Error	Wald Chi-Square	Pr > ChiSq
Intercept	1	-1.1431	0.0622	337.3241	<.0001
utilization	1	1.9713	0.0312	3999.3621	<.0001
Age	1	-0.0108	0.000832	167.2586	<.0001
Num_loans	1	0.2464	0.0106	538.4191	<.0001
Num_dependents	1	0.1941	0.00960	408.6747	<.0001
MonthlyIncome	1	-0.00043	5.983E-6	5078.5897	<.0001
Num_Savings_Acccts	1	0.00945	0.00517	3.3408	0.0676
DebtRatio	1	1.1702	0.0510	526.4205	<.0001

Odds Ratio Estimates

Effect	Point Estimate	95% Wald Confidence Limits	
utilization	7.180	6.755	7.633
Age	0.989	0.988	0.991
Num_loans	1.279	1.253	1.306
Num_dependents	1.214	1.192	1.237
MonthlyIncome	1.000	1.000	1.000
Num_Savings_Acccts	1.009	0.999	1.020
DebtRatio	3.223	2.916	3.561

Association of Predicted Probabilities and Observed Responses			
Percent Concordant	82.8	Somers' D	0.664
Percent Discordant	16.5	Gamma	0.668
Percent Tied	0.7	Tau-a	0.083
Pairs	1403379324	c	0.832

Now you can observe that the note has changed in the SAS output. If a customer scores high in this model, then she has a higher probability to default in the repayment of the loan. The following is the note, reproduced from the SAS output in Table 11-25.

Probability modeled is SeriousDlqin2yrs='1'.

Goodness of Fit Statistic

The concordance is 82.8 percent, and the discordance is 16.5 percent, which is really good for any logistic regression model. In real-life business problems, generally it is desirable to have a concordance of 75 percent or more. You can go ahead and use this model for prediction with greater confidence.

Insignificant Variables

Before you use this model for predictions, you have to see whether there are any independent variables that have no significant impact on the dependent variable. You can take a look at the P-value of individual independent variables (Table 11-26) to get an idea on their importance in the model. Any variable with a P-value of more than 5 percent in the Chi-square test is insignificant in the model. The predictions using this model will not be affected at all by keeping or removing such variables.

Table 11-26. *Variable Coefficient Estimates*

Analysis of Maximum Likelihood Estimates					
Parameter	DF	Estimate	Standard Error	Wald Chi-Square	Pr > ChiSq
Intercept	1	-1.1431	0.0622	337.3241	<.0001
utilization	1	1.9713	0.0312	3999.3621	<.0001
Age	1	-0.0108	0.000832	167.2586	<.0001
Num_loans	1	0.2464	0.0106	538.4191	<.0001
Num_dependents	1	0.1941	0.00960	408.6747	<.0001
MonthlyIncome	1	-0.00043	5.983E-6	5078.5897	<.0001
Num_Savings_Acccts	1	0.00945	0.00517	3.3408	0.0676
DebtRatio	1	1.1702	0.0510	526.4205	<.0001

In this data set, the number of savings accounts has no significant impact on the delinquency of a customer (P-value = 0.0676). Table 11-27 shows the goodness of fit statistics keeping the number of savings account variable in the model.

Table 11-27. *Goodness of Fit Statistics*

Percent Concordant	82.8
Percent Discordant	16.5

Let's see how the accuracy will change if you drop this variable from the final model. Ideally there should be no effect on the accuracy of the model because the variable is insignificant.

```
/* Final Model after Dropping Num_Savings_Acccts */
proc logistic data=loans_data descending;
model SeriousDlqin2yrs = utilization Age Num_loans Num_dependents MonthlyIncome DebtRatio;
run;
```

The following is the output, which indicates there is absolutely no change in accuracy even after dropping the number of savings accounts variable.

Percent Concordant	82.8
Percent Discordant	16.5

Final Model

The following is the code for creating the final logistic regression model:

```
proc logistic data=loans_data descending;
model SeriousDlqin2yrs = utilization Age Num_loans Num_dependents MonthlyIncome DebtRatio;
run;
```

Table 11-28 is the output for the previous code.

Table 11-28. *Output of proc logistic after Dropping Num_Savings_Acccts*

The SAS System			
The LOGISTIC Procedure			
Model Information			
Data Set	WORK.LOANS_DATA	**Response Variable**	SeriousDlqin2yrs
Number of Response Levels	2		
Model	binary logit		
Optimization Technique	Fisher's scoring		

Number of Observations Read	150000
Number of Observations Used	150000

Response Profile

Ordered Value	SeriousDlqin2yrs	Total Frequency
1	1	10026
2	0	139974

Probability modeled is SeriousDlqin2yrs='1'.

Model Convergence Status

Convergence criterion (GCONV=1E-8) satisfied.

Model Fit Statistics

Criterion	Intercept Only	Intercept and Covariates
AIC	73618.167	57642.594
SC	73628.085	57712.023
-2 Log L	73616.167	57628.594

Testing Global Null Hypothesis: BETA=0

Test	Chi-Square	DF	Pr > ChiSq
Likelihood Ratio	15987.5724	6	<.0001
Score	14075.3877	6	<.0001
Wald	11511.9253	6	<.0001

Analysis of Maximum Likelihood Estimates

Parameter	DF	Estimate	Standard Error	Wald Chi-Square	Pr > ChiSq
Intercept	1	-1.0930	0.0559	382.7749	<.0001
utilization	1	1.9707	0.0312	3997.7947	<.0001
Age	1	-0.0108	0.000831	167.3011	<.0001
Num_loans	1	0.2464	0.0106	538.4647	<.0001
Num_dependents	1	0.1941	0.00960	408.8748	<.0001
MonthlyIncome	1	-0.00043	5.983E-6	5078.5652	<.0001
DebtRatio	1	1.1706	0.0510	526.7798	<.0001

Odds Ratio Estimates

Effect	Point Estimate	95% Wald Confidence Limits	
utilization	7.175	6.750	7.627
Age	0.989	0.988	0.991
Num_loans	1.279	1.253	1.306
Num_dependents	1.214	1.192	1.237
MonthlyIncome	1.000	1.000	1.000
DebtRatio	3.224	2.917	3.563

Association of Predicted Probabilities and Observed Responses

Percent Concordant	82.8	Somers' D	0.664
Percent Discordant	16.5	Gamma	0.668
Percent Tied	0.7	Tau-a	0.083
Pairs	1403379324	c	0.832

The following is the final check list:

- *Applicability*: A linear regression model can't be applied for this data because the response variable is categorical (0 and 1). Logistic regression is a perfect fit for this dependent variable, which is binary in nature.

- *Chi-square test*: The P-value of the overall model Chi-square test is less than 5 percent, so the model can be termed significant.

- *VIF*: None of the variables has a VIF of more than 5 percent, so there is absolutely no multicollinearity.

- *Concordance*: The model concordance is 82.8 percent, which indicates that the resultant model is going to be a good predictor.

- *Individual variable impact*: The P-values for all the independent variables (finally included in the model) are less than 5 percent, which indicates there are no insignificant variables in the final model. One insignificant variable was already dropped.

Final Model Equation and Prediction Using the Model

Here is the final model after substituting all the coefficients:

$$P(y/x) = \frac{e^{\beta_0 + \beta_1 x_1 + \beta_2 x_2 + \beta_3 x_3 + \beta_4 x_4 + \ldots\ldots\ldots + \beta_p x_p}}{1 + e^{\beta_0 + \beta_1 x_1 + \beta_2 x_2 + \beta_3 x_3 + \beta_4 x_4 + \ldots\ldots\ldots + \beta_p x_p}}$$

$$P(Default) = \frac{e^{-1.09301.9707 + (-0.0108 * Age) + (0.2464 * Num\ Loans) + (0.1941 Num\ Dependent) + (-0.00043 * Income) + (1.1706 * \textbf{Debt Ratio})}}{1 + e^{-1.09301.9707 + (-0.0108 * Age) + (0.2464 * Num\ Loans) + (0.1941 Num\ Dependent) + (-0.00043 * Income) + (1.1706 * \textbf{Debt Ratio})}}$$

Predictions Using the Model

The bank can use the final model for quantifying the risk associated with each customer at the time of loan approval. The model gives the probability of default, so the bank can set a minimum limit, which may be that if any customer has a probability of default greater than 50 percent, as given by this model, the application will be rejected.

For example, consider the three customers shown in Table 11-29 with the details furnished in their loan applications.

Table 11-29. *Details in the Loan Application for Example Customers*

	Tom	David	Hanks
utilization	30%	70%	70%
Age	34	40	50
Num_loans	2	5	5
Num_dependents	3	4	4
MonthlyIncome	8000	5000	2500
DebtRatio	20%	45%	55%

You can get the calculated probability values (Table 11-30) by substituting these values in the logistic regression equation:

$$P\left(Default\right) = \frac{e^{-1.09301.9707+\left(-0.0108*Age\right)+\left(0.2464*Num\ Loans\right)+\left(0.1941Num\ Dependent\right)+\left(-0.00043*\ Income\right)+\left(1.1706*\textbf{Debt Ratio}\right)}}{1+e^{-1.09301.9707+\left(-0.0108*Age\right)+\left(0.2464*Num\ Loans\right)+\left(0.1941Num\ Dependent\right)+\left(-0.00043*\ Income\right)+\left(1.1706*\textbf{Debt Ratio}\right)}}$$

Table 11-30. *The Predicted Default Probabilities*

Customer	Predicted Default Probability	Decision
Tom	0.0047 (5%)	Approve the loan
David	0.56 (56%)	Reject the loan application
Hanks	0.79 (79%)	Reject the loan application

So, only Tom gets a loan. Generally, banks might not reject all the customers with a greater than 50 percent probability. They might approve lesser value loans for the risky customers or charge higher interest rates and reduce the loan amount. Table 11-31 might be an example of a moderated implementation of the final defaulter model.

Table 11-31. *Decision Making Using the Default Probability Range*

Default Probability Range	Decision
0%-45%	Approve loan application
45%-65%	Approve loan application with reduced loan amount
65%	Decline the loan application

Banks will send this model and the decision table to their front-end teams or install the model in a centralized system, which can be accessed by its loans processing team. It will be used as a primary tool for assessing any loan application. The decision methodology is scientific and reasonably accurate and uses analytics on the associated customer data.

Conclusion

In this chapter, we discussed that linear regression is not applicable for binary dependent variables. A logistic curve is best suited for such dependent variables. If the dependent variable has only two levels, you need to use logistic regression, which is also known as binomial logistic regression. If the response variable has more than two levels, such as good/bad/indeterminate, win/loss/draw, or satisfied/dissatisfied/neutral, then you need to use multinomial logistic regression.

You learned how to fit a logistic regression line using SAS and also observed some of the similarities and differences in the output (with linear regression). The logistic regression line also suffers from multicollinearity issues; however, the process of treating multicollinearity is the same as in linear regression. There are other nonlinear regressions as well. Logistics is not the only nonlinear regression. You have finished the regression-related topics with this chapter and will get into other methods of modeling such as ARIMA in the following chapters.

CHAPTER 12

■ ■ ■

Time-Series Analysis and Forecasting

In this chapter, we introduce the concept of time-series forecasting straight using a couple of examples. We'll talk about a popular bank, which has been offering credit cards to its customers for more than a decade. Every month the bank has some customers who file for bankruptcy. Some customers have defaulted on their loans for more than six months, and finally they need to be taken off the books. Obviously, this is a loss to the bank, in terms of dollars and accounts. Can the bank proactively estimate these losses for each of its portfolios? Can the bank proactively take necessary steps to control the loss and find out whether it is going to soar in the future? How does the bank forecast such losses based on historical data?

Let's consider another example. A contact center or business process outsourcing (BPO) process receives and processes thousands of calls every month. The number of calls is not the same every day. The call volume is different depending upon whether it's a weekend or a weekday and whether it's morning or evening. Can the BPO center forecast the call volumes so that it can better manage the resources and employee shifts? Again, one question arises: how do you forecast the future call volumes?

Another example is a laptop computer manufacturing company that has been in business for the last three decades. It has a couple of new models that will be launched in the next few months. Can the company determine the total addressable market? Can it estimate the sales of the next two quarters based on historical data?

Many companies track variables related to the economy such as gross domestic product (GDP), growth rate index, unemployment rate, and so on. Based on historical values, can you forecast the future values of these variables? If yes, everyone would like to have those numbers.

Similarly, can you forecast a stock price based on the past several years of data? Can you forecast the number of visitors for a web site? How do you forecast the number of viewers for a TV program?

There are many such business scenarios that involve forecasting variables that vary with time. The forecasting time lines for the variables can be days, months, quarters, years, or even seconds and milliseconds. Basically, it can be any unit that represents a time line. You can use time-series analysis and forecasting techniques for most of these business scenarios. These techniques are the topic of discussion in this chapter.

What Is a Time-Series Process?

Any process that varies over time is a time-series process provided the interval is fixed. If you are talking about monthly data and one month is the interval between two time points, then you should have one value of the variable every month for it to be defined as a time-series process.

Time-series data is a sequence of records collected from a process with equally spaced intervals in time. In simple terms, any metric that is measured over time is a time-series process, given that the time interval is the same between any two consecutive points. In the following text, you will consider some time-series plots as examples.

You denote a time series with Y_t, where Y is the value of the metric at time point t. Refer to Table 12-1, which is charted in Figure 12-1. It is an example of the daily call volume of the last two years for a voice-processing BPO.

Table 12-1. *Daily Call Volume of a Voice-Processing BPO*

Day	Denoted by	Call Volume
10/25/2014	Y_t	**99,552**
10/24/2014	Y_{t-1}	**113,014**
10/23/2014	Y_{t-2}	**118,502**
10/22/2014	Y_{t-3}	**119,437**
10/21/2014		**123,410**
10/20/2014		**129,368**
10/19/2014		**135,961**
10/18/2014		**131,698**
…	…	….
….	…	….
2/23/2014	Y_{t-p}	**113,214**

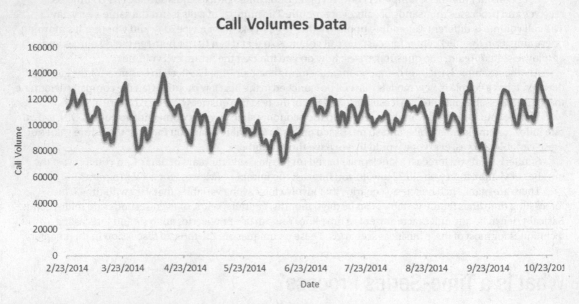

Figure 12-1. *Daily call volume for a voice-processing BPO*

Figure 12-2 shows an example of time-series data for monthly laptop computer sales.

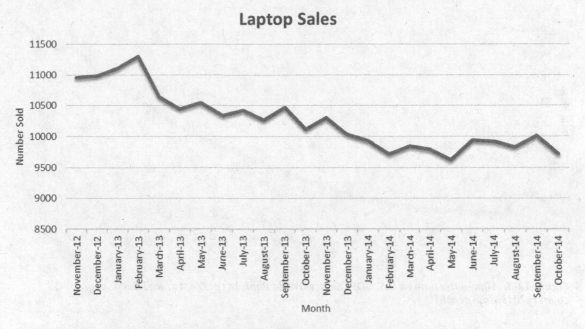

Figure 12-2. *Time-series data on monthly laptop sales*

Figure 12-3 shows the time-series chart of the United States' GDP for the past 50 years. This GDP is calculated at purchasers' prices. It is the sum of the gross value added by all resident producers in the economy. Product taxes and subsidies are treated appropriately while calculating the value of the products. It's not necessary to get into details of how GDP is actually calculated.

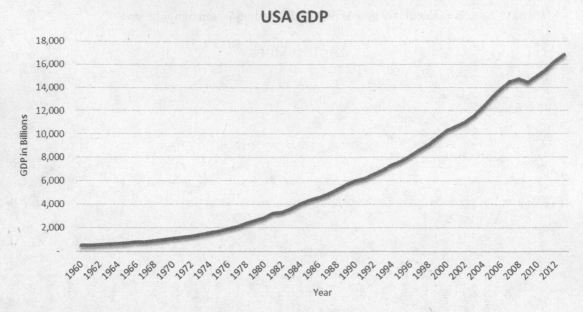

Figure 12-3. *Time-series data on U.S. GDP (Source: World Bank, http://data.worldbank.org/country?display=graph)*

Figure 12-4 shows the time-series chart of U.S. mobile cellular subscriptions (per 100 people).

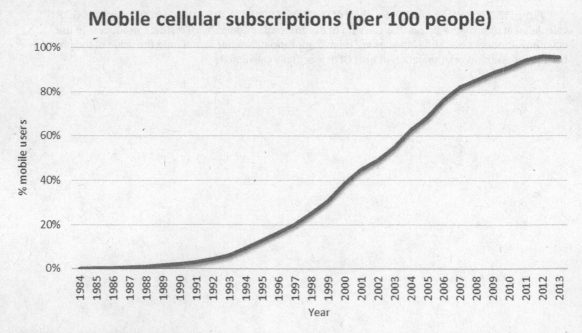

Figure 12-4. *U.S. mobile cellular subscriptions, per 100 people (Source: World Bank, http://data.worldbank.org/indicator/IT.CEL.SETS.P2)*

Main Phases of Time-Series Analysis

So far in this chapter, you have seen several examples of time-series data. The aim of time-series analysis is to comprehend the historical time line of data, analyze it to uncover hidden patterns, and finally model the patterns to use for forecasting. There are three main phases of time-series analysis (Figure 12-5).

1. Descriptive analysis

2. Modeling

3. Forecasting the future values

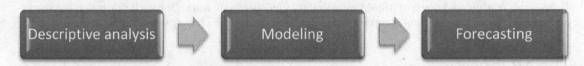

Figure 12-5. Main phases in time-series analysis

In the descriptive phase, you try to understand the nature of the time series. You try to answer questions about what the trend looks like, upward or downward, and whether there any seasonal trend in the process. Finding patterns in the data is a vital step. In the modeling or pattern discovery step, you model the inherent properties of the time-series data. There are several methods of finding patterns in the process of time-series analysis.

Once you have identified the patterns, forecasting is relatively easy. It is simple to use a model to forecast future values. In the next section, you will see some modeling techniques. Before applying these techniques, it helps if you get a basic feel of the data using descriptive statistics methodologies. You have already studied predictive modeling in earlier chapters. Normally, in predictive modeling, you predict the dependent variable Y using a set of X variable values. In time-series analysis, it's different. Here you predict Y using the previous values of Y. The following mathematical representation helps further clarify this difference:

Predictive modeling	**Y versus $X_1, X_2, \ldots X_p$**
Time-series analysis	Y versus Y (Y_t versus Y_{t-1})

Modeling Methodologies

There are several techniques for modeling a time series, both simple and heuristic. Mainly all these modeling techniques can be classified into two categories: heuristic and proper statistical methods. There are some major drawbacks in simple heuristic or instinctive methods. There are hardly any statistical tests to back up simplistic heuristic methods. That is why, in this book, the discussion is limited to more scientific statistical methods. The main focus will be on the Box–Jenkins approach (or ARIMA technique), which is probably one of the most commonly used techniques in the time-series domain.

Box–Jenkins Approach

George Box and Gwilym Jenkins popularized the auto-regressive integrated moving average (ARIMA) technique of modeling a time series. ARIMA methods were not actually invented by Box and Jenkins, but they made some significant contributions in simplifying the techniques. To understand model building using the Box–Jenkins approach, you first need to develop an understanding of terms like ARIMA, AR, MA, and ARMA process.

What Is ARIMA?

Time-series models are known as ARIMA models if they follow certain rules. One such rule is that the values of the time series should fall under certain limits. Later sections of the chapter will come back to ARIMA.

The AR Process

An auto-regressive (AR) process is where the current values of the series depend on their previous values. Auto-regression is also an indicative name, which specifies that it is regression on itself. The auto-regressive process is denoted by AR(p), where p is the order of the auto-regressive process. In the AR process, the current values of the series are a factor of previous values. P determines on how many previous values the current value of the series depends.

For example, consider Y_t, Y_{t-1}, Y_{t-2}, Y_{t-3}, Y_{t-4}, Y_{t-5}, Y_{t-6}, ... Y_{t-k}, which constitutes a time series. If this series is an AR(1) series, then the value of Y at the given time t will be a factor of its value at time t-1.

$$Y_t = a_1 * Y_{t-1} + \varepsilon_t$$

a_1 is the factor or the quantified impact Y_{t-1} on Y_t, and ε_t is the random error at time t. ε_t is also known as white noise. When I say that Y at any given time t is a factor of its value at time t-1, it may not exactly be equal to $a_1 * Y_{t-1}$. It may be slightly higher or lower than a1*Yt-1. But you can't precisely tell what will be the value of ε_t. It is just a small random value at time t, which may be positive or negative. This term (ε_t) can't be specifically quantified in an actual time-series model because it takes different values at different time points.

For an AR(2) series, the value of Y at time t will be a factor of the value at time t-1 and t-2.

$$Y_t = a_1 * Y_{t-1} + a_2 * Y_{t-2} + \varepsilon_t$$

a_1 is the factor or the quantified impact Y_{t-1} on Y_t, and a_2 is the quantified impact Y_{t-2} on Y_t.
Likewise, for a series to be AR(P), Y will depend on the previous Y values until the time point t-p.

$$Y_t = a_1 * Y_{t-1} + a_2 * Y_{t-2} + a_3 * Y_{t-3} + \ldots + a_p * Y_{t-p} + \varepsilon_t$$

a_1, a_2, a_3, ... a_p are the quantified impacts of Y_{t-1}, Y_{t-2}, ... Y_{t-p} on Y_t.
Let's look at an example in the sales domain. If you come across a sales series where the current month's sales depends upon the previous month's sales, then you call that an AR(1) series.

The following equation and the plot in Figure 12-6 represent an AR(1) process for time-series sales data:

$$Y_t = 0.76 * Y_{t-1} + \varepsilon_t$$

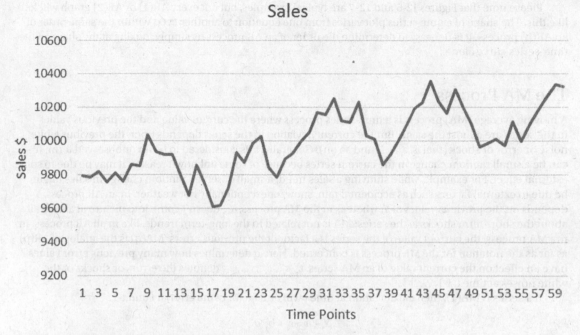

Figure 12-6. *A typical AR (1) time-series equation plot (product sales over time)*

The following equation and the plot in Figure 12-7 represent an AR(2) process of time-series sales data for a different product:

$$Y_t = 0.45 * Y_{t-1} + 0.40 * Y_{t-2} + \varepsilon_t$$

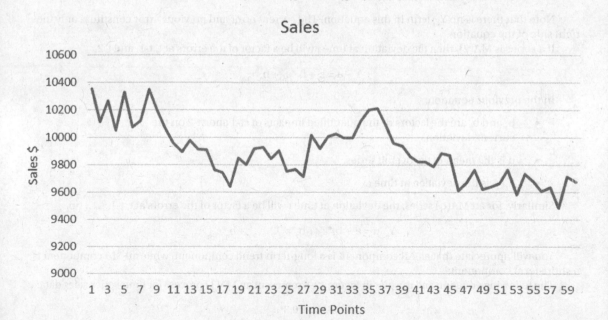

Figure 12-7. *A typical AR (2) time-series equation plot (product sales over time)*

Please note that Figures 12-6 and 12-7 are typical examples, but not every AR(1) or AR(2) graph will look like this. The shape of a time-series plot varies from one equation to another even within the same order of an AR(p) process. It is not easy to determine the order of an AR process by simply looking at the shape of the time-series curve alone.

The MA Process

A moving average (MA) process is a time-series process where the current value and the previous values in the series are almost the same. But the current deviation in the series depends upon the previous white noise or error or shock (that is, ε_t, ε_{t-1}, and so on). Generally, ε_t is considered to be an unobserved term. It can be a small random change in the current series because of some unknown reason; it may be due to some external effect. For example, while studying a sales trend, a small observed random change in sales might be due to external factors such as accidental rain, snow, or exceptionally hot weather. In an AR process, Y_t depends on the previous values of Y, whereas in the MA process, Y_{t-1} doesn't come into picture at all. It's all about the short-term shocks in the series; MA is not related to the long-term trends, like in an AR process. In the MA process, the current value of the series is a factor of the previous errors. MA(q) is the analog of AR(p) as far as the notation for the MA process is concerned. Here q determines how many previous error values have an effect on the current value of an MA series. ε_t, ε_{t-1}, ε_{t-3}, ..., ε_{t-k} denotes the errors or shocks or the white noises at time t, t-1, ... t-k.

If a series is MA(1), then the deviation at time t will be a factor of the error at t and t-1.

$$Y_t - \mu = \varepsilon_t + b_1{}^*\varepsilon_{t-1}$$

In the previous equation:

- b_1 is the factor or the quantified impact of ε_{t-1} on the current deviation.

- μ is the mean of the overall series.

- $Y_t - \mu$ is the deviation at time t.

Note that there is no Y_{t-1} term in this equation. The current error and previous error constitute only the right side of the equation.

If a series is MA(2), then the deviation at time t will be a factor of the errors at t, t-1, and t-2.

$$Y_t - \mu = \varepsilon_t + b_1{}^*\varepsilon_{t-1} + b_2{}^*\varepsilon_{t-2}$$

In the previous equation:

- b_1 and b_2 are the factors or the quantified impacts of εt-1 and εt-2 on the current deviation.

- μ is the mean of the overall series.

- $Y_t - \mu$ is the deviation at time t.

Similarly, for an MA(q) series, the deviation at time t will be a factor of the errors at t, t-1, ... t-p.

$$Y_t - \mu = \varepsilon_t + b_1{}^*\varepsilon_{t-1} + b_2{}^*\varepsilon_{t-2} \ldots \ldots + b_q{}^*\varepsilon_{t-q}$$

You will appreciate that an AR component is a long-term trend component, while an MA component is a short-term component.

The following equation and the plot in Figure 12-8 represent an MA(1) process for time-series sales data:

$$Y_t = \varepsilon_t + 0.3{}^*\varepsilon_{t-1}$$

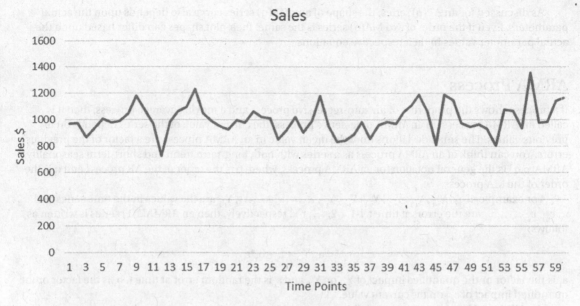

Figure 12-8. *A typical MA(1) time-series equation plot (product sales over time)*

The following equation and the plot in Figure 12-9 represent an MA(2) process for time-series sales data:

$$Y_t = \varepsilon_t + 0.1*\varepsilon_{t-1} + 0.3*\varepsilon_{t-2}$$

Figure 12-9. *A typical MA(2) time-series equation plot (product sales over time)*

As discussed for an AP(p) series, the shape of any MA(q) series curve also depends upon the actual parameters. Even if the order of two MA(q) series is the same, their plot shapes can differ based upon the actual parameter values in their respective equations.

ARMA Process

If a process shows the properties of an auto-regressive process and a moving average process, then it is called an ARMA process. In an ARMA time-series process, the current value of the series depends on its previous values. The small deviations from the mean value in an ARMA process are a factor of the previous errors. You can think of an ARMA process as a series with both long-term trend and short-term seasonality. ARMA(p,q) is the general notation for an ARMA process, where p is the order of the AR process and q is the order of the MA process.

For example, if Y_t, Y_{t-1}, Y_{t-2}, Y_{t-2}, Y_{t-4}, Y_{t-5}, Y_{t-6}, ..., Y_{t-k} are the values in the time series and ε_t, ε_{t-1}, ε_{t-2}, ..., ε_{t-k} are the errors at time t, t-1, t-2, ... t-k, respectively, then an ARMA(1,1) series is written as follows:

$$Y_t = a_1{}^*Y_{t-1} + \varepsilon_t + b_1{}^*\varepsilon_{t-1}$$

a_1 is the factor or the quantified impact of Y_{t-1} on Y_t, and ε_t is the random error at time t. b_1 is the factor or the quantified impact of ε_{t-1} on the current value.

An ARMA(2,1) series is written as follows:

$$Y_t = a_1{}^*Y_{t-1} + a_2{}^*Y_{t-2} + \varepsilon_t + b_1{}^*\varepsilon_{t-1}$$

a_1 and a_2 are the quantified impacts of Y_{t-1} and Y_{t-2} on Y_t, and ε_t is the random error at time t. b_1 is the factor or the quantified impact of ε_{t-1} on the current value.

In the same way, the generic ARMA (p,q) series is written as follows:

$$Y_t = a_1{}^*Y_{t-1} + a_2{}^*Y_{t-2} + a_3{}^*Y_{t-3} + \ldots + a_p{}^*Y_{t-p} + \varepsilon_t + b_1{}^*\varepsilon_{t-1} + b_2{}^*\varepsilon_{t-2}\ldots\ldots + b_q{}^*\varepsilon_{t-q}$$

a_1, a_2, a_3, ... a_p are the quantified impacts of Y_{t-1}, Y_{t-2}, ... Y_{t-p} on Y_t, and b_1, b_2, ... b_q are the factors or the quantified impacts of ε_{t-1} and ε_{t-2} ... ε_{t-q} on Y_t.

The equation and the plot in Figure 12-10 represent an ARMA(1,1) process of a typical sales time-series data:

$$Y_t = 0.75^* Y_{t-1} + \varepsilon_t + 0.2^* \varepsilon_{t-1}$$

Figure 12-10. *A typical ARMA(1,1) time-series equation plot (product sales over time)*

The following equation and the plot in Figure 12-11 represent an ARMA(2,1) process for time-series sales data:

$$Y_t = 0.35* Y_{t-1} + 0.45*Y\, Y_{t-2} + \varepsilon_t + 0.1* \varepsilon_{t-1}$$

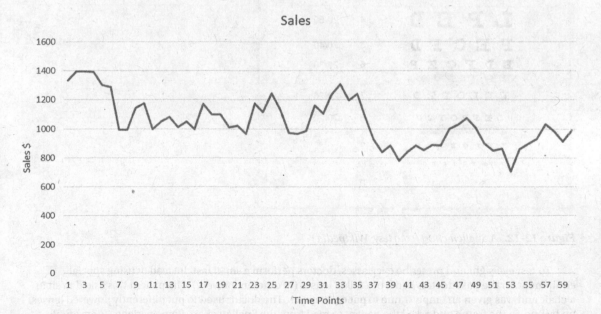

Figure 12-11. *A typical ARMA(2,1) time-series equation plot (product sales over time)*

Now you know what an ARMA process is. And you are also in a position to envisage the equation of a general ARMA(p,q) process. You are now ready to learn the Box–Jenkins ARIMA approach of time-series modeling.

Understanding ARIMA Using an Eyesight Measurement Analogy

I'll use an analogy related to eyesight measurement and prescription eyeglasses as an example to explain a serious concept. Nowadays there is a lot complicated equipment to measure human eyesight. It is relatively easy to accurately measure eyesight and get the right prescription for eyeglasses. But a few decades back, before today's sophisticated and computerized eye testing machines, doctors accomplished this task manually. Today, almost everyone who has visited an eye doctor may easily recognize the picture presented in the Figure 12-12. It is known as a Snellen chart (though this name is not as common as the chart itself).

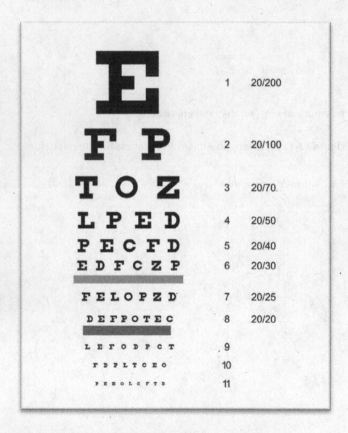

Figure 12-12. A Snellen chart (courtesy: Wikipedia)

To test eye sight and prescribe eyeglasses, doctors perform a small test. Instead of using special equipment, in the past doctors had a box full of lenses (of different powers). The patient was asked to sit in a chair and was given an empty frame to put on her eyes. The doctor used to put differently powered lenses, on by one, in the frame and asked the patient to read from the Snellen chart. Some patients, for example, read the top seven rows and struggled with the lower ones. The doctor then removed the first lens and put another. After much such iteration, the doctor used to finalize on the exact lenses to be used in the patient's glass. Some patients got diagnosed as nearsighted and some with farsightedness. The basic assumption in this process was that all the patients are literate. But what if the patient was illiterate and couldn't read any letters?

So, the major steps in the process of eyesight determination and prescription can be listed as follows:

1. Assume that patient is literate.

2. Based on some tests, identify nearsightedness or farsightedness and get a rough estimate of eyesight.

3. Estimate the exact eyesight by trying various lenses.

4. Use the test results to give the prescription.

A similar analogy can be used with Box–Jenkins approach. Say you have a time series in your hands and you want to forecast some future values in this series. You first need to identify whether the time-series process is an AR process or an MA process or an ARMA process. You also need to identify the orders, p and q, of AR(p) or MA(q) or ARMA(p,q) processes as applicable. Once you identify the type of series and the orders, you can attempt to write the series equation. You are already familiar with the AR(p), MA(q), and ARMA(p,q) equations. The next step is to find parameters such as a_1, a_2, ... a_p, and b_1, b_2, ... b_q as applicable. Before you move on to identifying the model, there is an assumption to be made; the series has to be stationary (in the coming sections, I explain more about stationary time series). This is to simplify the overall model identification process. Table 12-2 uses the eyeglasses prescription analogy to illustrate the Box–Jenkins approach. After that I discuss various steps involved in the Box–Jenkins approach.

Table 12-2. *An Analogy Between a Vision Test and the Box–Jenkins Approach*

Vision Test	Box–Jenkins Approach
Assume the patient is literate.	Assume that the time series is stationary; otherwise, make it stationary.
Based on some tests, identify nearsightedness or farsightedness and get a rough estimate of eyesight.	Based on plots (ACF and PACF functions, explained later in this chapter), identify whether the model is an AR or MA or ARMA process.
Estimate the exact eyesight by trying various lenses.	Estimate the parameters such as a_1, a_2, ... a_p, and b_1, b_2, ... b_q.
Use the test results to give the prescription.	Use the final model for forecasting.

Steps in the Box–Jenkins Approach

Once again I'll show a time-series forecasting problem. Consider some time-series data of a premium stock over a period of time. Let's assume you have the stock price data for the past year. Also assume that you want to predict the stock prices for the next week. You will use the Box–Jenkins approach for this forecasting. First you need to make sure that the stock price time-series process is stationary. Then you need to identify the type of process, which approximates the pattern followed by the stock price data. Is it an AR process or an MA process or an ARMA process? Once you have the basic model equation in place, you can estimate the parameters. Successfully completing all these steps concludes the model-building process. You now have the final equation that can be used for forecasting the future values of the stock under consideration. You also need to take a look at the model accuracy or the error rate before you can continue with the final forecasting and model deployment. What follows is a detailed explanation of each step.

Step 1: Testing Whether the Time Series Is Stationary

If a time-series process is stationary, it's much easier to build the model using the Box–Jenkins methodology.

What Is a Stationary Time Series?

"A time series is said to be stationary if there is no systematic change in mean (no trend), if there is no systematic change in variance, and if strictly periodic variations have been removed," according to Dr. Chris Chatfield in *The Analysis of Time Series: An Introduction* (Chapman and Hall, 2003). A stationary time series is in a state of statistical equilibrium. If the series is varying within certain limits and if the previous or future values of the mean and variance of the series are not going to be exceptionally high, then the series can be understood as a stationary series. In the stationary time-series process, the mean and variance hover around a single value. With the growth of a series, if the mean and variance of the series also tend to grow extremely high, then the series is considered to be nonstationary. It's difficult to identify patterns in a nonstationary time series, and model building can be tough. A nonconstant mean or nonconstant variance is a sign of a nonstationary time-series process (Figure 12-13). These processes are also called *explosive*.

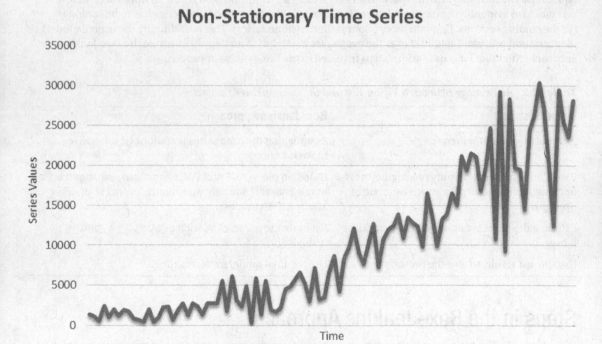

Figure 12-13. *An example of a nonstationary time series*

In some cases, it is easy to identify a nonstationary time series. You need to ask two questions:

- What will the series mean be at time ∞ (infinity)?
- What will the series variance be at time ∞ (infinity)?

If the answer to either one of these questions is ∞, then you can safely conclude that the series under consideration is nonstationary. As a nonstationary series grows, you can see that the values of the series start inflating too much; in fact, they tend to go out of proportion soon. It becomes really tough to predict the series values using the Box–Jenkins approach in such a case. Real-life scenarios may encounter many nonstationary series. You may have to transform or differentiate to make them stationary. Later sections will cover the details of series transformation and differentiation.

Figure 12-14 shows an example plot of a stationary time series.

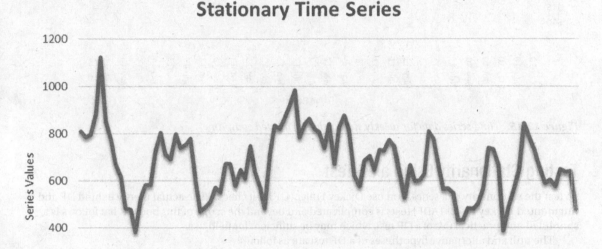

Figure 12-14. *An example of a stationary time series*

In Figure 12-14, some steadiness is apparent in the series plot. It's not inflating too much from its mean value line. In this case, the value of variance or mean will also show steadiness or a state of equilibrium. By now I have discussed many times that simply by looking at the curve shape, it is not a good idea to come to a conclusion. You need a scientifically proven statistical approach to test whether a series is stationary. In fact, curve shapes can be misleading at times and can lead to wrong conclusions. For example, let's take the series shown in Figure 12-15, which represents weekly fluctuations in Microsoft's stock price. The data shown in this plot is Microsoft's stock prices from January 2010 to September 2014 (Source: http://finance.yahoo.com/echarts?s=MSFT). Just looking at the curve shape, you can't undoubtedly tell whether it is a stationary or nonstationary series. You need a solid test for testing stationarity.

Figure 12-15. *Time-series data for weekly stock prices of a listed company*

Testing Stationarity Using a DF Test

To test the stationarity of a series, you use Dickey Fuller (DF) test checks. The actual theory behind DF and Augmented Dickey Fuller (ADF) tests is complicated and beyond the scope of this book. What follows is a simplified and practical use of a DF test, which may be sufficient for field use.

The null and alternative hypotheses of a DF test are as follows:

- H_0: The series is not stationary.

- H_1: The series is stationary.

So, you perform a DF test and take note of the P-value. Looking at the P-value of the DF test, if the P-value is less than 5 percent (0.05), you reject the null hypothesis, which is equivalent to rejecting the hypothesis that the series is not stationary. So, a series is concluded as stationary when the P-value of a DF test is less than 0.05. On the other hand, if the P-value is more than 0.05, then you go ahead and accept the null hypothesis, which means that the series is not stationary.

SAS Code for Testing Stationarity

The following is the SAS code for testing the stationarity of a series. Let's assume that the Microsoft stock price data is imported into the SAS environment and stored in a data set called ms.

```
proc arima data= ms;
identify var= stock_price  stationarity=(DICKEY);
run;
```

The following is an explanation of this code:

- proc arima: arima is the procedure to fit ARIMA models.

- data= ms;: The data set name is ms.

- identify var= stock_price: This mentions the time-series variable, which is Adj_close in this case.

- stationarity=(DICKEY): This is the keyword for the DF test.

Table 12-3 shows the output of this code.

Table 12-3. *Output of PROC ARIMA on the Data Set ms (DF Test)*

The ARIMA Procedure		
Name of Variable = stock_price		

Mean of Working Series	29.20438	
Standard Deviation	6.539502	
Number of Observations	251	

Autocorrelations

Lag	Covariance	Correlation	-1 9 8 7 6 5 4 3 2 1 0 1 2 3 4 5 6 7 8 9 1	Std Error
0	42.765085	1.00000	| |********************|	0
1	41.916461	0.98016	| .|*******************|	0.063119
2	41.181417	0.96297	| . |******************* |	0.107885
3	40.429605	0.94539	| .|****************** |	0.137942
4	39.522281	0.92417	| . |***************** |	0.161708
5	38.561047	0.90169	| .|***************** |	0.181535
6	37.544646	0.87793	| . |***************** |	0.198579
7	36.563693	0.85499	| . |**************** |	0.213483
8	35.643849	0.83348	| . |**************** |	0.226715
9	34.680732	0.81096	| . |*************** |	0.238611
10	33.800014	0.79036	| . |*************** |	0.249350
11	33.004501	0.77176	| . |************** |	0.259139
12	32.306690	0.75545	| . |************** |	0.268140
13	31.663022	0.74039	| . |************** |	0.276489
14	30.943593	0.72357	| . |************* |	0.284279
15	30.128160	0.70450	| . |************* |	0.291524
16	29.475695	0.68925	| . |************* |	0.298230
17	28.873215	0.67516	| . |************* |	0.304510
18	28.196376	0.65933	| . |************ |	0.310417
19	27.582788	0.64498	| . |************ |	0.315947
20	27.015599	0.63172	| . |************ |	0.321150
21	26.431044	0.61805	| . |************. |	0.326063
22	25.838137	0.60419	| . |************. |	0.330697
23	25.265838	0.59081	| . |************. |	0.335066
24	24.735666	0.57841	| . |************. |	0.339191

"." marks two standard errors

Inverse Autocorrelations

Lag	Correlation	-1 9 8 7 6 5 4 3 2 1 0 1 2 3 4 5 6 7 8 9 1
1	-0.43201	| *********| . | .
2	-0.03023	| .*| . |
3	-0.06586	| .*| . |
4	0.01759	| . | . |
5	-0.02020	| . | . |
6	0.02871	| . |* . |
7	0.02352	| . | . |
8	-0.06275	| .*| . |
9	0.03719	| . |* . |
10	0.00328	| . | . |
11	0.03405	| . |* . |
12	0.01159	| . | . |
13	-0.04483	| .*| . |
14	-0.07979	| .**| . |
15	0.09398	| . |** . |
16	0.01896	| . | . |
17	-0.05763	| .*| . |
18	0.03131	| . |* . |
19	0.01035	| . | . |
20	-0.02261	| . | . |
21	-0.00559	| . | . |
22	0.01411	| . | . |
23	-0.00034	| . | . |
24	-0.00268	| . | . |

Partial Autocorrelations

Lag	Correlation	-1 9 8 7 6 5 4 3 2 1 0 1 2 3 4 5 6 7 8 9 1
1	0.98016	.|*******************|
2	0.05757	. |* .	
3	-0.01432	. | .	
4	-0.10292	.**| .	
5	-0.05609	.*| .	
6	-0.04894	.*| .	
7	0.01308	. | .	
8	0.03603	. |* .	
9	-0.02412	. | .	
10	0.03221	. |* .	
11	0.03873	. |* .	
12	0.05418	. |* .	
13	0.02240	. | .	
14	-0.06212	. *| .	
15	-0.09624	.**| .	
16	0.06381	. |* .	
17	0.04601	. |* .	
18	-0.02960	. *| .	
19	0.02061	. | .	
20	0.01711	. | .	
21	-0.01757	. | .	
22	-0.01194	. | .	
23	0.00998	. | .	
24	0.00511	. | .	

Autocorrelation Check for White Noise

To Lag	Chi-Square	DF	Pr > ChiSq	Autocorrelations					
6	1338.63	6	<.0001	0.980	0.963	0.945	0.924	0.902	0.878
12	2356.83	12	<.0001	0.855	0.833	0.811	0.790	0.772	0.755
18	3147.49	18	<.0001	0.740	0.724	0.705	0.689	0.675	0.659
24	3768.57	24	<.0001	0.645	0.632	0.618	0.604	0.591	0.578

Augmented Dickey-Fuller Unit Root Tests

Type	Lags	Rho	Pr < Rho	Tau	Pr < Tau	F	Pr > F
Zero Mean	0	0.5872	0.8278	1.27	0.9484		
	1	0.5975	0.8302	1.41	0.9610		
	2	0.6488	0.8424	1.61	0.9740		
Single Mean	0	0.4367	0.9749	0.20	0.9725	0.81	0.8654
	1	0.8208	0.9852	0.42	0.9835	1.00	0.8151
	2	0.9180	0.9872	0.49	0.9862	1.30	0.7389
Trend	0	-9.5408	0.4616	-2.35	0.4071	4.13	0.3520
	1	-8.4784	0.5410	-2.25	0.4603	4.23	0.3321
	2	-7.0116	0.6587	-1.93	0.6390	3.34	0.5093

The output in Table 12-3 includes several tables, but the one that matters here is the last one named Augmented Dickey-Fuller Unit Root Tests. Within that table you need to focus on the section named Single Mean. The Zero Mean portion of the table is calculated by assuming the mean of the series is zero. The Trend portion of the table is calculated by assuming the null hypothesis of a DF test is true, meaning that the series is stationary with some trend component. You can safely ignore these two portions of the output from your inference. So, to keep it simple, let's concentrate only on Table 12-4, which is taken from the complete table, Augmented Dickey-Fuller Unit Root Tests.

Table 12-4. *The Single Mean Portion of Augmented Dickey-Fuller Unit Root Tests*

Augmented Dickey-Fuller Unit Root Tests

Type	Lags	Rho	Pr < Rho	Tau	Pr < Tau	F	Pr > F
Single Mean	0	0.4367	0.9749	0.20	0.9725	0.81	0.8654
	1	0.8208	0.9852	0.42	0.9835	1.00	0.8151
	2	0.9180	0.9872	0.49	0.9862	1.30	0.7389

You need to look at the P-value of a single mean for an augmented Dickey Fuller test.

- Pr<Rho values are more than 5 percent for lags 0, 1, and 2.

- Similarly, Pr<Tau values are more than 5 percent for lags0, 1, and 2.

So, you safely conclude here that the series under consideration is not stationary.

■ **Note**   As discussed earlier, the exact explanation of a DF test is complicated and is beyond the scope of this book. The following are some notes for statistics enthusiasts.

The DF test has two sets of P-values:

- The first set of P-values (Pr<Rho) is for testing the unit roots directly. Looking at these P-values (Table 12-4), you can conclude that the series is not stationary.

- The second set of P-values (Pr<Tau) is for testing the regression of a differentiated Y_t on a constant, time t, Y_{t-1}, and differentiated Y_{t-1}.

Achieving Stationarity

If a series is not stationary like the way you saw in the previous case (Figure 12-15), you can differentiate it to make it stationary. If Y_t is the original series, then $\Delta Y_t = Y_t - Y_{t-1}$. Instead of taking the original series values, you consider the differentiated series, as shown in Table 12-5.

Table 12-5. Differentiated Series Data

Time	Y_t (Original Series)	Y_{t-1} (Lag of Original Series)	$\Delta Y_t = Y_t - Y_{t-1}$ (Differentiated Series)
Jan	200	-	-
Feb	250	200	50
Mar	230	250	-20
Apr	245	230	15
May	280	245	35
Jun	256	280	-24

In the previous Microsoft example, the stock price time series was not stationary. Let's create a differentiated time series using the following SAS code. It is pretty straightforward.

```
data ms1;
set ms;
diff_stock_price=stock_price-lag1(stock_price);
run;
```

Y_{t-1}

ms1 is the new data set, Lag1 denotes the Y_{t-1} values, and diff_stock_price is the new differentiated series. Table 12-6 is a snapshot of the differentiated series.

Table 12-6. *A Snapshot of the Differentiated Stock Price Time Series*

Obs	stock_price	diff_stock_price
1	27.05	.
2	27.23	0.18
3	25.55	-1.68
4	24.86	-0.69
5	24.72	-0.14
6	24.64	-0.08
7	25.5	0.86
8	25.41	-0.09
9	25.34	-0.07
10	25.94	0.60

Note that some series may not be stationary even after the first differentiation. You then need to go for the second differentiation. If even the second differentiation doesn't work, you may have to try some other transformation logarithm.

Now instead of considering the original series, in other words, the stock price (Y_t), you now consider the differentiated stock price and test it for stationarity. The following is the code to perform this test:

```
proc arima data= ms1;
identify var=diff_stock_price stationarity=(DICKEY);
run;
```

Table 12-7 shows the output of this code. Again, you concentrate only on the Augmented Dickey-Fuller Unit Root Tests table of the output. The focus will be on the Single Mean portion.

Table 12-7. *Output of PROC ARIMA on ms1 (DF Test)*

Augmented Dickey-Fuller Unit Root Tests							
Type	Lags	Rho	Pr < Rho	Tau	Pr < Tau	F	Pr > F
Zero Mean	0	-270.879	0.0001	-17.17	<.0001		
	1	-289.824	0.0001	-12.06	<.0001		
	2	-265.730	0.0001	-9.31	<.0001		
Single Mean	0	-272.508	0.0001	-17.26	<.0001	148.89	0.0010
	1	-296.605	0.0001	-12.19	<.0001	74.30	0.0010
	2	-280.934	0.0001	-9.46	<.0001	44.80	0.0010
Trend	0	-275.376	0.0001	-17.43	<.0001	151.98	0.0010
	1	-306.994	0.0001	-12.36	<.0001	76.36	0.0010
	2	-303.187	0.0001	-9.62	<.0001	46.34	0.0010

Both the sets of P-values on Rho and Tau are well below 5 percent. Hence, you conclude that the series diff_stock_price is stationary.

The Integration Component in the ARIMA Process

In the previous section, you saw how to make a nonstationary process stationary by differentiating it. The original Microsoft price was in tens of dollars, whereas the differentiated value (the new series, which is stationary) is in fractions. Please refer to Table 12-6.

The differentiated series is stationary, and it can be used for analysis using the Box–Jenkins approach. The new series can be any ARMA(p,q) process. But the forecasted values of this series will be in fractions only. As a matter of fact, you have no interest in forecasting the differentiated time series. Our main interest is in forecasting actual Microsoft stock prices but obviously not the differentiated series. You can still derive the original series values from the differentiated series. You need to perform an integration act on the differentiated series, which is the mathematical opposite of differentiation. You made the series stationary by differentiating it. Now you apply the integration to the transformed series to bring it back into its original form. Here the order of the integration (d) is 1. The order of differentiation, which you did to the stock prices series, was also 1. An ARMA series with integration is also known as an ARIMA series. It is denoted by ARIMA (p,d,q), where p is the order of the AR component, d is the order of integration, and finally q is the order of the MA component.

The order of integration required in the current example process is 1. This is simply because to make it stationary, you did the differentiation once. Some series are stationary without any need of differentiation. For such time series, the order of integration is 0. The following are some examples of the ARIMA processes:

The ARIMA(1,1, 0) series is written as follows:

$$\Delta Y_t = a_1 {}^* \Delta Y_{t-1} + \varepsilon_t$$

a_1 is the factor or the quantified impact ΔY_{t-1} on ΔY_t, and ε_t is the random error at time t, in other words, $\Delta Y_t = Y_t - Y_{t-1}$.

The ARIMA(2,1, 0) series is written as follows:

$$\Delta Y_t = a_1 {}^* \Delta Y_{t-1} + a2 {}^* \Delta Y_{t-2} + \varepsilon_t$$

a_1 and a_2 are the factors or the quantified impacts of ΔY_{t-1} and ΔY_{t-2} on ΔY_t, and ε_t is the random error at time t, in other words, $\Delta Y_t = Y_t - Y_{t-1}$.

The ARIMA(1,1, 1) series is written as follows:

$$\Delta Y_t = a_1 {}^* \Delta Y_{t-1} + \varepsilon_t + b_1 {}^* \varepsilon_{t-1}$$

a_1 is the factor or the quantified impact of ΔY_{t-1} on ΔY_t, and ε_t and ε_{t-1} are the random errors at time t and t-1, in other words, $\Delta Y_t = Y_t - Y_{t-1}$.

Integration After Forecasting

You must be wondering how to integrate after forecasting. Table 12-6 shows the form of the series after differentiating the original series in the Microsoft stock price example. In this example, you found that the stock price was not stationary, but `diff_stock_price` is stationary.

Now let's assume you have gone ahead with the `diff_stock_price` series and produced three forecasted values after the analysis. Refer to observations 11, 12, and 13 in Table 12-8.

Table 12-8. *Forecasted Values in the Time Series diff_stock_price*

Obs	stock_price	diff_stock_price
1	27.05	.
2	27.23	0.18
3	25.55	-1.68
4	24.86	-0.69
5	24.72	-0.14
6	24.64	-0.08
7	25.5	0.86
8	25.41	-0.09
9	25.34	-0.07
10	25.94	0.60
11		-0.10
12		0.21
13		-0.11

Having done this, how do you know what the stock prices will be for observations 11, 12, and 13? In other words, the question here is, how do you bring back the forecasted series in its original units?

You know that the differentiated stock prices at any given time t can be calculated by the following formula:

Diff_stock_price at time t = (Stock price at time t) – (Stock price at time t-1)

Before forecasting:

Diff_stock_price at time 10 = (Stock price at time 10) – (Stock price at time 9)

Diff_stock_price at time 10 = 25.94 – 25.34

Diff_stock_price at time 10 = 0.60

After forecasting:

Diff_stock_price at time 11 = (Stock price at time 11) – (Stock price at time 10)

Inserting the values in this equation from Table 12-8, you get the following:

-0.10 = Stock price at time 11-25.94

Stock price at time 11 = 25.94 -0.10

Stock price at time 11 = 25.84

Similarly,

Diff_stock_price at time 12 = (Stock price at time 12) – (Stock price at time 11)

0.21 = Stock price at time 12 – 25.84

Stock price at time 12 = 25.84 + 0.21

Stock price at time 12 = 26.05

Repeating the same calculations, you get the stock price at time point 13 as 25.94 (Table 12-9).

Table 12-9. *Forecasted Values in the Time Series stock_price and diff_stock_price (After Integration)*

Obs	stock_price	diff_stock_price
1	27.05	.
2	27.23	0.18
3	25.55	-1.68
4	24.86	-0.69
5	24.72	-0.14
6	24.64	-0.08
7	25.5	0.86
8	25.41	-0.09
9	25.34	-0.07
10	25.94	0.60
11	25.84	-0.10
12	26.05	0.21
13	25.94	-0.11

The rule of thumb is that you subtract the previous lag values while differentiating; you add previous lag values while integrating. Once the series is stationary, it's ready for analysis, and forecasting becomes relatively easy. The next step is to identify the type of series. Is it AR, MA, or ARMA?

For the current example series, you have already identified the order of integration. Now you need to identify p and q. This is the next step.

Step 2: Identifying the Model

Model identification is one of the important steps in time-series modeling using the Box–Jenkins methodology. The final forecast accuracy will depend on how precisely you identify the model. The model identification includes finding out the type of time-series process, whether you are dealing with AR(1) or AR(2) or MA(1) or MA(2) or ARMA(1,1) or ARMA(2,1), and so on. By identifying the type of series, you will get a basic idea of the data. If it is AR, then you know that the current values of the series are dependent on the previous values. An AR series has a long-term trend. If it is MA, then the errors will have some impact. An MA series has some short-term seasonal effects. In fact, you can write the basic form of the series for ARMA(p,q) as follows:

$$Y_t = a_1{}^*Y_{t-1} + a_2{}^*Y_{t-2} + a_3{}^*Y_{t-3} + \ldots + a_p{}^*Y_{t-p} + \varepsilon_t + b_1{}^*\varepsilon_{t-1} + b_2{}^*\varepsilon_{t-2}\ldots\ldots + b_q{}^*\varepsilon_{t-q}$$

Identifying the model is definitely not easy. It is not enough to look at the series graph and conclude the type and order of the series. You need some concrete metrics for identifying the type and order of a series. You need to use the auto correlation function (ACF) and partial auto correlation function (PACF) functions and plots to identify the type of the series. I will go into details of this in the following sections. First, for a better comprehension, let's look at some time-series equations and their plots.

465

Time-Series Examples

To understand the identification of the series, you will consider some simulated time series. You will see plots for about 15 graphs with their equations (Figures 12-16 through 12-30). The following are some time-series graphs. Let's try to guess what their series will be.

Time Series Plot – 1 (Figure 12-16): $y_t = 0.66y_{t-1} + \varepsilon_t$

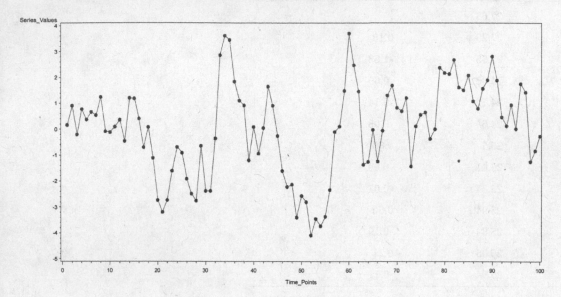

Figure 12-16. *A time-series plot for $y_t = 0.66y_{t-1} + \varepsilon_t$*

Time Series Plot – 2 (Figure 12-17): $y_t = 0.83y_{t-1} + \varepsilon_t$

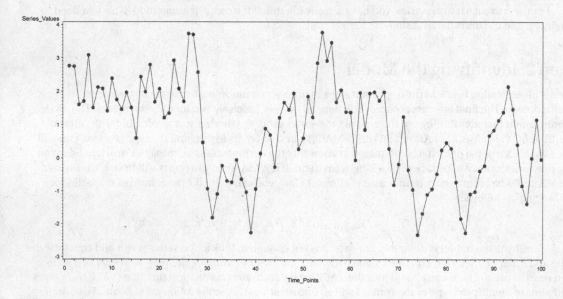

Figure 12-17. *A time-series plot for $y_t = 0.83y_{t-1} + \varepsilon_t$*

Time Series Plot – 3 (Figure 12-18): $y_t = -0.95y_{t-1} + \varepsilon_t$

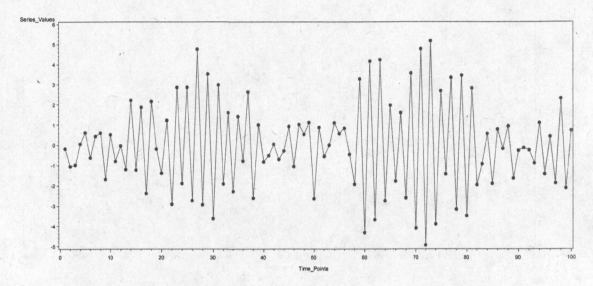

Figure 12-18. *A time-series plot for $y_t = -0.95y_{t-1} + \varepsilon_t$*

Time Series Plot – 4 (Figure 12-19): $y_t = 0.4y_{t-1} + 0.48y_{t-2} + \varepsilon_t$

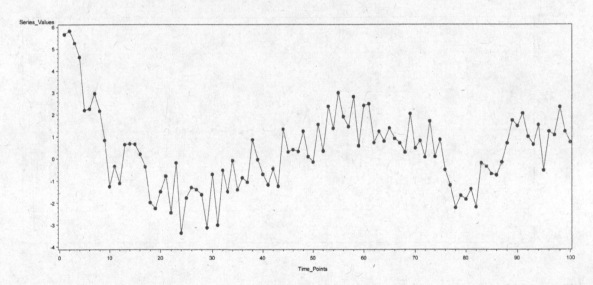

Figure 12-19. *A time-series plot for $y_t = 0.4y_{t-1} + 0.48y_{t-2} + \varepsilon_t$*

Time Series Plot – 5 (Figure 12-20): $y_t = 0.5y_{t-1} + 0.4y_{t-2} + \varepsilon_t$

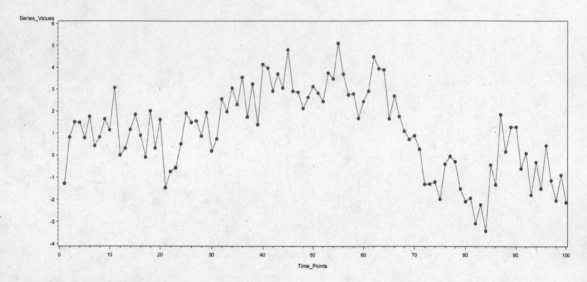

Figure 12-20. *A time-series plot for $y_t = 0.5y_{t-1} + 0.4y_{t-2} + \varepsilon_t$*

Time Series Plot – 6 (Figure 12-21): $y_t = 0.3y_{t-1} + 0.33y_{t-2} + 0.26y_{t-3} + \varepsilon_t$

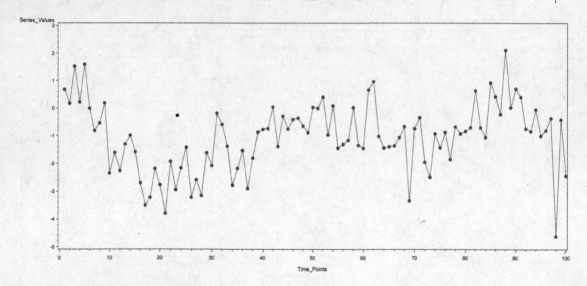

Figure 12-21. *A time-series plot for $y_t = 0.3y_{t-1} + 0.33y_{t-2} + 0.26y_{t-3} + \varepsilon_t$*

Time Series Plot – 7 (Figure 12-22): $y_t = \varepsilon_t - 0.8\,\varepsilon_{t-1}$

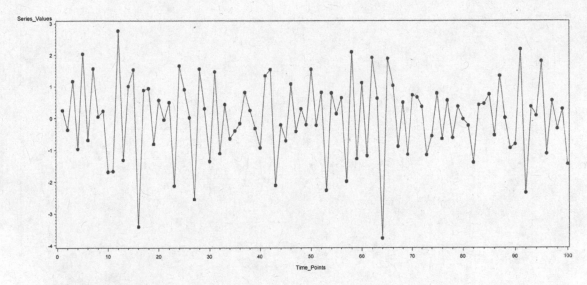

Figure 12-22. *A time-series plot for $y_t = \varepsilon_t - 0.8\,\varepsilon_{t-1}$*

Time Series Plot – 8 (Figure 12-23): $y_t = \varepsilon_t + 0.66\,\varepsilon_{t-1}$

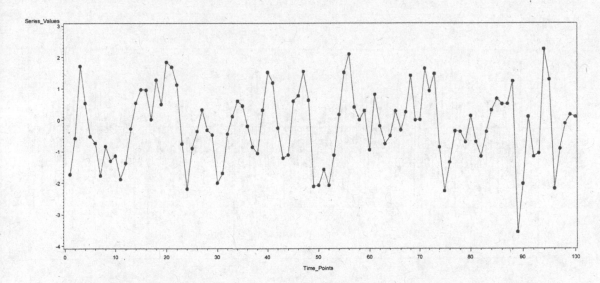

Figure 12-23. *A time-series plot for $y_t = \varepsilon_t + 0.66\,\varepsilon_{t-1}$*

Time Series Plot – 9 (Figure 12-24): $y_t = \varepsilon_t + 0.95\,\varepsilon_{t-1}$

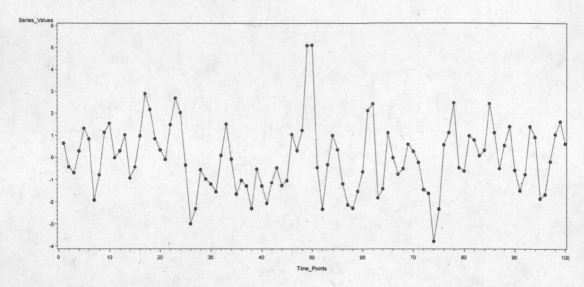

Figure 12-24. *A time-series plot for $y_t = \varepsilon_t + 0.95\,\varepsilon_{t-1}$*

Time Series Plot – 10 (Figure 12-25): $y_t = \varepsilon_t - 0.4\,\varepsilon_{t-1} - 0.38\,\varepsilon_{t-2}$

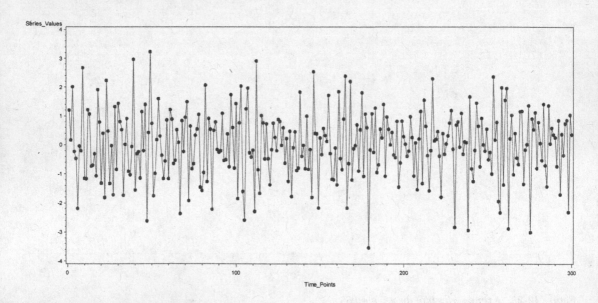

Figure 12-25. *A time-series plot for $y_t = \varepsilon_t - 0.4\,e_{t-1} - 0.38\,\varepsilon_{t-2}$*

Time Series Plot – 11 (Figure 12-26): $y_t = \varepsilon_t - 0.6\, e_{t-1} - 0.3\, \varepsilon_{t-2}$

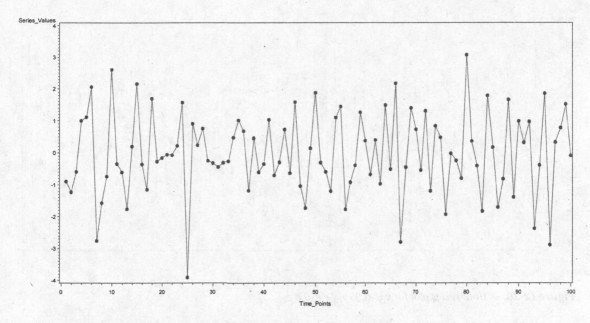

Figure 12-26. *A time-series plot for $y_t = \varepsilon_t - 0.6\, \varepsilon_{t-1} - 0.3\, \varepsilon_{t-2}$*

Time Series Plot – 12 (Figure 12-27): $y_t = \varepsilon_t - 0.55\, \varepsilon_{t-1} - 0.35\, \varepsilon_{t-2} - 0.15\, \varepsilon_{t-3}$

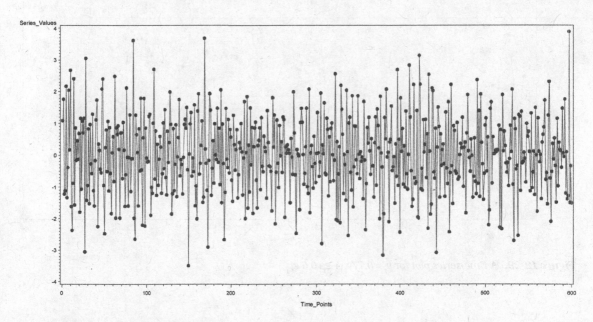

Figure 12-27. *A time-series plot for $y_t = \varepsilon_t - 0.55\, \varepsilon_{t-1} - 0.35\, \varepsilon_{t-2} - 0.15\, \varepsilon_{t-3}$*

Time Series Plot – 13 (Figure 12-28): $y_t = -0.95\, y_{t-1} + \varepsilon_t - 0.8\, \varepsilon_{t-1}$

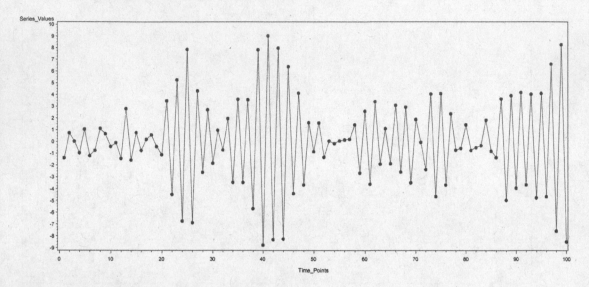

Figure 12-28. *A time-series plot for* $y_t = -0.95\, y_{t-1} + \varepsilon_t - 0.8\, \varepsilon_{t-1}$

Time Series Plot – 14 (Figure 12-29): $y_t = 0.77\, y_{t-1} + \varepsilon_t + 0.6\, \varepsilon_{t-1}$

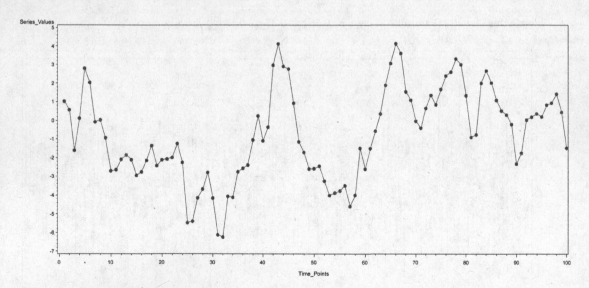

Figure 12-29. *A time-series plot for* $y_t = 0.77 y_{t-1} + \varepsilon_t + 0.6\, \varepsilon_{t-1}$

Time Series Plot – 15 (Figure 12-30): $y_t = 0.55y_{t-1} + 0.35y_{t-2} + \varepsilon_t - 0.5\,\varepsilon_{t-1}$

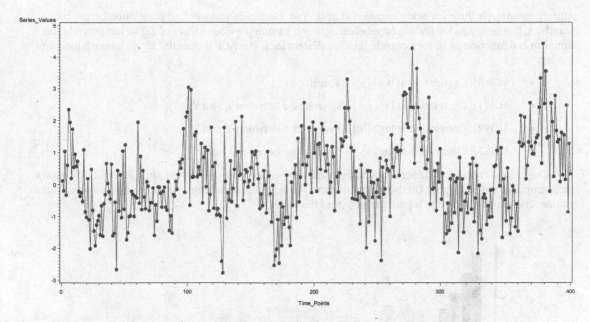

Figure 12-30. *A time-series plot for $y_t = 0.55y_{t-1} + 0.35y_{t-2} + \varepsilon_t - 0.5\,\varepsilon_{t-1}$*

Was a simple observation enough to judge the type of series? In the time-series plots from Figures 12-16 through 12-30, some of them are AR processes. There are examples of AR(1), AR(2), AR(3). Some are pure MA processes. There are examples of MA(1), MA(2), and MA(3). Some series are ARMA processes.

In the preceding examples, I simulated the data; hence, I could write the equations. But in practical scenarios you will not have the model equations. You will just have the series values and the series graph. You know that it's not easy to classify any of these series as AR, MA, or ARMA just by looking at the series values. You need some concrete metrics for identifying the type of series. You need to use ACF and PACF functions and plots to identify the type of the series. I discuss these topics in the following sections. The following lines are the 15 time-series equations put at one place so that you can compare them:

TS - 1: $y_t = 0.66\ y_{t-1} + \varepsilon_t$
TS - 2: $y_t = 0.83\ y_{t-1} + \varepsilon_t$
TS - 3: $y_t = -0.95\ y_{t-1} + \varepsilon_t$
TS - 4: $y_t = 0.4\ y_{t-1} + 0.48\ y_{t-2} + \varepsilon_t$
TS - 5: $y_t = 0.5\ y_{t-1} + 0.4\ y_{t-2} + \varepsilon_t$
TS - 6: $y_t = 0.3\ y_{t-1} + 0.33\ y_{t-2} + 0.26\ y_{t-3} + \varepsilon_t$
TS - 7: $y_t = \varepsilon_t - 0.8\ \varepsilon_{t-1}$
TS - 8: $y_t = \varepsilon_t + 0.66\ \varepsilon_{t-1}$
TS - 9: $y_t = \varepsilon_t + 0.95\ \varepsilon_{t-1}$
TS - 10: $y_t = \varepsilon_t - 0.4\ \varepsilon_{t-1} - 0.38\ \varepsilon_{t-2}$
TS - 11: $y_t = \varepsilon_t - 0.6\ \varepsilon_{t-1} - 0.3\ \varepsilon_{t-2}$
TS - 12: $y_t = \varepsilon_t - 0.55\ \varepsilon_{t-1} - 0.35\ \varepsilon_{t-2} - 0.15\ \varepsilon_{t-3}$
TS - 13: $y_t = -0.95\ y_{t-1} + e_t - 0.8\ \varepsilon_{t-1}$
TS - 14: $y_t = 0.77\ y_{t-1} + \varepsilon_t + 0.6\ \varepsilon_{t-1}$
TS - 15: $y_t = 0.55\ y_{t-1} + 0.35\ y_{t-2} + \varepsilon_t - 0.5\ \varepsilon_{t-1}$

Auto Correlation Function

Auto correlation is the correlation between Yt upon Yt-1. Generally you will find correlation between two variables, but here you are finding correlation between Y upon previous values of Y. The auto correlation function is a function of all such correlations at different lags. The ACF is denoted by ρh, where h indicates the lag.

- ACF(0): Correlation at lag0 (ρ_0) = Y_t and Y_t =1

- ACF(1): Correlation at lag1 (ρ_1) = Correlation between Y_t and Y_{t-1}

- ACF(2): Correlation at lag2 (ρ_2) = Correlation between Y_t and Y_{t-2}

- ACF(3): Correlation at lag3 (ρ_3) = Correlation between Y_t and Y_{t-3}

The graphs created using auto correlation values are called *auto correlation plots*. Figure 12-31 shows an example of one such plot. On the x-axis you have the lag values, while the y-axis has the auto correlation values. The graph might vary based on the type of the series.

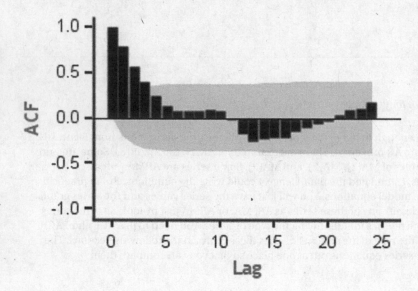

Figure 12-31. *An example auto correlation plot*

In AR models, the current values of the series depend on its previous values. You can expect the auto correlation to chart a pattern in AR models.

Partial Auto Correlation Function

The partial auto correlation function is the partial correlations between Y and its previous values calculated at different lags. The partial correlation is the correlation after removing the effect of the other variables. You can visualize partial correlation as the correlation coefficient after removing or excluding the effect of the other independent variables. You can also treat it is as an exclusive correlation. Let's see an example. You can expect a credit score to be dependent upon the number of existing loans, income, payment dues, length of credit history, and number of dependents. Trying to find a simple correlation between the number of loans and credit score will give you the association between the two variables, which will also include the effect of all other variables on the credit score. To find a partial correlation, you need to first eliminate the effect of the other impacting variables. For all practical purposes, you can envisage the partial correlation coefficient as a regression coefficient.

For the credit score example, here's the equation:

Credit score = β_0 + β_1 (number of loans)+ β_2 (payment due)+ β_3 (length of credit history) + β_4 (number of dependents)

The partial correlation between the number of loans and the credit score is β_1, assuming that there is no multicolliniarity. β_1 is the exclusive impact of the variable number of loans on the credit score. Now you can appreciate that a partial coefficient is the quantified association between two variables, which is not explained by any of their mutual associations with any other variable.

In the context of time-series analysis, partial auto correlation is found by regressing the old values of Y on the current value. The PACF is denoted by θ_h, where h indicates the lag.

- PACF(0): Partial correlation at lag0 (θ_0) = Y_t and Y_t = 1

- PACF(1): Partial correlation at lag1 (θ_1) = Regression coefficient of Y_{t-1} when Y_{t-1} is regressed up on Y_t

- PACF(2): Partial correlation at lag2 (θ_2) = Regression coefficient of Y_{t-2} when Y_{t-1} and Y_{t-2} are regressed up on Y_t

- PACF(3): Partial correlation at lag3 (θ_3)= Regression coefficient of Y_{t-3} when Y_{t-1}, Y_{t-2} and Y_{t-3} are regressed up on Y_t

The graphs created using partial auto correlation values are called *partial auto correlation plots*. Figure 12-32 shows an example of one such plot. On the x-axis you have the lag values, while the y- axis has the partial auto correlation values. As is the case with ACF, a PACF graph might vary based on the type of series.

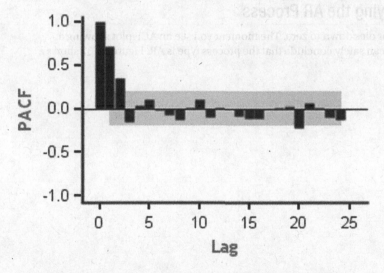

Figure 12-32. An example of a partial auto correlation plot

In MA models, the current errors in the series depend on its previous error values. You expect the partial auto correlation to follow a pattern in MA models.

Creating ACF and PACF Plots in SAS

Using the ACF and PACF plots, you can get an idea about the type of series. This is the second step in the time-series analysis process. You have to first make sure that the time series is stationary. You may have to differentiate the series at times to make it stationary; a stationary series is ready for analysis. Once the series is ready, you want to categorize what type of series it is. Just to recap, to identify the type of series, you need to study the ACF and PACF plots generated for the series. What follows next is the code for creating ACF and PACF plots in SAS:

```
proc arima data=ts15 plots=all;
identify var=Series_Values ;
run;
```

You need to use the plots=all option to generate ACF and PACF plots. The SAS output gives four plots.

- The original series graph
- ACF plot
- PACF plot
- IACF plot

■ **Note**   IACF stands for inverse auto correlation. It will be discussed in later sections.

Rules of Thumb for Identifying the AR Process

The ACF for the AR process tails off or dies down to zero. The moment you see an ACF plot showing a reducing tendency toward zero, you can safely conclude that the process type is AR. Figure 12-33 shows a few examples.

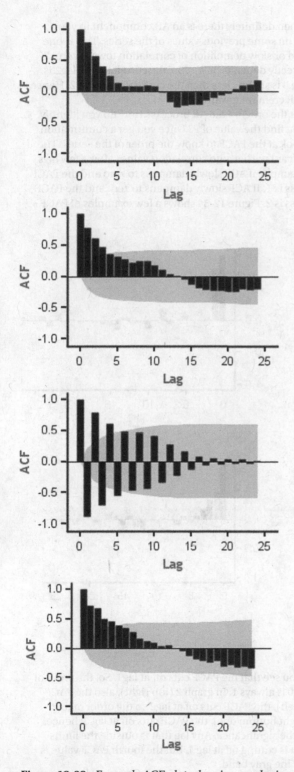

Figure 12-33. Example ACF plots showing a reducing tendency toward zero

If you observe any such behaviors in your series, then definitely there is an AR component in your series. It means the current values of the series depend on some previous values of the series. This is one of the reasons why the ACF function shows a dampened or slow demolition of correlation toward zero. Y_t depends on Y_{t-1}, and Y_{t-1} depends on Y_{t-2}; hence, Y_t indirectly depends on Y_{t-2}. So, the correlation at lag2 is slightly less than what it is at lag1. The correlation at lag 3 is slightly less than the correlation at lag 2. Hence, the ACF of the AR process shows this behavior of slow decrement toward zero.

But this is just the first half of the story. What about the order of the AR process? How do you identify whether a series is AR(1) or AR(2) or AR(p)? How do you find the value of P? Once you get a confirmation from the ACF plot that the process is AR, you need to look at the PACF to know the order of the series. The PACF cuts off for an AR process. There will be no significant partial auto correlation values after a few lags. The lag number indicates the order of the series. For example, if ACF slowly dampens to zero and the PACF function cuts off at lag1, then the order of the AR process is 1. If ACF slowly dampens to zero and the PACF function cuts off at lag2, then the order of the AR process is 2. Figure 12-34 shows a few examples of PACF plots for an AR process.

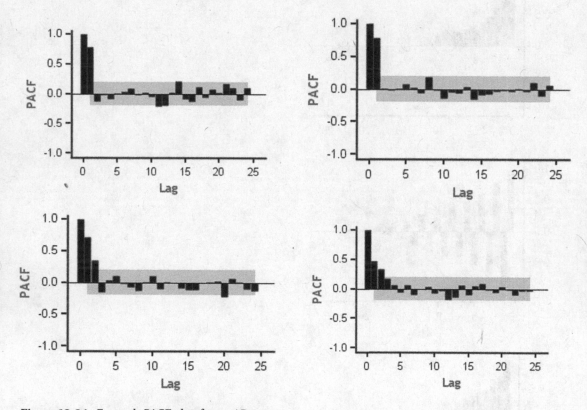

Figure 12-34. *Example PACF plots for an AR process*

Counting clockwise from the top left, in graph 1, you see that the PACF cuts off at lag1. So, the order of that AR process is 1. The partial correlation value at lag0 is always 1. In graph 2 (top right), also the PACF cuts off at lag1. In graph 3 (in other words, the bottom left), the PACF cuts off at lag2 so the order of that AR process is 2. In the last graph (in other words, the right bottom one), the PACF cuts off at lag 3; hence, the order of that AR process is 3. The gray band shows the significance. Any lag that is outside the limits of that band is significant. The last graph appears as if it is cutting off at lag 4, but the fourth PACF value is nonsignificant enough because it is within the limits of the gray band.

Here is the justification for the cutoffs in the PACF graph for the AR process. Imagine an AR process of order 1; the Y_{t-1} will have maximum impact on Y_t. If you remove the previous impacts from Y_{t-1} and Y_t and calculate the correlation, then you will get a high correlation only for lag1, and the rest of the lags will be negligible for the AR(1) process. Similarly, for an AR(2) process, the Y_{t-1}, Y_{t-2} will have a maximum impact on Y_t. If you remove all the previous effects from Y_{t-1}, Y_{t-2}, and Y_t, then the correlation will be high for lag1 and lag2 only. All the remaining correlations, beyond lag 2, will be negligible for an AR(2) process. Please refer to Table 12-10 for a summary.

Table 12-10. *The Rules of Thumb to Identify an AR(P) Process*

AR Process	Behavior
ACF	Slowly tails off or diminishes to zero. Either reduces in one direction or reduces in a sinusoidal (sine wave) passion.
PACF	Cuts off. The cutoff lag indicates the order of the AR process.

In the simulated list of time series in Figure 12-16 through Figure 12-30, the first six plots belong to the AR series. Let's take a look at their ACF and PACF plots (Figure 12-35 through Figure 12-40). Once again, I would like to remind you that for any new series, in an actual business situation, the series equation will not be available. This is just an exercise to get the series equation.

Time Series Plot 1 (Figure 12-16): $y_t = 0.66y_{t-1} + \varepsilon_t$

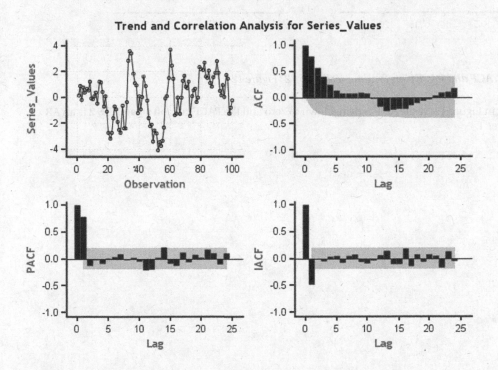

Figure 12-35. *ACF and PACF plots for time-series plot 1 (Figure 12-16)*

Note that in Figure 12-35 the ACF is dying down to zero and the PACF cuts off at lag1. TS 1 is an AR process of order 1.

Time Series Plot 2 (Figure 12-17): $y_t = 0.83y_{t-1} + \varepsilon_t$

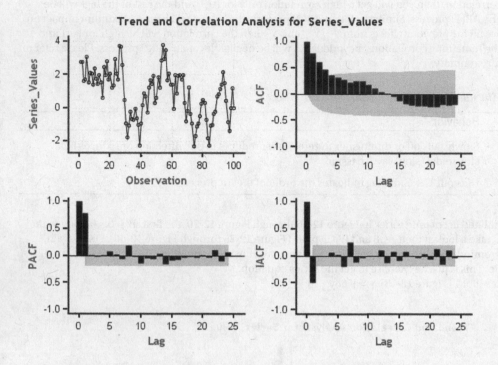

Figure 12-36. *ACF and PACF plots for time-series plot 2 (Figure 12-17)*

Note that in Figure 12-36 the ACF is dying down to zero and the PACF cuts off at lag 1. TS 2 is an AR process of order 1.

Time Series Plot 3 (Figure 12-18): $y_t = -0.95y_{t-1} + \varepsilon_t$

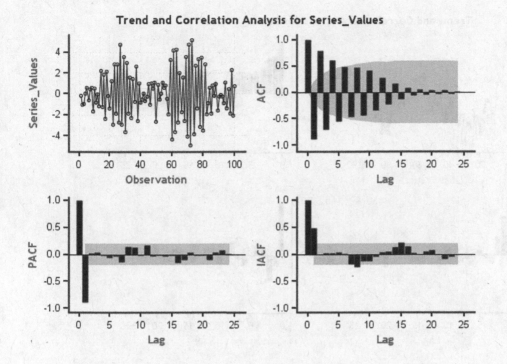

Figure 12-37. *ACF and PACF plots for time-series plot 3 (Figure 12-18)*

Note that in Figure 12-37 the ACF is dying down to zero and the PACF cuts off at lag 1. TS 3 is an AR process of order 1.

Time Series Plot 4 (Figure 12-19): $y_t = 0.4y_{t-1} + 0.48y_{t-2} + \varepsilon_t$

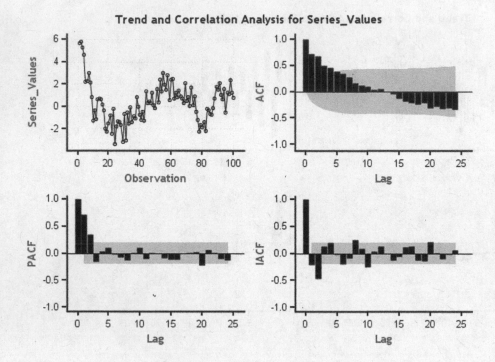

Figure 12-38. *ACF and PACF plots for time-series plot 4 (Figure 12-19)*

Note that in Figure 12-38 the ACF is dying down to zero and the PACF cuts off at lag 2. TS 4 is an AR process of order 2.

Time Series Plot 5 (Figure 12-20): $y_t = 0.5y_{t-1} + 0.4y_{t-2} + \varepsilon_t$

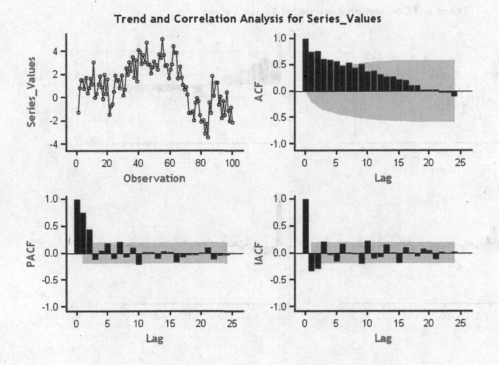

Figure 12-39. *ACF and PACF plots for time-series plot 5 (Figure 12-20)*

Note that in Figure 12-39 the ACF is dying down to zero and the PACF cuts off at lag 2. TS 5 is an AR process of order 2.

Time Series Plot 6 (Figure 12-21): $y_t = 0.3y_{t-1} + 0.33y_{t-2} + 0.26y_{t-3} + \varepsilon_t$

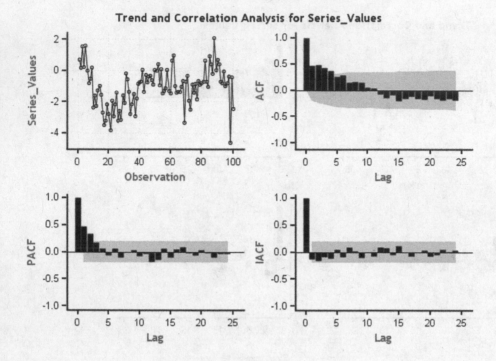

Figure 12-40. *ACF and PACF plots for time-series plot 6 (Figure 12-21)*

Note that in Figure 12-40 the ACF is dying down to zero and the PACF cuts off at lag 3. TS 6 is an AR process of order 3.

Rules of Thumb for Identifying the MA Process

The method of identifying the MA process is almost the same as the AR process. For an MA process, the PACF process slowly dies down, and the ACF cuts off. Since, in an MA process, the current value of the series has no direct dependency on the past values, the ACF cuts off. The errors of the series depend on the previous errors. This is the cause for the PACF to dampen slowly. Table 12-11 lists the rules of thumb for identifying an MA process.

Table 12-11. *The Rules of Thumb to Identify MA Process*

MA Process	Behavior
ACF	Cuts off. The cutoff lag indicates the order of the MA process.
PACF	Slowly tails off or diminishes to zero. Either reduces in one direction or reduces in a sinusoidal or sine wave pattern.

There is a small issue with the PACF plots. Since you are dealing with errors, which are generally small in magnitude, the PACF plot is not clear, and it's hard to interpret. You can make use of an inverse auto correlation function (IACF) plot as a substitute to a PACF plot.

Inverse Auto Correlation Function

The inverse auto correlation function is an auto correlation function wherein you interchange the roles of errors and actual series values. The ACF plot for ARMA (2,0) looks the same as an IACF plot for ARMA(0,2). You can use IACF as a substitute to PACF in your analysis. For some models, an IACF graph shows a clear trend and is hence easy to interpret. You can comprehend IACF as an auto correlation function calculated between the errors. Table 12-12 lists the modified and final rules of thumb for identifying an MA process.

Table 12-12. *The Rules of Thumb to Identify MA Process with IACF*

MA process	Behavior
ACF	Cuts off. The cutoff lag indicates the order of the MA process.
PACF/IACF	Slowly tails off or diminishes to zero. Either reduces in one direction or reduces in a sinusoidal or a sine wave pattern

Figures 12-41 through 12-46 show some of the MA series from the earlier examples with their ACF, PACF, and IACF plots.

Time Series Plot 7 (Figure 12-22): $y_t = \varepsilon_t - 0.8\, \varepsilon_{t-1}$

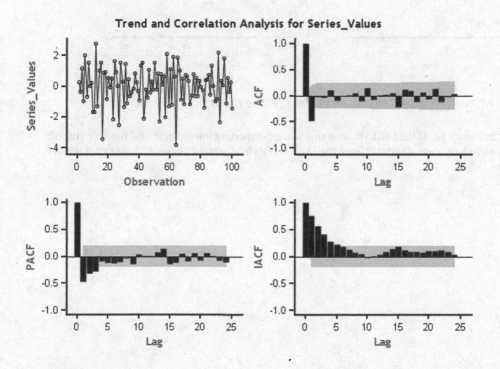

Figure 12-41. *ACF and PACF plots for time-series plot 7 (Figure 12-22)*

Note that in Figure 12-41 the IACF shows a dampening trend toward zero and the ACF cuts off at lag1. TS 7 is an MA process of order 1.

Time Series Plot 8 (Figure 12-23): $y_t = \varepsilon_t + 0.66\,\varepsilon_{t-1}$

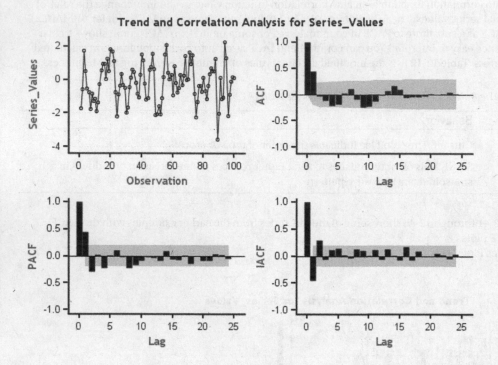

Figure 12-42. *ACF and PACF plots for time-series plot 8 (Figure 12-23)*

Note that in Figure 12-42 the IACF shows a sine wave dampening toward zero and the ACF cuts off at lag1. TS 8 is an MA process of order1. Since the IACF is not clear, you may expect a lower coefficient for previous errors.

Time Series Plot 9 (Figure 12-24): $y_t = \varepsilon_t + 0.95\,\varepsilon_{t-1}$

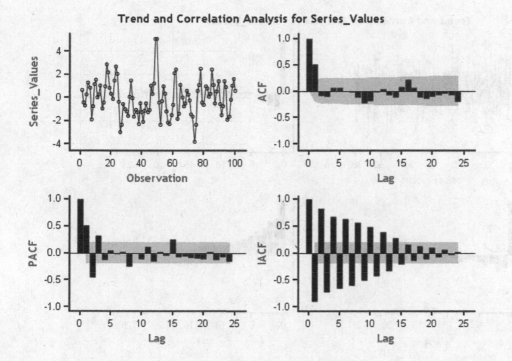

Figure 12-43. *ACF and PACF plots for time-series plot 9 (Figure 12-24)*

Note that in Figure 12-43 the IACF shows a dampening toward zero and the ACF cuts off at lag1. TS 9 is an MA process of order1.

Time Series Plot 10 (Figure 12-25): $y_t = \varepsilon_t - 0.4\,\varepsilon_{t-1} - 0.38\,\varepsilon_{t-2}$

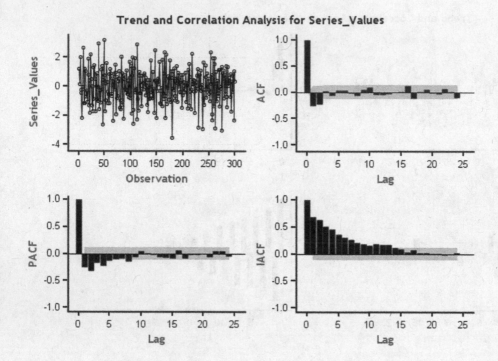

Figure 12-44. *ACF and PACF plots for time-series plot 10 (Figure 12-25)*

Note that in Figure 12-44 the IACF shows a dampening toward zero and the ACF cuts off at lag2. TS 10 is an MA process of order2.

Time Series Plot 11 (Figure 12-26): $y_t = \varepsilon_t - 0.6\,\varepsilon_{t-1} - 0.3\,\varepsilon_{t-2}$

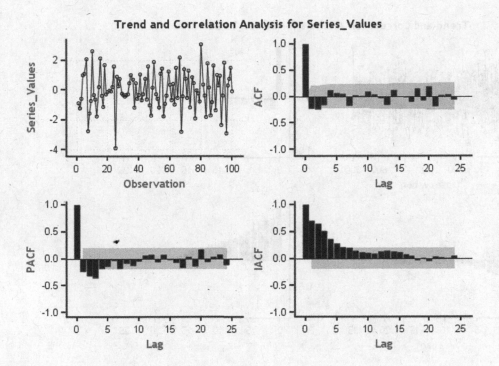

Figure 12-45. *ACF and PACF plots for time-series plot 11 (Figure 12-26)*

Note that in Figure 12-45 the IACF shows a dampening toward zero and the ACF cuts off at lag2. TS 11 is an MA process of order2.

Time Series Plot 12 (Figure 12-27): $y_t = \varepsilon_t -0.55\ \varepsilon_{t-1} -0.35\ \varepsilon_{t-2} -0.15\ \varepsilon_{t-3}$

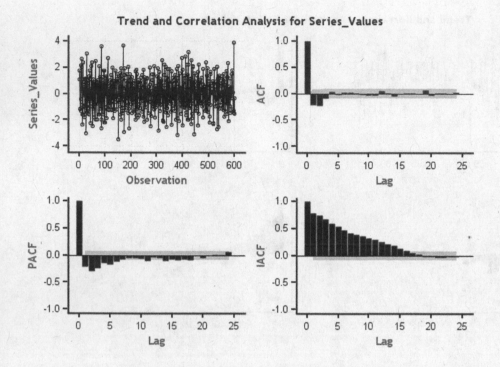

Figure 12-46. *ACF and PACF plots for time-series plot 12 (Figure 12-27)*

Note that in Figure 12-46 the IACF shows a dampening toward zero and the ACF cuts off at lag3. TS 12 is an MA process of order 3.

Rules of Thumb for Identifying the ARMA Process

When compared with pure AR and pure MA processes, it's difficult to identify the orders of an ARMA process. If the ACF dampens to zero, then it's an AR process, and the PACF gives the order of the AR process. If the IACF dampens to zero, then it's an MA process, and the order is decided by looking at the ACF plot. But for an ARMA process, the ACF, PACF, and IACF will all behave the same way. Though you can quickly identify that the process is an ARMA process, you will not be able to decide the orders. So, when the series is an ARMA process, then you will not be able to identify the orders by looking at the ACF, PACF, and IACF graphs. Table 12-13 summarizes the rule of thumb. Figures 12-47 through Figure 12-49 give the example plots of ACF, PACF, and IACF for the ARMA processes discussed in earlier sections.

Table 12-13. *The Rule of Thumb to Identify an ARMA Process*

ARMA Process	Behavior
ACF/PACF/IACF	All of them damp down to zero.

Time Series Plot 13 (Figure 12-28): $y_t = -0.95y_{t-1} + \varepsilon_t - 0.8\,\varepsilon_{t-1}$

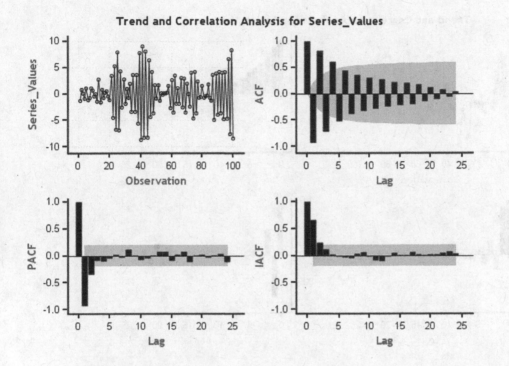

Figure 12-47. *ACF and PACF plots for time-series plot 13 (Figure 12-28)*

Note that in Figure 12-47 the ACF, PACF, and IACF damp down to zero. TS 13 is an ARMA process.

Time Series Plot 14 (Figure 12-29): $y_t = 0.77y_{t-1} + \varepsilon_t + 0.6\,\varepsilon_{t-1}$

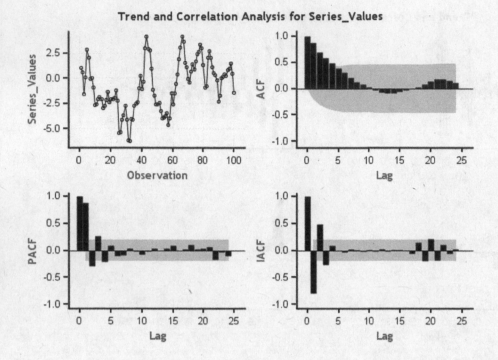

Figure 12-48. *ACF and PACF plots for time-series plot 14 (Figure 12-29)*

Note that in Figure 12-48 the ACF, PACF, and IACF damp down to zero. TS 14 is an ARMA process.

Time Series Plot 15 (Figure 12-30): $y_t = 0.55y_{t-1} + 0.35y_{t-2} + \varepsilon_t - 0.5\,\varepsilon_{t-1}$

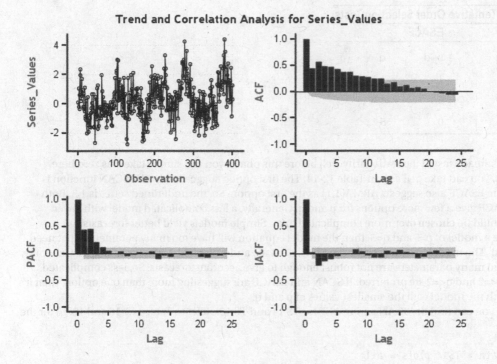

Figure 12-49. *ACF and PACF plots for time-series plot 15 (Figure 12-30)*

Note that in Figure 12-49 the ACF, PACF, and IACF damp down to zero. TS 15 is an ARMA process.

Finding Out ARMA Orders (p,q) Using SCAN and ESACF

You have already learned that just by looking at the ACF, PACF, and IACF graphs, you will not be able to identify the orders p and q in an ARMA process. Fortunately, you have some advanced techniques for determining the orders of ARMA models. To determine the orders, you need to calculate two more functions: the smallest canonical correlation (SCAN) and extended sample auto correlation function (ESACF). They were developed by Ruey Tsay and George Tiao. According to their research, if a series is ARMA (p,q), then the ESACF plot is a triangle with (p,q) as the vertex. And the SCAN plot is a rectangle with (p,q) as the vertex. The details of this theory are beyond the scope of this book; however, as usual, you can use SAS to calculate the ESACF and SCAN function values and arrive at the values of p and q. What follows is the SAS code to calculate the ESACF and SCAN function values:

```
proc arima data= TS13 plots=all;
identify var=x    SCAN ESACF ;
run;
```

You need to add just two options: SCAN and ESACF. This code gives much more elaborate output. You can directly look at the final consolidated recommendations by SCAN and ESACF (Table 12-14).

Table 12-14. *Finding the Order of an ARMA Process Using SCAN and ESACF, SAS Output*

ARMA(p+d,q) Tentative Order Selection Tests			
SCAN		**ESACF**	
p+d	q	p+d	q
1	1	1	1
4	0	5	4
0	5	0	5

Since you already tested the stationarity well before this phase, you can simply take d as zero here. In other words, you can take p+d as p in Table 12-14. The first option suggestion by the SCAN function is ARMA(1,1). The ESACF also suggests ARMA(1,1) as the first option. So, the underlined series is 1,3. Both SCAN and ESACF give a few more options for p and q. Generally, a less complicated model with fewer parameters should be chosen over more complicated ones. Simple models yield better forecasts.

If you take a model of p=5 and q=5, then the model equation will have too many parameters that need to be estimated. The model (p=5,q=5) will have a1,a2,a3,a4,a5, and b1,b2,b3,b4,b5 as parameters. Generally, models with too many parameters are not robust enough to give accurate forecasts. So, less complicated models with p<=2 and q<=2 are preferred. If SCAN and ESACF are suggesting more than one option, then it's better to go with the model with the smallest values of p and q.

Similarly, you can find the SCAN and ESACF for TS 14 and TS 15. Refer to Tables 12-15 and 12-16 for the output.

```
proc arima data = TS14 plots = all;
identify var = x SCAN ESACF ;
run;
```

Table 12-15. *Finding the Order of an ARMA Process Using SCAN and ESACF, SAS Output*

ARMA(p+d,q) Tentative Order Selection Tests			
SCAN		**ESACF**	
p+d	q	p+d	q
1	1	1	1
3	0	2	1
0	4	3	1
		5	1
		0	4

Table 12-16. *Finding the Order of an ARMA Process Using SCAN and ESACF, SAS Output*

ARMA(p+d,q) Tentative Order Selection Tests			
SCAN		ESACF	
p+d	q	p+d	q
2	1	2	1
1	2	1	2
4	0		

Again, the first preference given by both SCAN and ESACF is ARMA(1,1). So, TS 14 is an ARMA(1,1) process.

```
proc arima data= TS15 plots=all;
identify var=x    SCAN ESACF ;
run;
```

You already took a stationary series, which required no further differentiation. This means that can be taken as d=0. Both SCAN and ESACF are giving the first preference as an ARMA(2,1) process. So, the time series TS 15 is an ARMA(2,1) process.

Let's recap what has happened up to now. You want to forecast the future values. You need the ARIMA technique for model building and forecasting. To build a model using the Box–Jenkins approach, you need the series to be stationary. As part of step 1, you tested the stationarity. If the series is not stationary, you make it stationary by differentiating it. That is where the value of d in ARIMA (p,d,q) is decided. Then in step 2 you identify the orders of p and q. Once you have p and q, you can write a skeleton equation of the model. For example, you identify a model as AR(1), and then you can write a skeleton equation as follows:

$$Y_t = a_1 * Y_{t-1} + \varepsilon_{t-1}$$

Now as a next step you will need to find an estimate of a_1. Depending on the type of series—AR(p), MA(q), or ARMA(p,q)—you need to estimate the number of parameters. If the series is AR(2), you need to estimate two parameters.

Identification SAS Example

An e-commerce web site wants to track the number of unique visitors per day and has collected data for nearly one year. And now the company wants to forecast the number of visitors for the coming week. Before forecasting, you need to identify what type of series this is.

The following is the code for identifying the number of unique visitors (per day). Refer to Table 12-17 and Figure 12-50 for the output.

```
ods graphics on; /* Option to display graphs */
proc arima data = web_views plots = all;
identify var = Visitors stationarity = (DICKEY) ;
run;
```

Table 12-17. *Output of PROC ARIMA on web_views Series*

Name of Variable = Visitors	
Mean of Working Series	8946.432
Standard Deviation	1215.248
Number of Observations	340

Autocorrelation Check for White Noise

To Lag	Chi-Square	DF	Pr > ChiSq	Autocorrelations					
6	379.84	6	<.0001	0.687	0.542	0.406	0.310	0.204	0.184
12	395.13	12	<.0001	0.159	0.106	0.069	0.039	0.025	0.010
18	410.19	18	<.0001	0.040	0.025	0.064	0.089	0.112	0.122
24	421.77	24	<.0001	0.135	0.089	0.062	0.018	0.031	0.024

Augmented Dickey-Fuller Unit Root Tests

Type	Lags	Rho	Pr < Rho	Tau	Pr < Tau	F	Pr > F
Zero Mean	0	-2.0950	0.3195	-1.07	0.2567		
	1	-1.2110	0.4378	-0.81	0.3664		
	2	-0.8499	0.4993	-0.62	0.4473		
Single Mean	0	-105.309	0.0001	-7.86	<.0001	30.86	0.0010
	1	-80.6368	0.0016	-6.30	<.0001	19.88	0.0010
	2	-84.4002	0.0016	-6.12	<.0001	18.74	0.0010
Trend	0	-115.577	0.0001	-8.28	<.0001	34.28	0.0010
	1	-91.2041	0.0007	-6.67	<.0001	22.25	0.0010
	2	-96.9292	0.0007	-6.43	<.0001	20.75	0.0010

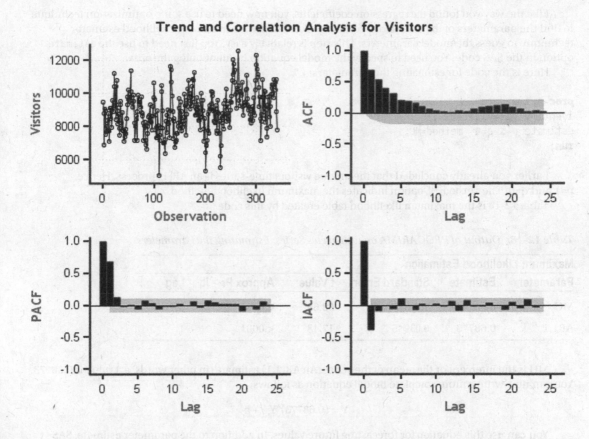

Figure 12-50. *ACF, PACF, and IACF plots from the output of PROC ARIMA on web_views series*

From Table 12-17, you can interpret that the process is stationary. The ACF is dampening to zero, so the process is an AR process. The PACF cuts off at lag1. A close observation of the PACF graph reveals that lag-2 is not significant. So, the unique visitors time series is an AR (1) process. The skeleton equation of the time series is as follows:

$$Y_t = a_1{}^*Y_{t-1} + \varepsilon_{t-1}$$

If you can estimate the value of a_1, then you can use this model for forecasting the number of unique visitors for the coming week, which is the ultimate goal.

Step 3: Estimating the Parameters

By now you have the model equation. You have already identified the time series as AR(p) or MA(q) or ARMA(p,q). You now have to find the model parameters. This model equation looks similar to the regression models discussed in Chapter 10 on multiple linear regressions. Just to recap, refer to the following analogy of equations.

Regression equation: $y = \beta_0 + \beta_1 x_1 + \beta_2 x_2 + \beta_3 x_3 + \beta_4 x_4 + \beta_5 x_5 + \beta_6 x_6 \dots \dots + \beta_p x_p$

Time-series equation: $Y_t = a_1{}^*Y_{t-1} + a_2{}^*Y_{t-2} + a_3{}^*Y_{t-3} + \dots + a_p{}^*Y_{t-p} + \varepsilon_t + b_1{}^*\varepsilon_{t-1} + b_2{}^*\varepsilon_{t-2} \dots \dots + b_q{}^*\varepsilon_{t-q}$

Like the way you found the regression coefficients, you now need to use some optimization technique to find the parameters of the time-series equation. You can use the maximum likelihood estimator technique to guess the model parameters. This step is relatively easy. You just need to use the estimate option in the SAS code. You need to specify the model equation by mentioning the order.

Here is the code for estimating the parameters:

```
proc arima data=web_views;
identify var=Visitors ;
estimate p=1 q=0  method=ML;
run;
```

Earlier you already concluded that the unique visitors time-series is an AR(1) process. Hence, p=1 and q=0. The method=ML option indicates the maximum likelihood method.

Table 12-18 is the maximum likelihood table created by this code.

Table 12-18. *Output of PROC ARIMA on web_views Series: Estimating the Parameters*

Maximum Likelihood Estimation					
Parameter	Estimate	Standard Error	t Value	Approx Pr > |t|	Lag
MU	8942.4	152.37003	58.69	<.0001	0
AR1,1	0.68773	0.03945	17.43	<.0001	1

MU is the intercept or the mean of the series. An AR(1,1) estimate (in other words, a_1) value is 0.68773. You can now write a more complete model equation as follows:

$$Y_t = (0.68773)*Y_{t-1} + \varepsilon_t$$

You can use this equation for forecasting future values. In addition to the parameter estimate, SAS produces detailed output. The following are some notes on how to interpret some of the unfamiliar yet important tables in the output. Refer to Table 12-19.

Table 12-19. *One Important Output Table of PROC ARIMA on web_views Series: Estimating the Parameters*

Constant Estimate	2792.49
Variance Estimate	781147.7
Std Error Estimate	883.8256
AIC	5580.809
SBC	5588.467
Number of Residuals	340

The AIC and SBC values help you to compare between two models. Low AIC and SBC value models are preferred. A low value of Akaike's information criteria (AIC) and Schwarz's Bayesian criterion (SBC) indicates that the model with the current estimates is the right fit for the data. This helps when you are trying to work with various orders such as AR(1) versus AR(2) versus ARMA(1,1), and so on. Refer to Table 12-20 and Figure 12-51. An explanation is given after Figure 12-51.

Table 12-20. *Autocorrelation Check of Residuals, an Output Table of PROC ARIMA on web_views Series: Estimating the Parameters*

Autocorrelation Check of Residuals

To Lag	Chi-Square	DF	Pr > ChiSq	Autocorrelations					
6	9.97	5	0.0761	-0.088	0.094	0.019	0.070	-0.075	0.038
12	12.67	11	0.3153	0.063	-0.002	0.004	-0.015	0.004	-0.059
18	17.03	17	0.4521	0.063	-0.069	0.035	0.016	0.041	0.018
24	24.50	23	0.3765	0.103	-0.015	0.037	-0.069	0.032	-0.051
30	28.42	29	0.4954	0.044	0.078	-0.045	-0.022	0.007	0.001
36	34.61	35	0.4869	0.026	0.051	0.110	0.002	-0.031	0.008
42	40.64	41	0.4865	0.056	-0.075	-0.003	0.061	-0.015	0.053
48	50.06	47	0.3529	-0.062	0.131	0.029	0.040	-0.018	0.016

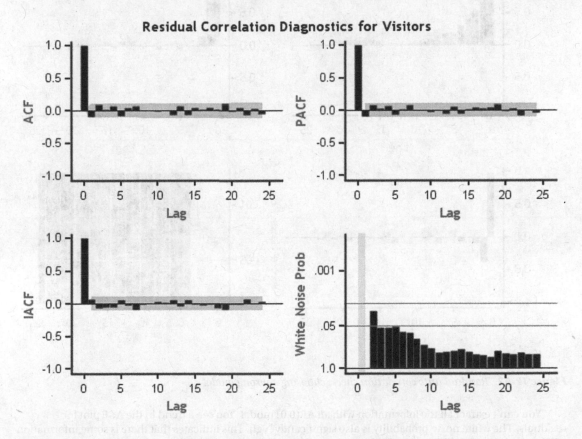

Figure 12-51. *Residual auto correlation check, output table of PROC ARIMA on web_views Series; estimating the parameters*

A residual auto correlation check helps you to identify whether there is any information left in the data after building the model. The model should be good enough to extract all the hidden patterns in the data. If there are still some trends left, then you observe them in residual plots. These residual plots show no trend in ACF, PACF, and IACF. You also observed that white noise probabilities (Figure 12-51, bottom right) are negligible.

For example, if you try to fit your model wrongly—in other words, instead of an AR(1) model, you try to fit an ARMA(0,0) model—then the residual plots show the trends (Figure 12-52). For a better understanding, you can compare the following code with the previous code. Both are the same except the values of p and q.

```
proc arima data=web_views;
identify var=Visitors ;
estimate p=0 q=0  method=ML;
run;
```

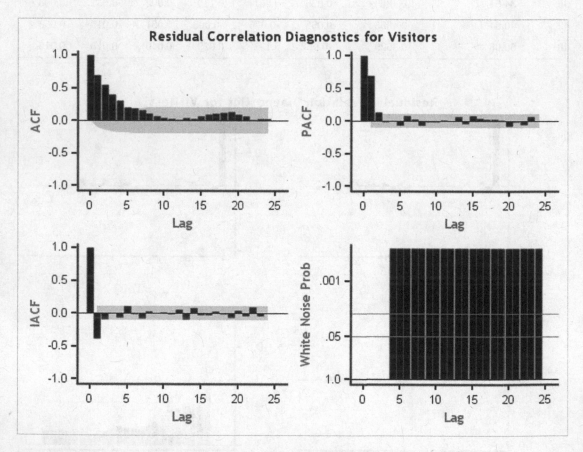

Figure 12-52. Residual auto correlation check, choosing a wrong model

You can't extract all the information with an AR(0,0) model. You see a trend in the ACF plot for residuals. The white noise probability is also significantly high. This indicates that there is some information still left in the model. This residual information needs to be extracted.

Once you are done with the model parameters estimation, you are left only with the forecasting. This is the last step in the model building.

Step 4: Forecasting Using the Model

Up to now you have worked really hard to find the type of the model, estimate the parameters, and create the final model equation. Once you have the final equation, what remains is to substitute the historical values to get the forecasts. SAS is there as usual to help you out. You just need to use a forecast option in the code.

```
proc arima data=web_views;
Identify var=Visitors;
Estimate p=1 q=0 method=ml;
Forecast lead=7;
run;
```

lead=7 indicates the next seven forecasted values, which, in this case, are the unique number of visitors that are likely to visit in the next seven days. This code also appends two new components to the output (forecast table and forecast graph). Table 12-21 is the forecast table, and Figure 12-53 charts the forecast graph from the SAS output.

Table 12-21. *Next Seven Days Forecasts for the Variable Visitors, SAS Output*

Forecasts for variable Visitors

Obs	Forecast	Std Error	95% Confidence Limits	
341	8139.5548	883.8256	6407.2885	9871.8211
342	8390.2684	1072.6622	6287.8892	10492.6476
343	8562.6905	1151.2361	6306.3092	10819.0718
344	8681.2696	1186.5879	6355.6000	11006.9391
345	8762.8194	1202.9463	6405.0880	11120.5508
346	8818.9033	1210.6063	6446.1586	11191.6480
347	8857.4736	1214.2123	6477.6612	11237.2861

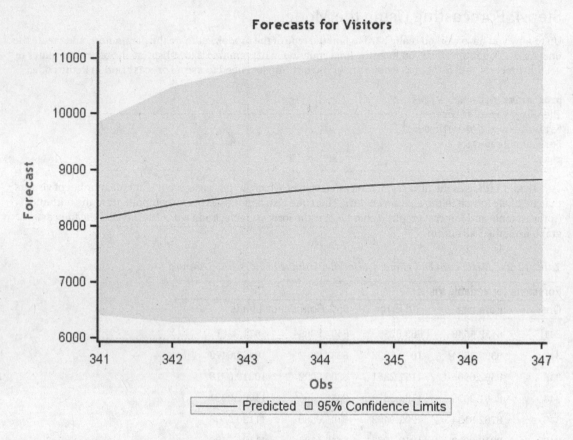

Figure 12-53. *Predicted forecast for the visitors, SAS output*

Data is available until observation 340. The values from 341 to 347 are forecasted and displayed along with the 95 percent confidence limits. There is a steady increment expected in the number of visitors to the web site. If you see a sharp dip, then it can serve as an alert message. If there is a huge increment in the forecasts, then you may need to step up the web site's IT infrastructure, including the number of servers, network bandwidth, and so on. Also, you need to make sure that there are no technical glitches. In the current case, everything seems to be normal; the forecasted flow of visitors shows a manageable, steady increase.

This is a logical end to our time-series analysis and forecasting process using the Box–Jenkins ARIMA approach. Every analyst wants to accurately forecast the future values by building the best model for the available historical data. This can be definitely achieved by using the Box–Jenkins approach. The following are some suggestions while building time-series models:

- Have sufficient historical data, at least 30 data points. Make sure you don't run into too much history. Only the historical values that will have an impact on future forecasts should be considered.

- Do not forecast too far into the future. With one year of data, forecasting the next two years is not a good idea. It may be that 10 percent or fewer data points into the future is recommended.

- Remove outliers before building the model.

Case Study: Time-Series Forecasting Using the SAS Example

Air Voice is a mobile network company that has more than 70 million customers. The customer care center of this company handles millions of calls per day. The company wants to increase customer satisfaction by resolving most of the customer queries within the agreed-upon service level agreements (SLAs). This can be done by better resource planning by forecasting the number of calls every week. The following is a typical call volume forecasting exercise.

The daily call volume data is available for the last 300 days. The volume is in multiples of thousands. The following is the procedure to analyze and forecast the future call volume values.

Step 1: Testing the Stationarity

The following is the SAS code to test the stationarity. It was already explained in the earlier sections. Refer to Table 12-22 for the SAS output relevant to this task.

```
proc arima data=call_volume;
identify var=call_volume stationarity=(DICKEY);
run;
```

Table 12-22. *Testing the Stationarity of call_volume Time Series, SAS Output*

Augmented Dickey-Fuller Unit Root Tests							
Type	Lags	Rho	Pr < Rho	Tau	Pr < Tau	F	Pr > F
Zero Mean	0	-0.3118	0.6116	-0.75	0.3904		
	1	-0.4327	0.5844	-0.70	0.4139		
	2	-0.3451	0.6040	-0.74	0.3964		
Single Mean	0	-17.2064	0.0217	-3.26	0.0181	5.44	0.0242
	1	-39.0108	0.0016	-4.71	0.0002	11.14	0.0010
	2	-25.7633	0.0026	-3.87	0.0027	7.59	0.0010
Trend	0	-17.4227	0.1110	-3.18	0.0899	5.31	0.1108
	1	-40.4434	0.0007	-4.70	0.0009	11.22	0.0010
	2	-26.2061	0.0169	-3.79	0.0183	7.48	0.0191

The process is stationary on the basis of a single mean P-value in the Augmented Dickey Fuller test.

Step 2: Identifying the Model Type Using ACF and PACF Plots

The following code creates the ACF and PACF plots to facilitate the identification of model type. Refer to Figure 12-54 for the SAS output relevant to this task.

```
ods graphics on;
proc arima data=call_volume plots=all;
identify var=call_volume stationarity=(DICKEY);
run;
```

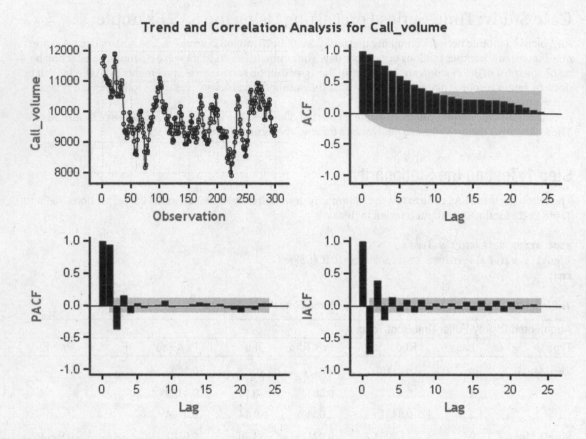

Figure 12-54. *ACF and PACF plots for identification of the model type, SAS output*

The ACF is dampening. The PACF and IACF are also decreasing to zero. This should be an ARMA process. You will use the SCAN and ESACF options to find the orders. Refer to the following SAS code and relevant output in Table 12-23.

```
proc arima data=call_volume plots=all;
identify var=call_volume SCAN ESACF;
run;
```

Table 12-23. *SCAN and ESACF table for finding ARMA process orders p and q, SAS output*

ARMA(p+d,q) Tentative Order Selection Tests			
SCAN		ESACF	
p+d	q	p+d	q
1	1	1	1
5	0		

The first option given by both SCAN and ESACF is ARMA(1,1). You will try to estimate the parameters a_1 and b_1.

Step 3: Estimating the Parameters

Refer to the following SAS code and the relevant SAS output in Table 12-24 to find out the model equation parameters.

```
proc arima data=call_volume plots=all;
identify var=call_volume SCAN ESACF;
estimate p=1 q=1 method=ML;
run;
```

Table 12-24. *Estimation of the Parameters, Maximum Likelihood Estimation Table from the SAS Output*

Maximum Likelihood Estimation					
Parameter	Estimate	Standard Error	t Value	Approx Pr > |t|	Lag
MU	9825.6	201.76908	48.70	<.0001	0
MA1,1	-0.68286	0.04442	-15.37	<.0001	1
AR1,1	0.90783	0.02424	37.46	<.0001	1

The AR and MA parameters values are now estimated. The model equation is as follows:

$$y_t = 9825.6 + 0.90783 y_{t-1} + \varepsilon_t - 0.68286\, \varepsilon_{t-1}$$

Step 4: Forecasting

You can now forecast the next week's call volumes. The applicable SAS code and its interpretation are left as an exercise for you. I have already discussed the process in detail, in the earlier sections, for other cases. Refer to Table 12-25 for the relevant portion of the SAS output.

Table 12-25. *Forecasts Data for Call Volume, SAS Output*

Forecasts for variable Call_volume				
Obs	Forecast	Std Error	95% Confidence Limits	
301	9876.9777	198.7791	9487.3779	10266.5776
302	9872.2397	373.4869	9140.2188	10604.2607
303	9867.9384	471.0529	8944.6917	10791.1852
304	9864.0336	538.3305	8808.9252	10919.1420
305	9860.4887	588.0196	8707.9914	11012.9860
306	9857.2705	626.0134	8630.3068	11084.2342
307	9854.3489	655.6732	8569.2530	11139.4449

As you can observe in Table 12-25, there are no major dips or increments expected over the next week. No major changes are required in the resource planning!

Checking the Model Accuracy

Testing the model accuracy is important in any model-building exercise. Residual graphs and AIC and SBC values tell you how well you have built the model. To test the forecasting accuracy, it might be a good idea to put aside some data for testing purposes and later use it to test the model accuracy. For example, in the call volume forecasting data, you have data available for 300 days. Of these 300 days of data, you assume that the data is available only for 295 days and use the next 5 days of data for testing the accuracy of forecasts. You forecast these 5 days of values using the model. You expect a good model to give the forecasts close to the actual values. Refer to Table 12-26.

Table 12-26. *Forecasted Values and Errors in the Previous Call Volume Example*

Day	Observed	Forecasted	Abs Error	Percent Error	Squared Deviation
296	9345.557	9407.4053	61.84803	0.7%	3825.179
297	9275.306	9444.8852	169.5787	1.8%	28756.94
298	9396.698	9478.9372	82.23937	0.9%	6763.314
299	9217.704	9509.8748	292.171	3.2%	85363.87
300	9533.522	9537.9828	4.461153	0.0%	19.90189
			122.1	1.3%	24945.8

Ideally you would like to have the error be zero or less than 5 percent. The following are some measures of errors or accuracy. In the following text, Y_i is the actual value and $\hat{Y}_i$ is the expected value.

1. Mean absolute deviation (MAD)

    - Average(absolute deviations)
    - $MAD = \sum_{i=1}^{n} |Y_i - \hat{Y}_i| \big/ n$

2. Mean absolute percent error(MAPE). This is the most widely used measure.

    - Average(percent deviations)
    - $MAPE = \dfrac{100}{n} \sum_{i=1}^{n} |Y_i - \hat{Y}_i| \big/ Y_i$

3. Mean square error (MSE)

    - Average(squared deviations)
    - $MSE = \sum_{i=1}^{n} (Y_i - \hat{Y}_i)^2 \big/ n$

4. Another related measure is root mean square error, which is $RMSE = \sqrt{MSE}$.

In the previous example, here are the values:

- MAD = 122.1
- MAPE = 1.3 percent
- MSE = 24,945.8
- RMSE = 157.9

It is generally a good practice to keep 5 to 10 percent of the sample data for validation purposes. The exact percentage will depend on the size of the data set.

Conclusion

In this chapter, you learned what a time-series process is. I discussed one of the most widely used time-series techniques, that is, the Box–Jenkins ARIMA technique. I also discussed the four major steps in ARIMA model building. The ARIMA models are one of the many types of models. There are sessional ARIMA models; there are ARIMAX models, which will use an independent variable X along with the historical values of Y; and there are heuristic TSI models. I suggest you research other time-series techniques only after developing a solid understanding of the basics of ARIMA model building. The model accuracy-testing methodologies discussed at the end of the chapter should be taken seriously. It's good practice for analysts to furnish the error rates along with the forecasted values.

CHAPTER 13

■ ■ ■

Introducing Big Data Analytics

By this time you have read 12 chapters on the topic of business analytics and can now appreciate the theories and practices that make up this domain. But the learning never ends, especially in a fast-growing area like this one. By now, you might already be aware that big data is catching up in popularity very quickly. A fast-growing number of organizations, notably online ones with a large amount of visitor traffic, are generating an unprecedented amount of data daily, which is almost impossible to handle with classic techniques.

We discussed this a bit in Chapter 1, and in this last chapter of the book, we expand your understanding so that you get a feel of the big data domain and understand what tools and techniques are used by the experts working in this field.

Traditional Data-Handling Tools

In your day-to-day work, you might use one or more of the following tools:

- *Text editor/Notepad*: For modifying small amounts of data

- *Microsoft Excel*: For simple and medium calculations

- *Microsoft Access/SQL*: For storing and querying relational data

- *SAS/R*: For advanced analytics

- *Tableau/QlikView*: For data visualization and reporting

Have you ever tried to open a 2GB file using Notepad? Have you ever worked with an Excel file of 4GB? Is it possible to create a 4GB Excel file in the first place? How much time does it take to query a 10GB database? Have you ever performed any analytics operation like regression on a 20GB data set? Can SAS or R handle 20GB data with equal efficiency as relatively smaller data sets? We'll present some examples to give you a feel of where we are coming from.

Walmart Customer Data

About 245 million active customers and members visit Walmart in 11,000 stores under 71 banners in 27 countries, which includes its e-commerce web sites (http://corporate.walmart.com/our-story/). Imagine that Walmart wants to know its customers' buying patterns for the last month. Let's try to get a rough estimate of how much data is generated by Walmart in a month. The following calculations may give a conservative estimate:

(245,000,000 customers) × (2 visits per month) × (20 items bought per visit) = 9,800,000,000 rows

There might be as many as 9.8 billion rows in one month. For each row, Walmart must be tracking details such as the date stamp, item ID, number of items, item price, net price, and so on. Again, using a conservative approach, let's imagine that there are a minimum of six columns in each such row. What will the size of the file be with 9.8 billion rows and 6 columns? Generally speaking, for a CSV or TXT file with 10,000 rows and 6 columns per row, the approximate file size may be around 210KB. Based on this, the size of a similar file with 9.8 billion rows will be approximately 196GB. And this is just one month's conservative estimate for only the billing transactions. Envisage the data for customers' previous purchases, demographics data, product details, store details, and so on. A basic file that needs be considered for any customer analytics will be much larger than our estimate of 160GB.

Facebook Data

Let's imagine that Facebook (or a social media network of that size) wants to build an effective online advertisement (ad) engine. Depending upon its user profiles, Facebook wants to display the most relevant ad on each web page. Or it wants to improve the Facebook feeds. Let's try to estimate the minimum amount of data generated by Facebook in a month.

Facebook has more than 1 billion users. There are 864 million daily active users. Just for the sake of simplicity, let's treat the user activities such as share, likes, status updates, picture uploads, login and logout, and so on, as one activity. Consider the following conservative calculations:

$$(864,000,000 \text{ customers}) \times (30 \text{ activities per month}) = 25,920,000,000 \text{ rows}$$

A file with 25.92 billion rows and 6 columns (for example) may be on the order of 510GB. Facebook may need to handle at least this much amount of data if it wants to do any meaningful analytics for its ad engine. And this is just a simulated example to give you a feel of the amount of data involved in the process.

Note that the Walmart and Facebook estimates are conservative as far as the amount of data involved. The actual data size involved might be much bigger. Some real-life business scenarios might require you to handle petabytes of data. What is a petabyte? Refer to Table 13-1.

Table 13-1. *Units of Data*

Unit Measure	Conversion
1GB Gigabyte	1024MB
1TB Terabyte	1024GB
1PB Petabyte	1024TB = 1024GB × 1024GB = 1,048,576GB

As an example, consider that a typical DVD stores about 4.7GB of data. To store 1PB of data, you would need 223,101 DVDs.

Examples of the Growing Size of Data

The real-time data produced and analyzed by some enterprises is sometimes even beyond our imagination.

- Walmart has a database with more than 2.5 petabytes of data.
- Facebook handles 300 petabytes of data daily.
- Google handles nearly 200 petabytes of data every day.

- NASA stores and processes around 32 petabytes of climate data and related simulations.

- Large Hadron Collider (LHC) at CERN produced 30TB of data in 2012.

By now, it may be clear that you can't handle these large volumes of data using traditional tools. Does big data really mean a large data set? What is the definition of big data?

What Is Big Data?

There is no standard definition for *big data*. The following are a few definitions that are floating around the Internet:

- Any data set that is difficult to handle with conversional or traditional data-processing tools is called *big data*.

- Data that is difficult to capture, store, process, search, query, transfer, analyze, and visualize is called *big data*.

- *Big data* is a complex, varied, rapidly growing, and huge data set that requires new architecture, techniques, and algorithms to analyze it.

- *Big data* is a data set that can be analyzed and mined for insights, but because of its size, complexity, and growth rate, traditional tools fail to process it.

The following two sections basically revise what we discussed in Chapter 1.

The Three Main Components of Big Data

Big data is not just about the size, though the volume of the data plays a big role in the failure of traditional tools while handling it. There are three main components of big data: volume, verity, and velocity. The traditional methods and tools have their own limitations in processing the data. The following bullets shed some more light on each of these characteristics:

- *Volume*: The sheer size of the data, in big data analytics, is so large that it becomes almost impossible to process. So far we have mentioned several examples, and to add some more examples, transactional data from credit card companies, call data from a cell phone company, and billing data from a big retailer like Amazon fall under this category. Size is considered the first characteristic of defining big data.

- *Variety*: The type of data that we are generating these days is limited not just to structured numerical tabular data. Examples are audio files from telecom companies, video files from numerous closed-circuit cameras, and blog data from many web sites. All these are nonconventional types of data. Other data in this category includes text, numerical data, sequences, time series, multidimensional arrays, XML, still images, and social media data. These types of data may come pooled with the conventional data. Preparing this type of data for analysis is obviously not simple.

- *Velocity*: Data is growing day by day at a rapid pace. In some situations, data keeps growing so fast that we do not have enough time to analyze it. Imagine a Formula One scenario. We invariably have some sensors placed in each super car. The data collected by these sensors is analyzed so as to give some suggestions for improvements for the next round. We can't take, for example, three months to transform the data, prepare it for analysis, build a model, and finally validate it so that some useful recommendations can be made based on this analysis. All this needs to happen well before the next lap starts. This type of analytics even uses real-time data streaming from car sensors. Table 13-2 compares big data to conventional data. It's reproduced from Chapter 1 for a ready reference.

Table 13-2. *Big Data vs. Conventional Data*

Big Data	Normal or Conventional Data
Huge data sets.	Data set size in control.
Unstructured data such as text, video, and audio.	Normally structured data such as numbers and categories, but it can take other forms as well.
Hard-to-perform queries and analysis.	Relatively easy-to-perform queries and analysis.
Needs a new methodology for analysis.	Data analysis can be achieved by using conventional methods.
Need tools such as Apache's Hadoop, Hive, HBase, Pig, Sqoop, and so on, to handle big data.	Tools such as SQL, SAS, R, and Excel alone may be sufficient to handle traditional data.
Raw transactional data.	The aggregated, sampled, or filtered data.
Used for reporting, basic analysis, and text mining. Advanced analytics is only in a starting stage in big data.	Used for reporting, advanced analysis, and predictive modeling.
Big data analysis needs both programming skills (such as Java) and analytical skills to perform analysis (as of today).	Analytical skills are sufficient for conventional data; advanced analysis tools don't require expert programing skills.
Petabytes/exabytes of data.	Megabytes/gigabytes of data.
Millions/billions of accounts.	Thousands/millions of accounts.
Billions/trillions of transactions.	Millions of transactions.
Generated by big financial institutions, Facebook, Google, Amazon, eBay, Walmart, and so on.	Mostly generated by small enterprises.

Now you know that big data is complex, huge, and rapidly growing. Also, it's almost impossible for the traditional tools to process this kind of data. Why do we really need to analyze big data? Can't we simply use aggregated data or sample data sets and stick to the traditional techniques and tools? What are the applications of big data analytics in real life? The following section answers all these questions.

Applications of Big Data Analytics

Not long ago, it was not a common practice to send soft copies of data. We used to mail all the data in the form of hard copies, arranged in files and bundles. With the advent of computers, smartphones, and other devices, we are able to store and process huge amounts of data. Now the data is growing in an exponential manner. There are already some smart companies that are using this vast amount of available data to their benefit. Here are some examples:

- *E-commerce retailers*: Online retailers such as Amazon and eBay need to handle the millions of customers online who are purchasing thousands of products. They analyze petabytes of data every day to recommend the right products, improve searches, and improve service quality. Performing big data analytics reveals so many customer insights such as customer lifetime value, customer loyalty, product popularity, and so on.

- *Online entertainment industry*: Sites such as Netflix are using big data analytics to understand customer sentiments. They are analyzing big data for movie recommendations and even for ticket pricing based on the demand.

- *Automobile industry*: Car manufacturers collect data from millions of vehicles worldwide using sensors. Because of the data's sheer size, its variety, and the pace with which it accumulates, it falls under the big data category. This data is used in understanding the maintenance issues with the models, strong and weak points in a particular model, customer driving habits, causes of accidents, suggestions on quality improvements, and so on.

- *Telecom industry*: In addition to the billing and complaints data, telecom companies also maintain call detail record data (CDR) and text messages data. These days a smartphone is an integral part of human life. By analyzing CDR data, one can get some vital analytics on people's habits such as sleeping hours, most active hours, office hours, driving time, and so on. This data can be utilized by the civic authorities in planning and managing cities by setting up the right infrastructure at places. Numerous such uses can be found once the analytics models are in place.

- *Online social media industry*: One of the most popular users of big data is the social media industry. Social networking sites such as Facebook, Google+, Twitter, and LinkedIn generate petabytes of data every day. They analyze this data for friends' recommendations, advertisement management, feed management, and so on.

These are just a few examples. In fact, big data analytics is also used in the healthcare industry in drug invention. The industry uses simulated data rather than conventional lab experiments. Another use of big data is video analytics for self-driving cars. In addition, search engines are using big data in indexing, searching, and querying. Big data is used in politics too. Many popular politicians have used big data analytics along with focused e-mail campaigns to get an estimate of their chances of winning. So, as you see, the uses and possibilities are beyond one's imagination.

Now that you are familiar with the big data introductory concepts, we will discuss some more applications in the form of the following questions, many of which were considered as far-fetched ideas just a few years ago:

- How about getting some real-time recommendations and discounts (maybe on a cell phone) when we drop an item into a cart in a supermarket?

- How about a city with 100 percent self-driven cars?

- Can we wear some glasses and get complete information of the objects that we see through those glasses?

- Can we predict the result and change in the winning probability of a football match as it happens?

- In real time can we control the traffic and better manage it during the peak hours?

- Can we create a robot that can understand the human language and reply in real time?

- Can we have a smartphone that realizes our blood pressure levels? Can we have an app that rings an alarm at the time of a heart attack?

- Can we analyze a call center's historic voice data and automatically resolve customer queries instead of an agent physically answering a question?

All this looks too advanced to be true. But these are all possibilities with big data. Unfortunately, big data can't be processed using traditional tools, old techniques, or typical algorithms; we need an alternative solution.

The Solution for Big Data Problems

The reason for the failure of traditional data-processing techniques when analyzing big data is apparent. The conventional tools take the complete data set as one unit. In that way it would be painful to handle the data in the orders of petabytes. Sometimes it's not even possible to save big data files on one machine. We need gigantic supercomputers with multiple processors to handle big data using one machine. A supercomputer is not a simple machine; it costs millions of dollars. The solution to the big data challenge is to connect multiple computers using computer networks (such as LANs), writing the code in such a way that it is separately understood and executed at each computer and the results are finally assembled as one output. This technique is called *distributed computing*.

Distributed Computing

Even today, the need of high-computing power is sometimes listed as a major limitation in dealing with big data. Distributed computing appears to work well in these situations. It works on the principle of "divide and rule." Instead of considering a huge data set as a single unit, you can it cut into pieces or small data chunks. You can then save these relatively smaller pieces of data on an array or network of computing devices. Each computer in this network is called a *node*. Distributing the data onto a cluster setup is the first part of distributed computing. The second part is to adapt the code for this kind of cluster of computers. The following points summarize the distributed computing approach for dealing with big data:

1. Modify the overall problem into smaller tasks and execute smaller tasks on an array of individual computers. That translates to dividing the data into smaller blocks and storing the blocks on individual nodes in the cluster.

2. Collate the results from all the individual machines and deliver a consolidated output.

Let's take the same example of Facebook or a similar networking site. In a day there are billions of activities on this social networking web site. Say you want to get the number of status updates in a day. Imagine that the overall status updates file is 500GB. The following could be the simple SQL code to calculate the total count of status updates:

```
Select count(*) from feb_status /* feb_status is the status updates data set name */
```

This code will treat the whole 500GB file as one unit, and it might take an unusually long time. What if you take a network of 20 computers and divide this 500GB file into 10,000 pieces of 50MB each? You can store these 10,000 individual data chunks on 20 computers. By doing so, you are distributing the one big task of computation onto 20 machines. Each computer can locally calculate the number of status updates, and then the count data can be totaled to get the desired consolidated number of status updates. All 20 machines are working in parallel on independent chunks of data. If the overall computing task takes 20 hours using the traditional method and a single machine, it might be natural to expect that with this divide-and-rule policy, it will take only an hour or less to complete. In practice, using distributed computing, it will take much less than an hour to complete the whole task.

A supercomputer also could have done the same task efficiently. But, generally speaking, even medium and small enterprises can afford a cluster of 20 PCs when compared to the prohibitive costs of one large supercomputer. Moreover, 20 PCs can be put to other commercial use when there are no big data files to be analyzed. A supercomputer, on the other hand, will make sense only for high-end tasks, which might not be available all the time.

The programing model wherein you divide a big computing task into smaller tasks and collate the intermediate results to generate the ultimate result is called the *MapReduce programming model*.

What Is MapReduce?

In distributed computing, the aim is to solve a global problem. You write a program to divide this global task into smaller ones and finally assign the pieces to individual machines forming a cluster. This divide program is called a *map function*. Once you are done with the local machine tasks, you then write another program to sum up the results from the individual map functions to generate a final consolidated result. This consolidation function is termed a *reduce function*. We discuss each function in the following sections.

Map Function

The map function is locally executed on all individual chunks of data. The result of a map function comes in the form of a key-value pair. These key-value pairs, from different map functions, carry the intermediate results. Generally any individual map function is similar to the overall task. The only difference is the size, which is considerably smaller for the map function. The outputs of map functions are used as the inputs of the reduce function.

Reduce Function

The reduce function takes the intermediate results from the map functions and creates the final output. The key-value pairs generated by the map functions is sorted and aggregated in the reduce function. The reduce function doesn't act on the individual pieces of data. In fact, it has no interaction with the original input data, which was fed only to the map functions.

Figure 13-1 shows a reduce diagram that explains the overall MapReduce programming model that is used for handling big data. It's a cost-effective and efficient model of processing big data.

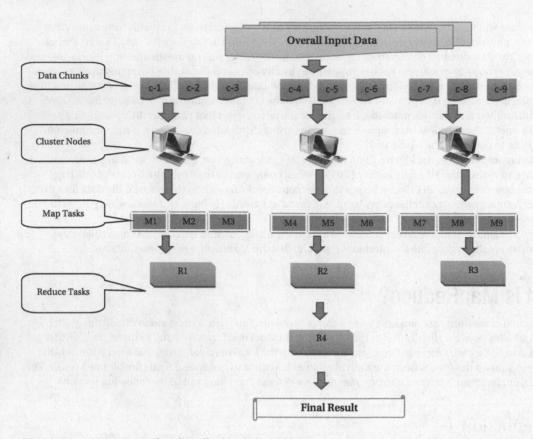

Figure 13-1. *How MapReduce handles big data problems*

In addition to the MapReduce programming model, the following processes also need your attention:

- *Cluster setup*: You set up a cluster by creating a network of individual machines. You may need to get the help of a system admin for setting up a cluster.

- *Work assignment*: Choose some nodes as slaves and assign the map tasks followed by the reduce task. You need to write a separate function to assign the work to individual nodes.

- *Task scheduling*: Create a master node, which will keep track of all the MapReduce tasks, meaning the scheduling. You can't start with a reduce task. That doesn't mean you need to wait until all the map tasks are completed, though. You may need to write some separate code for scheduling the tasks.

- *Load balancing*: There will be challenges; say one system in the cluster is slow. Some nodes in the cluster may have a good configuration. To achieve the desired efficiency, you need to assign the tasks based on the capacity of nodes. This may require you to write functions for load balancing as well.

- *Fault tolerance*: What if one of the nodes shuts down suddenly? Is there any way you can retrieve the data in that node? Maybe it is a good idea to replicate that specific load chunk onto a different machine and use it as a backup. So, you may need to take care of replication also while distributing the data.

It appears that MapReduce is a solution to big data problems, but, as you see, there are so many complicated interconnected tasks that make it difficult for an average programmer to try. You need multiple skills to solve a big data problem using the MapReduce programming model. It would be great if you have a readymade framework that takes care of setting up a cluster, adding and deleting nodes, assigning work, scheduling tasks, load balancing, fault tolerance, and so on. Apache Hadoop is the one such popular framework. Hadoop is open source, and it's a windfall for those who want to focus on MapReduce.

What Is Apache Hadoop?

The solution for the challenges of several issues of distributed computing using the MapReduce programing model is the Hadoop framework. The Hadoop framework is an open source tool, built on the Linux operating system.

Hadoop is a framework. It is a simplified platform to write and execute MapReduce and other big data tasks. Hadoop has two major components: the Hadoop distributed file system (HDFS) and MapReduce. We go into the details of these two components in the following sections.

Let's be clear on a few facts.

- Hadoop is *not* a database.

- Hadoop is *not* a synonym of big data.

- Hadoop is *not* a programing language.

Hadoop Distributed File System

In the world of distributed computing and MapReduce, there should not be any abnormally big files. All big files will be cut into pieces. The file system in Hadoop is called the Hadoop distributed file system. Every file in that system will be 64MB or less. If you transfer any file from outside to HDFS, it will be broken into pieces of 64MB. In the case of a cluster, HDFS will cut the files into pieces and distribute them to the different nodes within the cluster. You can imagine HDFS as a chopper that cuts the data into small chunks. With it, developers do not need to worry about the division and distribution of data. All that you need to do is copy the file from the local file system to HDFS. In fact, HDFS will even take care of details such as the data block locations. There will be replication of data blocks; if one system goes down, then replicated data blocks will be used. Figure 13-2 shows how HDFS distributes the data.

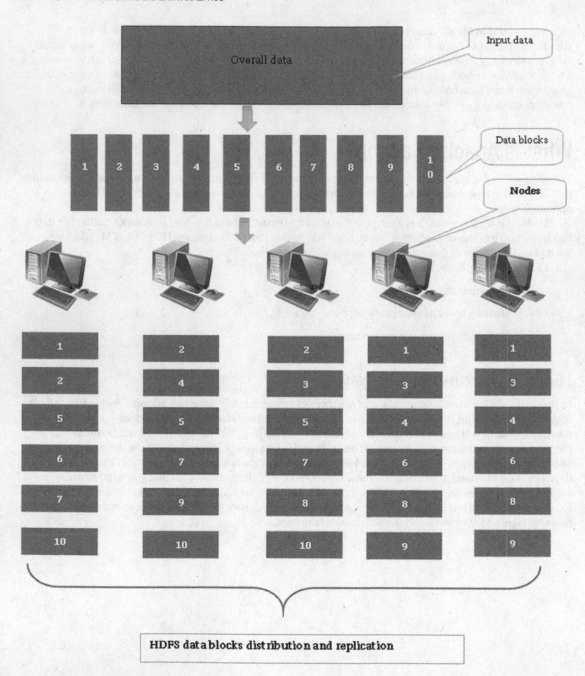

Figure 13-2. How HDFS distributes data

In Figure 13-2, each block has been replicated three times. By default each block size will be 64MB, and it will be replicated three times on the network.

MapReduce

MapReduce is a parallel processing programing model. In partnership with HDFS, MapReduce code blocks are executed at local locations. HDFS has all the data blocks and related information. MapReduce has all the task-related information. MapReduce and HDFS together process big data almost effortlessly. The Hadoop framework takes care of coordination between the MapReduce code and the HDFS data blocks. In Hadoop, the MapReduce code blocks need to be written in the Java language. It is the map code for each of the local data blocks and the reduce code for consolidation.

In addition to these two main components, the Hadoop framework plays a key role in the following processes:

- *Clusters made easy*: Hadoop allows users to quickly set up a cluster network. Hadoop makes it extremely easy to add and remove a node without disturbing the rest of the network. On the Hadoop framework, it is as easy as adding or deleting an IP address. Without Hadoop, setting up a computer cluster and maintaining it for coherent processing may not be easy.

- *Storing made easy*: Hadoop picks one node as the master node, which is known as a *name node*. Remaining nodes are slave nodes, known as *data nodes*. The data nodes contain the actual data blocks, and the name nodes contain the data about the data nodes and data block locations (basically the metadata). Hadoop gives an option to choose which machines will be the master and which the slaves.

- *Job scheduling made easy*: Hadoop automatically takes care of job orchestration. MapReduce will have several map and reduce tasks that need to be executed at the data nodes. To manage all the resources, Hadoop has a job tracker and a task tracker. The job tracker works along with the master node, and its primary function is to allocate the jobs to multiple task trackers. Task trackers run at different data nodes. Task trackers present the results and report the status of the jobs to the job tracker.

- *Fault tolerance*: Task trackers regularly send some pulses to the job tracker. It helps the job tracker to understand the status at each task tracker, whether it is alive or dead. If by any chance the job tracker doesn't receive heartbeat pulses from any task tracker, then it quickly allocates the task to a different task tracker on the replicated data block.

- *Load balance*: Hadoop automatically runs some backup jobs. If a node is slow and it is slowing down the entire MapReduce process, then the job tracker will assign that particular task to other high-speed nodes and thereby reduce the number of tasks assigned to the slow machine.

You can see that many bits and pieces related to big data processing are automatically taken care of in Hadoop. All this is handled in a proficient manner. Developers don't need to worry about cluster management, scalability, fault tolerance, job scheduling, load balancing, data distribution, and so on. Moreover, Hadoop is free and open source. Refer to Table 13-3 for some useful resources for Hadoop.

Table 13-3. *Some Resources for Hadoop*

Hadoop logo	
Hadoop home page	`http://hadoop.apache.org/`
Hadoop documentation	`http://hadoop.apache.org/docs/current/`
Hadoop download page	`http://hadoop.apache.org/releases.html`

The Hadoop framework makes several tasks easy. It automatically takes care of many key components in distributed computing. Writing, debugging, and maintaining key code components can be tough even for an expert programmer. The core MapReduce programming and other interactions with Hadoop are in the Java programing language; you need to be reasonably comfortable with Java programming to effectively work with Hadoop. The logic of the code algorithms need to be adjusted for the MapReduce programing model. The usual Java code that produces a result in a traditional way might not be the same with HDFS and MapReduce. Wouldn't it be great if some tools were available that could automatically convert your conventional SQL scripts or commands into what is required for MapReduce? Fortunately, some tools available in the Hadoop ecosystem can help you to simplify MapReduce programing. Apache's Hive, Pig, Sqoop, and Mahout are a few important ones. The following sections have the details.

Apache Hive

Hive is data warehouse software. You can install it on the top of Hadoop. Hive automatically converts conventional SQL-like queries to MapReduce code. The queries are written in the Hive query language (HiveQL). HiveQL is similar to SQL, and it is extremely useful for data analysts for producing some basic business intelligence (BI) reports on big data. Hive can't work independently without HDFS and MapReduce. Hive was originally developed by Facebook and later made public. Many data analysts use Hive for data summarization, reporting, and analytics. Refer to Table 13-4 for some basic resources for Hive.

Table 13-4. *Some Resources for Hive*

Hive logo	
Hive home page	`http://hive.apache.org/`
Hive documentation	`https://cwiki.apache.org/confluence/display/Hive/LanguageManual`
Hive download page	`http://hive.apache.org/downloads.html`

Apache Pig

Pig is software that is mainly used for extract, transform, and load (ETL) operations. Pig provides an engine for executing data flows in parallel. Like Hive, Pig also runs on the top of Hadoop. You need to write Pig Latin scripts to interact with the Pig tool. Pig Latin scripts are much easier than Java MapReduce code. Pig Latin scripts can be written for data-processing operations such as join, sort, filter, load, transform, and so on. Pig is less of a data analytics tool; its major use is for data handling and manipulation. Pig Latin can be extended to use customized user-defined functions. Pig was originally developed by Yahoo, and later it was made public. Like the way all Hive queries are automatically converted to MapReduce code before execution, Pig Latin scripts are also internally converted to MapReduce. The job of writing MapReduce code blocks for ETL is considerably simplified by Pig. Refer to Table 13-5 for some basic resources for Pig.

Table 13-5. *Resources for Pig*

Pig logo	
Pig home page	https://pig.apache.org/
Pig documentation	http://pig.apache.org/docs/r0.12.1/index.html
Pig download page	http://pig.apache.org/

There are several other utility tools to help big data developers. Interacting with Hadoop, writing MapReduce code, and performing analytics on big data are all getting easier every day with these tools. The following are a few more useful tools in the Hadoop ecosystem.

Other Tools in the Hadoop Ecosystem

The Hadoop ecosystem is like an app store on a smartphone. There is a tool in the Hadoop ecosystem for almost every need related to big data handling. Refer to Table 13-6 for more details.

Table 13-6. *Other Tools in the Hadoop Ecosystem*

Tool	Logo	Usage
Apache Sqoop		Used for efficient transferring of bulk data between Hadoop and structured data stores such as a relational database management system (RDBMS). An example is a data transfer between Hadoop and MySQL.
Apache Flume		Transferring large amounts of log data, that is, streaming data from Hadoop. Flume is used for loading real-time streaming data.
Apache Avro		A serialization system. It is used for modeling and serializing.
Apache Mahout		Advanced analytics on big data. Mahout gives access to some predictive modeling and machine learning algorithms.
Apache ZooKeeper		Coordination service. ZooKeeper is used for managing large clusters.
Apache Oozie		A workflow scheduler system to manage Hadoop jobs.
Apache HBase		Provides a fault-tolerant way of storing large quantities of sparse data. It is a NoSQL database, that is, a database stored in nontabular format for Hadoop.

Figure 13-3 is a simplified version of the Hadoop core ecosystem.

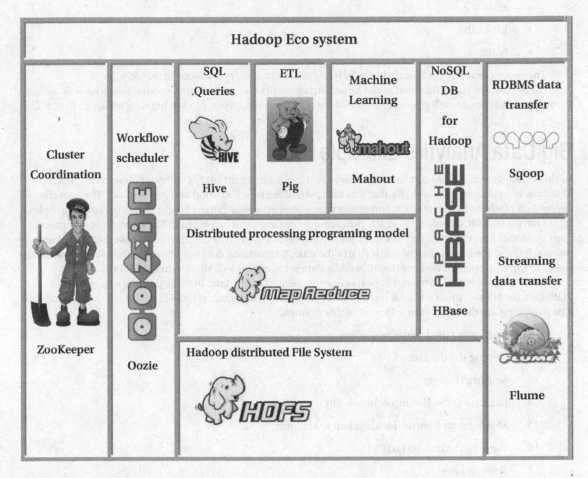

Figure 13-3. *The Hadoop ecosystem*

Companies That Use Hadoop

Hadoop is a widely used tool. The following are some major companies using Hadoop:

- Facebook
- Google
- IBM
- Yahoo
- Amazon
- AOL
- Fox Interactive Media
- New York Times

- Adobe

- eBay

- LinkedIn

- Ning

The more complete list is available at `http://wiki.apache.org/hadoop/PoweredBy`.

Up to now, we have discussed some important concepts and tools about big data. As we have done throughout this book, we'll give an example of big data analytics so you can see big data analytics in action.

Big Data Analytics Example

In this example, we will use data from the Stack Overflow web site (`http://stackoverflow.com/`). Stack Overflow is a privately held web site that was created in 2008 by Jeff Atwood and Joel Spolsky. The web site serves as a platform for users to ask and answer questions on a wide range of computer programming topics. If you have a specific question about any programing language, you can log in to Stack Overflow and submit your question. You can even tag your questions using keywords such as C++, Java, database, SQL, PHP, and so on. Stack Overflow has a portal where it shares the data. Enthusiastic data scientists can log in and download the data from this portal. We have downloaded a data set from this web site for this big data example.

This example is to understand various concepts related to big data. In this example, we are using a 7GB data set. It's too small a size for big data, but practical considerations limit us to this size in this example. The following are the steps to be followed in this example:

1. Examining the business problem

2. Getting the data set

3. Starting Hadoop

4. Looking at the Hadoop components

5. Moving data from the local system to Hadoop

6. Viewing the data on HDFS

7. Starting Hive

8. Creating a table using Hive

9. Executing a program using Hive

10. Viewing the MapReduce status

11. Seeing the final result

Let's start with the business problem.

Examining the Business Problem

The business wants to extract some basic descriptive statistics such as the total number of questions and the top ten topics from the given file. Data such as the total number of unique users, the top ten users, and the top ten users' questions is also desired. You could even go to the extent of building a model that detects and suggests the tags automatically when a user enters any question on Stack Overflow web site. For the demo purposes, we only try to find the total number of questions mentioned in the input file.

Getting the Data Set

The data set comes from Stack Overflow. The size of the data set is 7GB, and it's a text file. This input file needs to be analyzed for finding the basic descriptive statistics.

Starting Hadoop

Hadoop is installed on a Linux machine. In this example, we are using a single-node, or a pseudocluster, Hadoop setup. Basically we are working on one machine only. Since the operating system is Linux, all the operations will be in the form of Linux commands executed in a terminal. The first step is to open a terminal in Linux. The username of our Linux machine is Hadoop. If the username were neo, then you would have neo@local host home. Refer to Figure 13-4.

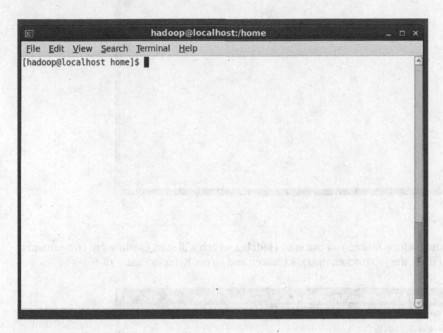

Figure 13-4. *Open terminal in Linux*

To start Hadoop, you need to go to the Hadoop folder. The path of Hadoop may differ from system to system. Refer to Figure 13-5.

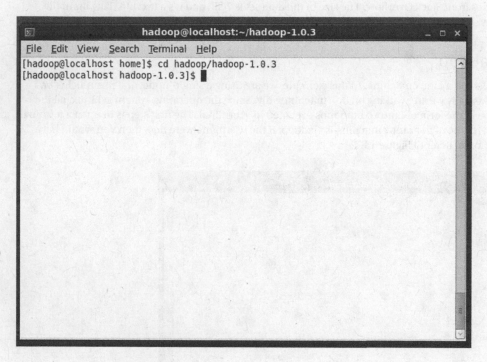

Figure 13-5. The Hadoop folder

Now, you are inside the Hadoop folder. You can start Hadoop, which will start the different components in the ecosystem such as HDFS, the job tracker, the task tracker, and so on. Refer to Figure 13-6.

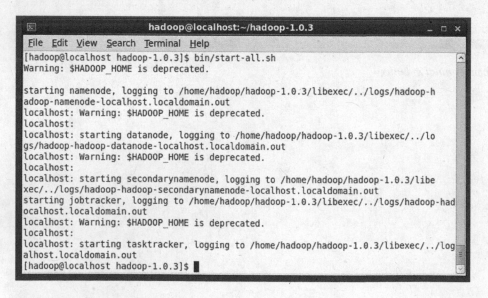

Figure 13-6. Starting Hadoop

start-all.sh is the shell script that will start Hadoop. If you get a warning that Hadoop is deprecated, you can ignore it. The warning will not appear if you are using the latest version of Hadoop. The following are some important messages when you start Hadoop. These messages will give the status of the Hadoop component's initiation.

- starting namenode, logging to /home/hadoop/hadoop-1.0.3/libexec/../logs/ hadoop-hadoop-namenode-localhost.localdomain.out

- starting datanode, logging to /home/hadoop/hadoop-1.0.3/libexec/../logs/ hadoop-hadoop-datanode-localhost.localdomain.out

- starting secondarynamenode, logging to /home/hadoop/hadoop-1.0.3/ libexec/../logs/hadoop-hadoop-secondarynamenode-localhost.localdomain.out

- starting jobtracker, logging to /home/hadoop/hadoop-1.0.3/libexec/../ logs/hadoop-hadoop-jobtracker-localhost.localdomain.out

- localhost: starting tasktracker, logging to /home/hadoop/hadoop-1.0.3/ libexec/../logs/hadoop-hadoop-tasktracker-localhost.localdomain.out

Looking at the Hadoop Components

You can now observe that all the Hadoop components have started. You can check on what ports the name node, data, and secondary name node (the backup for the name node) are running. Refer to Figure 13-7. The Jps option shows the ports on which these processes (such as the name node, job tracker, task tracker, and so on) are running.

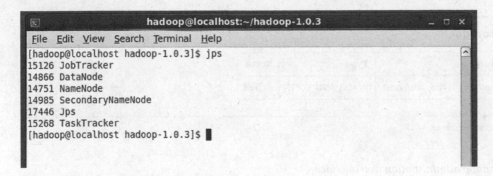

Figure 13-7. Jps option to see the processes

You can also check and manage Hadoop using the Hadoop administration web interface. To do this, you need to use a web browser and go to http://localhost:50070/dfshealth.jsp. Refer to Figure 13-8.

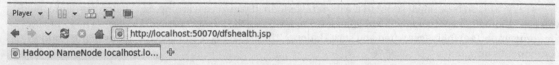

NameNode 'localhost.localdomain:54310'

Started: Wed Dec 03 10:37:20 PST 2014
Version: 1.0.3, r1335192
Compiled: Tue May 8 20:31:25 UTC 2012 by hortonfo
Upgrades: There are no upgrades in progress.

Browse the filesystem
Namenode Logs

Cluster Summary

450 files and directories, 172 blocks = 622 total. Heap Size is 31.57 MB / 966.69 MB (3%)

Configured Capacity	:	35.18 GB
DFS Used	:	37.36 MB
Non DFS Used	:	16.99 GB
DFS Remaining	:	18.15 GB
DFS Used%	:	0.1 %
DFS Remaining%	:	51.6 %
Live Nodes	:	1
Dead Nodes	:	0
Decommissioning Nodes	:	0
Number of Under-Replicated Blocks	:	5

NameNode Storage:

Storage Directory	Type	State
/home/hadoop/hadoop-hadoop/dfs/name	IMAGE_AND_EDITS	Active

This is Apache Hadoop release 1.0.3

Figure 13-8. *Hadoop administration web interface*

By clicking the "Browse the filesystem" link in Figure 13-8, you get the file system details shown in Figure 13-9.

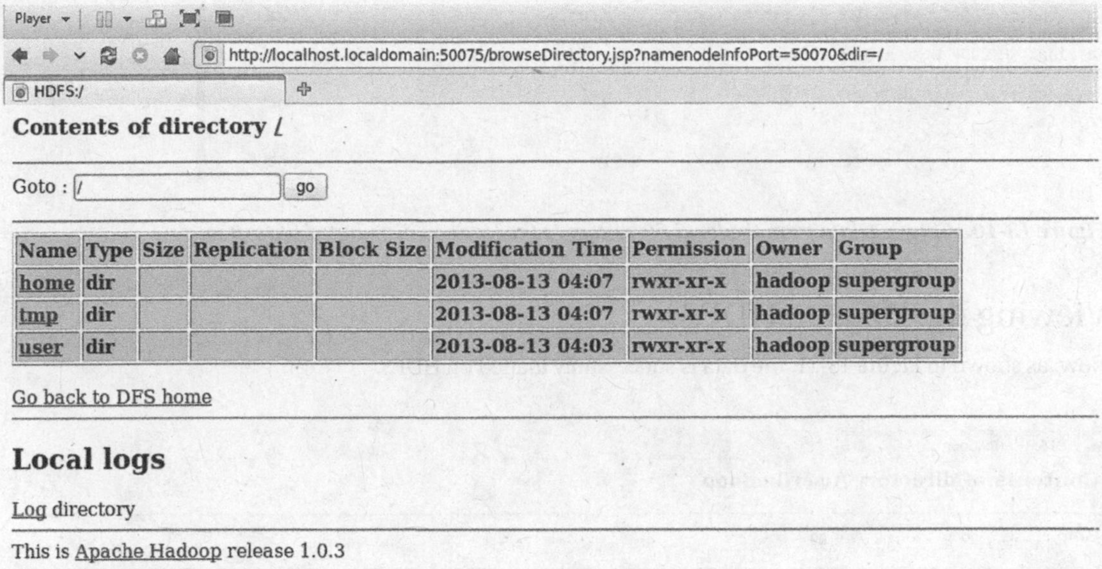
Player
http://localhost.localdomain:50075/browseDirectory.jsp?namenodeInfoPort=50070&dir=/
HDFS:/
Contents of directory /
Goto : / go
Name Type Size Replication Block Size Modification Time Permission Owner Group
home dir 2013-08-13 04:07 rwxr-xr-x hadoop supergroup
tmp dir 2013-08-13 04:07 rwxr-xr-x hadoop supergroup
user dir 2013-08-13 04:03 rwxr-xr-x hadoop supergroup
Go back to DFS home
Local logs
Log directory
This is Apache Hadoop release 1.0.3

Figure 13-9. *The Hadoop distributed file system*

The following is the cluster summary (Figure 13-8) before you get the data inside Hadoop:

- 450 files and directories and 171 blocks = 621 total
- Heap size: 31.57MB / 966.69MB (3 percent)
- Configured capacity: 35.18GB
- Distributed file system (DFS) used: 37.36MB
- Non-DFS used: 21.33GB
- DFS remaining: 13.81GB
- DFS used: 0.1 percent
- DFS remaining: 39.27 percent
- Live nodes: 1
- Dead nodes: 0
- Decommissioning nodes: 0
- Number of under-replicated blocks: 5

Moving Data from the Local System to Hadoop

On the DFS file system, you need to move the data file into the HDFS location. The command is simple.

```
bin/hadoop fs -copyFromLocal /home/final_stack_data.txt stack_data
```

Type this command and refer to Figure 13-10. This single command will cut the data into many 64MB blocks and distribute them on HDFS.

```
[hadoop@localhost hadoop-1.0.3]$ bin/hadoop fs -copyFromLocal  /home/hadoop/Desktop/final_st
ack_data.txt  stack_data
Warning: $HADOOP_HOME is deprecated.

[hadoop@localhost hadoop-1.0.3]$ █
```

Figure 13-10. *Copying data from the local file system to the Hadoop distributed file system*

Viewing the Data on HDFS

Now, as shown in Figure 13-11, the data is successfully loaded on HDFS.

| HDFS:/user/hadoop | ⊕ |

Contents of directory /user/hadoop

Goto : [/user/hadoop] [go]

Go to parent directory

Name	Type	Size	Replication	Block Size	Modification Time	Permission	Owner	Group
stack_data	file	6.99 GB	1	64 MB	2014-12-03 11:16	rw-r--r--	hadoop	supergroup
tmp	dir				2013-08-07 04:42	rwxrwxrwx	hadoop	supergroup
warehouse	dir				2013-08-13 04:03	rwxrwxrwx	hadoop	supergroup

Go back to DFS home

Local logs

Log directory

This is Apache Hadoop release 1.0.3

Figure 13-11. *Data transferred to HDFS*

In Figure 13-11 you can see a new data set named stack_data in HDFS. By clicking it, you can see the content in that data set. Refer to Figure 13-12.

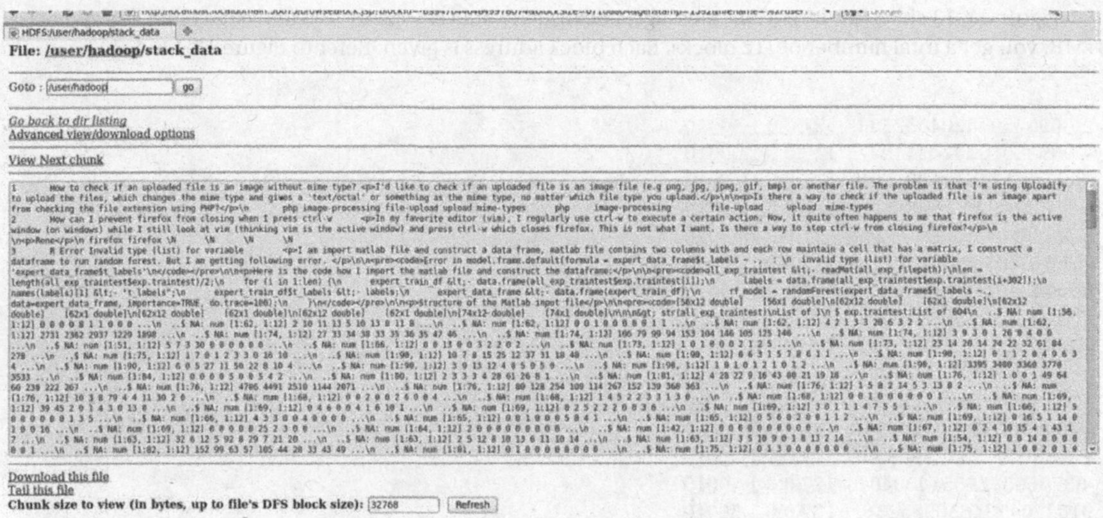

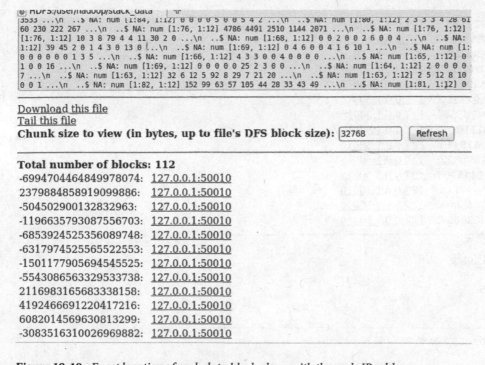

Figure 13-12. *Snapshot of a data block*

By clicking stack_data, you can see the data in the first chunk, as well as information about other data blocks. Refer to Figure 13-13.

Figure 13-13. *Exact location of each data block along with the node IP address*

Figure 13-13 shows an important message; because the data set size is 7GB and each block is sized at 64MB, you get a total number of 112 blocks. Each block address is given. Refer to Figure 13-14.

```
- - - - - - - - - - - - - - - - - - -    - - - - - - - - - - - - - -
-2609655914204062309:    127.0.0.1:50010
-7005085182424244977:    127.0.0.1:50010
-6667541116129086501:    127.0.0.1:50010
5484256018119500820:     127.0.0.1:50010
8208839047628257524:     127.0.0.1:50010
2766109055267620894:     127.0.0.1:50010
8560439098249530113:     127.0.0.1:50010
-8154540198588547564:    127.0.0.1:50010
-4801738609267090161:    127.0.0.1:50010
-9015278876832924082:    127.0.0.1:50010
1085565467444628722:     127.0.0.1:50010
3798046273410056334:     127.0.0.1:50010
7296897902937293398:     127.0.0.1:50010
-8370668727851400401:    127.0.0.1:50010
9104204734650985236:     127.0.0.1:50010
6311620234711798174:     127.0.0.1:50010
6160663678924941510:     127.0.0.1:50010
4644802008316034569:     127.0.0.1:50010
-8153836824628051043:    127.0.0.1:50010
8324023128208491095:     127.0.0.1:50010
-5038372401501817231:    127.0.0.1:50010
Done
```

```
325115968952222180:      127.0.0.1:50010
-3729272779043017207:    127.0.0.1:50010
5644059051038803629:     127.0.0.1:50010
8096145564457288508:     127.0.0.1:50010
4092415142371904364:     127.0.0.1:50010
1919613947396927412:     127.0.0.1:50010
-4339549546894785584:    127.0.0.1:50010
-8710325722631310689:    127.0.0.1:50010
5183222670479419349:     127.0.0.1:50010
1152771480347890522:     127.0.0.1:50010
-1727689516010445994:    127.0.0.1:50010
3216459131957652775:     127.0.0.1:50010
1119923936463448805:     127.0.0.1:50010
3946101514503684637:     127.0.0.1:50010
```

Go back to DFS home

Local logs

Log directory

This is Apache Hadoop release 1.0.3

Figure 13-14. *Block addresses*

Recall that we are not using a multinode cluster. So, you will have only one machine IP address for all the blocks of data. You can see all the blocks are stored at the same IP address (Figure 13-14). If the blocks were stored on different nodes, then you would see different IP addresses for each block. If you click a particular data block, you see its contents. Refer to Figure 13-15.

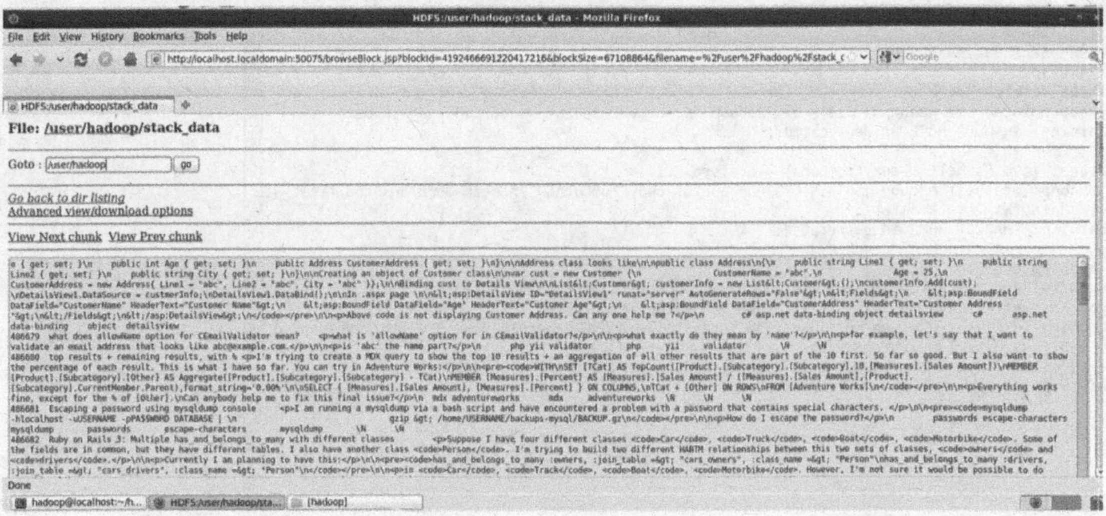

Figure 13-15. *Data inside a block*

Now that you have the data on a distributed system, you can get started with analytics. As discussed earlier, the task here is to just get some basic descriptive statistics on the data. We used Hive to perform our operations.

Starting Hive

To start Hive, you simply need to type the command hive. Refer to Figure 13-16.

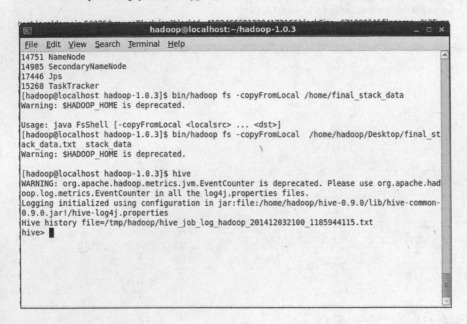

Figure 13-16. *Starting Hive*

The show tables command will show the existing tables in Hive. The default and previous table names in Hive will appear with this command. Refer to Figure 13-17.

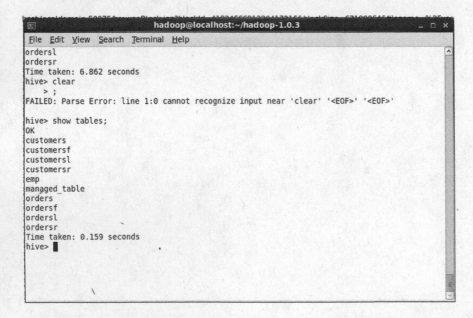

Figure 13-17. *Existing tables in Hive*

Creating a Table Using Hive

By now you know that Hive is like a SQL engine on top of Hadoop. In the next step, just as you do in SQL, you need to design a table and populate it with data. Refer to Figure 13-18 for the code used to create the table.

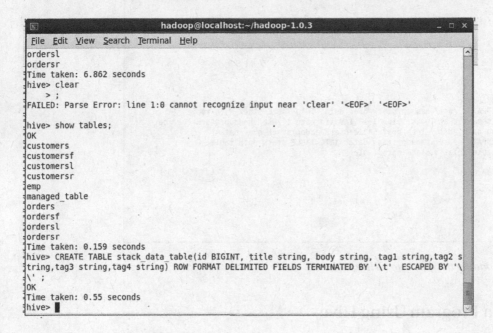

Figure 13-18. *Creating a table in Hive*

Now that the table is created, the following is the command to populate the table with the data on HDFS. Refer to Figure 13-19.

```
LOAD DATA INPATH '/user/hadoop/stack_over_data' INTO TABLE stack_data_table;
```

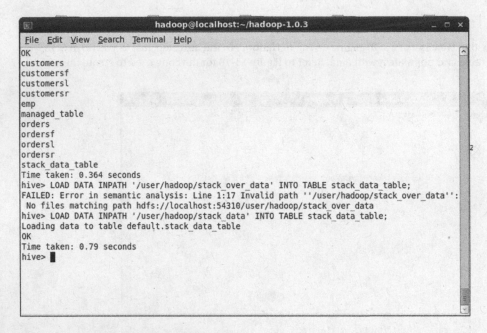

```
                        hadoop@localhost:~/hadoop-1.0.3                    _ □ ×
 File  Edit  View  Search  Terminal  Help
OK
customers
customersf
customersl
customersr
emp
managed_table
orders
ordersf
ordersl
ordersr
stack_data_table
Time taken: 0.364 seconds
hive> LOAD DATA INPATH '/user/hadoop/stack_over_data' INTO TABLE stack_data_table;
FAILED: Error in semantic analysis: Line 1:17 Invalid path ''/user/hadoop/stack_over_data'':
 No files matching path hdfs://localhost:54310/user/hadoop/stack_over_data
hive> LOAD DATA INPATH '/user/hadoop/stack_data' INTO TABLE stack_data_table;
Loading data to table default.stack_data_table
OK
Time taken: 0.79 seconds
hive> █
```

Figure 13-19. *Loading the data into a Hive table*

Executing a Program Using Hive

You have the table in Hive with the data loaded. Now you need to write the HiveSQL code, and it will automatically get converted into MapReduce code. Refer to Figure 13-20. Execute the following HiveSQL command:

```
Select count(*) from stack_data_table ;
```

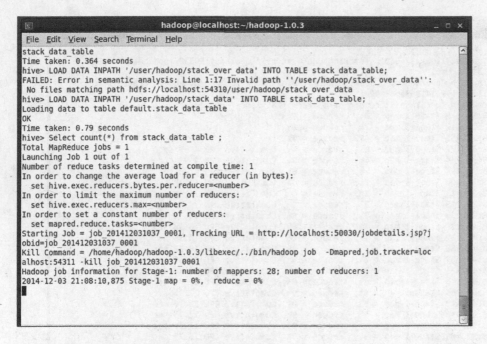

Figure 13-20. *Executing the HiveSQL command*

Viewing the MapReduce Status

Figure 13-21 shows the status of map and reduce. In this figure, note that the map status is 28 percent and the reduce job is completed by 7 percent.

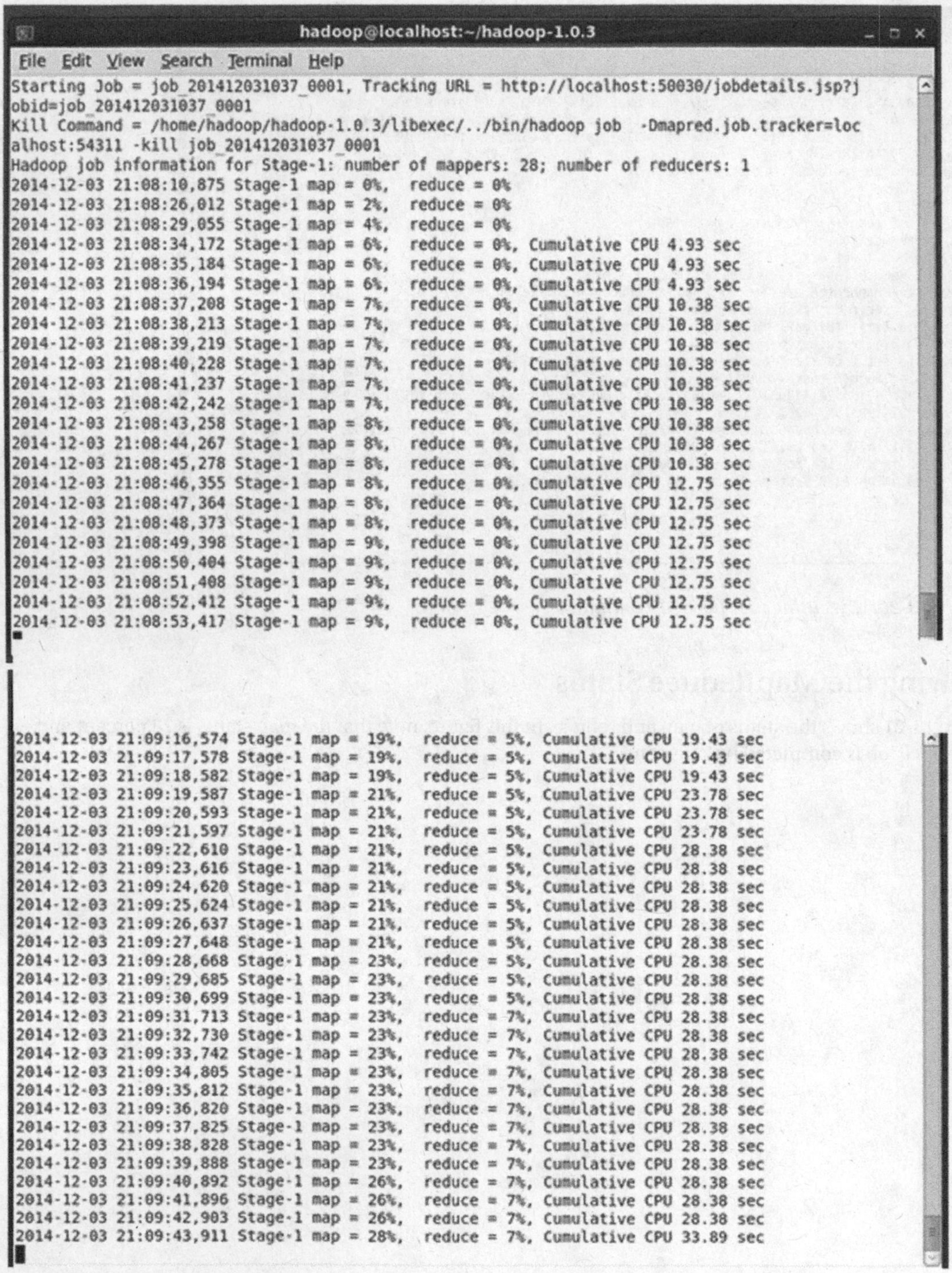

```
                        hadoop@localhost:~/hadoop-1.0.3                    _ □ ×

File  Edit  View  Search  Terminal  Help
Starting Job = job_201412031037_0001, Tracking URL = http://localhost:50030/jobdetails.jsp?j
obid=job_201412031037_0001
Kill Command = /home/hadoop/hadoop-1.0.3/libexec/../bin/hadoop job  -Dmapred.job.tracker=loc
alhost:54311 -kill job_201412031037_0001
Hadoop job information for Stage-1: number of mappers: 28; number of reducers: 1
2014-12-03 21:08:10,875 Stage-1 map = 0%,  reduce = 0%
2014-12-03 21:08:26,012 Stage-1 map = 2%,  reduce = 0%
2014-12-03 21:08:29,055 Stage-1 map = 4%,  reduce = 0%
2014-12-03 21:08:34,172 Stage-1 map = 6%,  reduce = 0%, Cumulative CPU 4.93 sec
2014-12-03 21:08:35,184 Stage-1 map = 6%,  reduce = 0%, Cumulative CPU 4.93 sec
2014-12-03 21:08:36,194 Stage-1 map = 6%,  reduce = 0%, Cumulative CPU 4.93 sec
2014-12-03 21:08:37,208 Stage-1 map = 7%,  reduce = 0%, Cumulative CPU 10.38 sec
2014-12-03 21:08:38,213 Stage-1 map = 7%,  reduce = 0%, Cumulative CPU 10.38 sec
2014-12-03 21:08:39,219 Stage-1 map = 7%,  reduce = 0%, Cumulative CPU 10.38 sec
2014-12-03 21:08:40,228 Stage-1 map = 7%,  reduce = 0%, Cumulative CPU 10.38 sec
2014-12-03 21:08:41,237 Stage-1 map = 7%,  reduce = 0%, Cumulative CPU 10.38 sec
2014-12-03 21:08:42,242 Stage-1 map = 7%,  reduce = 0%, Cumulative CPU 10.38 sec
2014-12-03 21:08:43,258 Stage-1 map = 8%,  reduce = 0%, Cumulative CPU 10.38 sec
2014-12-03 21:08:44,267 Stage-1 map = 8%,  reduce = 0%, Cumulative CPU 10.38 sec
2014-12-03 21:08:45,278 Stage-1 map = 8%,  reduce = 0%, Cumulative CPU 10.38 sec
2014-12-03 21:08:46,355 Stage-1 map = 8%,  reduce = 0%, Cumulative CPU 12.75 sec
2014-12-03 21:08:47,364 Stage-1 map = 8%,  reduce = 0%, Cumulative CPU 12.75 sec
2014-12-03 21:08:48,373 Stage-1 map = 8%,  reduce = 0%, Cumulative CPU 12.75 sec
2014-12-03 21:08:49,398 Stage-1 map = 9%,  reduce = 0%, Cumulative CPU 12.75 sec
2014-12-03 21:08:50,404 Stage-1 map = 9%,  reduce = 0%, Cumulative CPU 12.75 sec
2014-12-03 21:08:51,408 Stage-1 map = 9%,  reduce = 0%, Cumulative CPU 12.75 sec
2014-12-03 21:08:52,412 Stage-1 map = 9%,  reduce = 0%, Cumulative CPU 12.75 sec
2014-12-03 21:08:53,417 Stage-1 map = 9%,  reduce = 0%, Cumulative CPU 12.75 sec


2014-12-03 21:09:16,574 Stage-1 map = 19%,  reduce = 5%, Cumulative CPU 19.43 sec
2014-12-03 21:09:17,578 Stage-1 map = 19%,  reduce = 5%, Cumulative CPU 19.43 sec
2014-12-03 21:09:18,582 Stage-1 map = 19%,  reduce = 5%, Cumulative CPU 19.43 sec
2014-12-03 21:09:19,587 Stage-1 map = 21%,  reduce = 5%, Cumulative CPU 23.78 sec
2014-12-03 21:09:20,593 Stage-1 map = 21%,  reduce = 5%, Cumulative CPU 23.78 sec
2014-12-03 21:09:21,597 Stage-1 map = 21%,  reduce = 5%, Cumulative CPU 23.78 sec
2014-12-03 21:09:22,610 Stage-1 map = 21%,  reduce = 5%, Cumulative CPU 28.38 sec
2014-12-03 21:09:23,616 Stage-1 map = 21%,  reduce = 5%, Cumulative CPU 28.38 sec
2014-12-03 21:09:24,620 Stage-1 map = 21%,  reduce = 5%, Cumulative CPU 28.38 sec
2014-12-03 21:09:25,624 Stage-1 map = 21%,  reduce = 5%, Cumulative CPU 28.38 sec
2014-12-03 21:09:26,637 Stage-1 map = 21%,  reduce = 5%, Cumulative CPU 28.38 sec
2014-12-03 21:09:27,648 Stage-1 map = 21%,  reduce = 5%, Cumulative CPU 28.38 sec
2014-12-03 21:09:28,668 Stage-1 map = 23%,  reduce = 5%, Cumulative CPU 28.38 sec
2014-12-03 21:09:29,685 Stage-1 map = 23%,  reduce = 5%, Cumulative CPU 28.38 sec
2014-12-03 21:09:30,699 Stage-1 map = 23%,  reduce = 5%, Cumulative CPU 28.38 sec
2014-12-03 21:09:31,713 Stage-1 map = 23%,  reduce = 7%, Cumulative CPU 28.38 sec
2014-12-03 21:09:32,730 Stage-1 map = 23%,  reduce = 7%, Cumulative CPU 28.38 sec
2014-12-03 21:09:33,742 Stage-1 map = 23%,  reduce = 7%, Cumulative CPU 28.38 sec
2014-12-03 21:09:34,805 Stage-1 map = 23%,  reduce = 7%, Cumulative CPU 28.38 sec
2014-12-03 21:09:35,812 Stage-1 map = 23%,  reduce = 7%, Cumulative CPU 28.38 sec
2014-12-03 21:09:36,820 Stage-1 map = 23%,  reduce = 7%, Cumulative CPU 28.38 sec
2014-12-03 21:09:37,825 Stage-1 map = 23%,  reduce = 7%, Cumulative CPU 28.38 sec
2014-12-03 21:09:38,828 Stage-1 map = 23%,  reduce = 7%, Cumulative CPU 28.38 sec
2014-12-03 21:09:39,888 Stage-1 map = 23%,  reduce = 7%, Cumulative CPU 28.38 sec
2014-12-03 21:09:40,892 Stage-1 map = 26%,  reduce = 7%, Cumulative CPU 28.38 sec
2014-12-03 21:09:41,896 Stage-1 map = 26%,  reduce = 7%, Cumulative CPU 28.38 sec
2014-12-03 21:09:42,903 Stage-1 map = 26%,  reduce = 7%, Cumulative CPU 28.38 sec
2014-12-03 21:09:43,911 Stage-1 map = 28%,  reduce = 7%, Cumulative CPU 33.89 sec
```

Figure 13-21. *The MapReduce intermediate status*

Once the MapReduce is completed, you get the results. The total time taken to process this data is about five minutes.

You are simply trying to find the total count of rows, which is the business problem chosen for this demo. When you execute a HiveSQL command, you see that the map and reduce jobs start automatically. The map jobs will be counting rows in each block of data; reduce is for totaling the counts.

The Final Result

The final result shows that there are 6,034,195 records (Figure 13-22). Nearly 6 million questions are included in the original data input file.

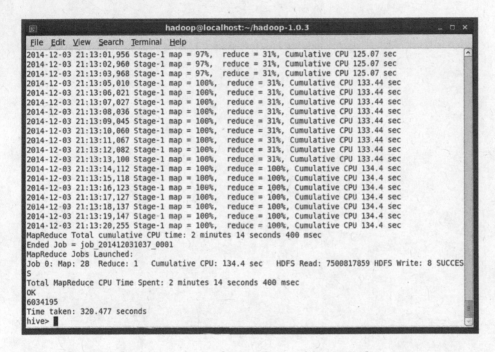

Figure 13-22. The final count of rows; the completion of the MapReduce process

Once you understand the basics and see how you got the basic statistics of counting the number of rows, you can write the respective HiveQL commands to get other results such as the top ten topics, the frequency of each topic, and so on.

Conclusion

In this chapter, you learned how a basic big data problem is handled. Big data analytics is the latest buzzword in the market. Big data today is at the beginning stages of its development. A lot of development is expected in this field, particularly in the domains of predictive modeling, text mining, video analytics, and machine learning. As of now, not every algorithm is converted into a MapReduce program. But in the near future, we may have a tool that works on a distributed platform and can perform all the advanced analytics tasks for us. This is a rapidly growing field, so you may see several new tools in the near future. For now, as an exercise, we suggest you do some research on the other components in the Hadoop ecosystem.

Index

Get the eBook for only $10!

> Now you can take the weightless companion with you anywhere, anytime. Your purchase of this book entitles you to 3 electronic versions for only $10.

This Apress title will prove so indispensible that you'll want to carry it with you everywhere, which is why we are offering the eBook in 3 formats for only $10 if you have already purchased the print book.

Convenient and fully searchable, the PDF version enables you to easily find and copy code—or perform examples by quickly toggling between instructions and applications. The MOBI format is ideal for your Kindle, while the ePUB can be utilized on a variety of mobile devices.

Go to www.apress.com/promo/tendollars to purchase your companion eBook.

Apress®
THE EXPERT'S VOICE™

Printed in the United States
By Bookmasters